Modeling and Simulation in Science, Engineering and Technology

Series Editor

Nicola Bellomo
Politecnico di Torino
Italy

Advisory Editorial Board

Andreas Deutsch
Sabine Dormann

Cellular Automaton Modeling of Biological Pattern Formation

Characterization, Applications, and Analysis

Foreword by Philip K. Maini

Birkhäuser
Boston • Basel • Berlin

Andreas Deutsch
Dresden University of Technology
Center for High Performance Computing
D-01062 Dresden
Germany

Sabine Dormann
University of Osnabrück
Department of Mathematics
D-49069 Osnabrück
Germany

AMS Subject Classifications: 00A72, 0001, 0002, 9208, 92B05, 92B20

Library of Congress Cataloging-in-Publication Data
Deutsch, Andreas, 1960-
Cellular automaton modeling of biological pattern formation : characterization, applications, and analysis / Andreas Deutsch, Sabine Dormann.
p. cm — (Modeling and simulation in science, engineering & technology)
Includes bibliographical references and index.
ISBN 0-8176-4281-1 (alk. paper)
1. Pattern formation (Biology) 2. Cellular automata–Mathematical models. I. Dormann, Sabine. II. Title. III. Series.

QH491.D42 2004
571.3–dc22

20030638990
CIP

ISBN 0-8176-4281-1 Printed on acid-free paper.

Printed in the United States of America. (TXQ/MV)

9 8 7 6 5 4 3 2 1 SPIN 10855758

Birkhäuser is a part of *Springer Science+Business Media*
www.birkhauser.com

To our parents

Foreword

The recent dramatic advances in biotechnology have led to an explosion of data in the life sciences at the molecular level as well as more detailed observation and characterization at the cellular and tissue levels. It is now absolutely clear that one needs a theoretical framework in which to place this data to gain from it as much information as possible. Mathematical and computational modelling approaches are the obvious way to do this. Heeding lessons from the physical sciences, one might expect that all areas in the life sciences would be actively pursuing quantitative methods to consolidate the vast bodies of data that exist and to integrate rapidly accumulating new information. Remarkably, with a few notable exceptions, quite the contrary situation exists. However, things are now beginning to change and there is the sense that we are at the beginning of an exciting new era of research in which the novel problems posed by biologists will challenge the mathematicians and computer scientists, who, in turn, will use their tools to inform the experimentalists, who will verify model predictions. Only through such a tight interaction among disciplines will we have the opportunity to solve many of the major problems in the life sciences.

One such problem, central to developmental biology, is the understanding of how various processes interact to produce spatio-temporal patterns in the embryo. From an apparently almost homogeneous mass of dividing cells in the very early stages of development emerges the vast and sometimes spectacular array of patterns and structures observed in animals. The mechanisms underlying the coordination required for cells to produce patterns on a spatial scale much larger than a single cell are still largely a mystery, despite a huge amount of experimental and theoretical research. There is positional information inherent in oocytes, which must guide patterns, but cells that are completely dissociated and randomly mixed can recombine to form periodic spatial structures. This leads to the intriguing possibility that at least some aspects of spatio-temporal patterning in the embryo arise from the process of self-organization. Spatial patterns also arise via self-organization in other populations of individuals, such as the swarming behaviour of bacteria, and in chemical systems, so that it is a widespread phenomenon.

Modelling in this area takes many forms, depending on the spatio-temporal scale and detail one wishes (or is able) to capture. At one extreme are coupled systems of ordinary differential equations, in which one assumes that the system is well stirred so that all spatial information is lost and all individuals (for example, molecules) are assumed to have identical states. At the other extreme are cellular automata models, in which each element may represent an individual (or a collection of individuals) with assigned characteristics (for example, age) that can vary from one individual to the next. This approach allows for population behaviour to evolve in response to individual-level interactions. In hybrid cellular automata, one can model intracellular phenomena by ordinary differential equations, while global signalling may be modelled by partial differential equations. In this way, one can begin to address the crucial issue of modelling at different scales. There are many modelling levels between these extremes and each one has its own strengths and weaknesses.

Andreas Deutsch and Sabine Dormann bring to bear on this subject a depth and breadth of experience that few can match. In this book they present many different modelling approaches and show the appropriate conditions under which each can be used. After an introduction to pattern formation in general, this book develops the cellular automaton approach and shows how, under certain conditions, one can take the continuum limit, leading to the classical partial differential equation models. Along the way, many interesting pattern formation applications are presented. Simple rules are suggested for various elementary cellular interactions and it is demonstrated how spatio-temporal pattern formation in corresponding automaton models can be analyzed. In addition, suggestions for future research projects are included. It is also shown that the model framework developed can be used more generally to tackle problems in other areas, such as tumour growth, one of the most rapidly growing areas in mathematical biology at the present time. The accompanying website (www.biomodeling.info) allows the reader to perform online simulations of some of the models presented.

This book, aimed at undergraduates and graduate students as well as experienced researchers in mathematical biology, is very timely and ranges from the classical approaches right up to present-day research applications. For the experimentalist, the book may serve as an introduction to mathematical modelling topics, while the theoretician will particularly profit from the description of key problems in the context of biological pattern formation. The book provides the perfect background for researchers wishing to pursue the goal of multiscale modelling in the life sciences, perhaps one of the most challenging and important tasks facing researchers this century.

Philip K. Maini
Centre for Mathematical Biology
Oxford, GB
January 2004

Contents

List of Figures

List of Notation

CA	cellular automaton
LGCA	lattice-gas cellular automaton/automata 67

Greek symbols

$\beta \in \mathbb{N}_0$	number of rest (zero-velocity) channels 72
$\Gamma(q)$	Boltzmann propagator 95
δ	length of temporal unit 107
ϵ	length of spatial unit 107
$\boldsymbol{\eta}(r) = (\eta_i(r))_{i=1}^{\tilde{b}} \in \{0,1\}^{\tilde{b}}$	node configuration in a LGCA 72
$\boldsymbol{\eta}_{\mathcal{N}(r)}$	local configuration in a LGCA 72
Λ_M	spectrum of the matrix M 91
μ	spectral radius 91
$\mu(q)$	spectral radius according to wave number q 96
$\nu = \lvert\mathcal{N}_b^I\rvert$, $\nu_o = \lvert\mathcal{N}_{bo}^I\rvert$	number of neighbors in the interaction neighborhood 70
$\rho(r,k) \in [0,1]$	local particle density of node r at time k 86
$\rho(k) \in [0,1]$	total particle density in the lattice at time k 86
$\varrho(r,k) \in [0,\tilde{b}]$	local mass of node r at time k 86
$\varrho(k) \in [0,\tilde{b}]$	total mass in the lattice at time k 86

$\sigma \in \{1, \ldots, \varsigma\}$	single component in a model with ς components . . . 69
$\Psi_a(\boldsymbol{\eta}(r,k)) \in \{0,1\}$	indicator function . . . 139

Further symbols

$\lvert\cdot\rvert$	cardinality of a set
$[y]$	integer closest to $y \in \mathbb{R}^+$
$\lceil y \rceil$	smallest integer greater than or equal to $y \in \mathbb{R}^+$ . . . 222
$\boldsymbol{y}^T$	transpose of the vector $\boldsymbol{y}$
$A_j \in \mathcal{A}_{\tilde{b}}$	permutation matrix . . . 116
b	coordination number; number of nearest neighbors on the lattice $\mathcal{L}$ . . . 68
$\tilde{b} = b + \beta$	total number of channels at each node . . . 72
$c_i \in \mathcal{N}_b, i = 1, \ldots, b$	nearest neighborhood connections of the lattice $\mathcal{L}$ . . . 68
$\mathcal{C}_i\left(\boldsymbol{\eta}_{\mathcal{N}(r)}(k)\right) \in \{-1,0,1\}$	change in occupation numbers . . . 78
$\tilde{\mathcal{C}}_i\left(\boldsymbol{f}_{\mathcal{N}(r)}(k)\right) \in [0,1]$	change of the average number of particles . . . 88
$\boldsymbol{D}\left(\boldsymbol{\eta}_{\mathcal{N}(r)}\right)$	director field . . . 164
$\mathcal{E} = \{z^1, \ldots, z^{\lvert\mathcal{E}\rvert}\}$	(finite) set of elementary states . . . 71
$\boldsymbol{f}(r) = (f_i(r))_{i=1}^{\tilde{b}} \in [0,1]^{\tilde{b}}$	vector of single particle distribution functions; average occupation numbers . . . 86
$\boldsymbol{f}^s(k)$	vector of spatially averaged occupation numbers . . . 220
$\boldsymbol{F}(q) = (F_i(q))_{i=1}^{\tilde{b}}$	Fourier-transformed value . . . 94
$\mathtt{I}$	(general) interaction operator . . . 77
$\boldsymbol{J}(\boldsymbol{\eta}(r))$	flux of particles at node r . . . 164
$k = 0, 1, 2, \ldots$	time step . . . 75
$\mathcal{L} \subset \mathbb{R}^d$	d-dimensional regular lattice . . . 67
$L_i, i = 1, \ldots, d$	number of cells in space direction i . . . 68
$\mathtt{M}$	shuffling (mixing) operator . . . 115

$\mathrm{n}(r) \in \{0, \ldots, \tilde{b}\}$	total number of particles present at node r 72
$\mathcal{N}_b$	neighborhood template 68
$\mathcal{N}_b^I$	interaction neighborhood template 70
$\mathcal{N}_{bo}^I$	outer interaction neighborhood template 70
$\mathtt{P}$	propagation operator 77
$q \in \{0, \ldots, L\}$	(discrete) wave number 93
q^s	wave number observed in simulation 220
$q_* \in Q^c$	dominant critical wave number 96
Q^c	set of critical wave numbers 96
Q^+, Q^-	subsets of critical wave numbers 96
$r \in \mathcal{L}$	spatial coordinate, cell, node, site 67
(r, c_i)	channel with direction c_i at node r 72
$\mathtt{R}$	reactive interaction operator ("Turing" model) 212
$\mathcal{R} : \mathcal{E}^\nu \to \mathcal{E}$	cellular automaton rule 73
$s(r) \in \mathcal{E}$ $s : \mathcal{L} \to \mathcal{E}$	state value at node r 71
$\boldsymbol{s} = (s(r_i))_{r_i \in \mathcal{L}} \in \mathcal{S}$	global configuration 71
$\boldsymbol{s}_{\mathcal{M}} = (s(r_i))_{r_i \in \mathcal{M}}$, $\mathcal{M} \subset \mathcal{L}$	local configuration 72
$\mathcal{S} = \mathcal{E}^{\lvert\mathcal{L}\rvert}$	state space 72
$W : \mathcal{E}^\nu \to [0, 1]$	time-independent transition probability 74
X_σ	particle of "species" σ 78

Cellular Automaton Modeling of Biological Pattern Formation

Part I

General Principles, Theories, and Models of Pattern Formation

1 Introduction and Outline

"Things should be made as simple as possible, but not any simpler."
(A. Einstein)

This book deals with the problem of biological pattern formation. What are the mechanisms according to which individual organisms develop and biological patterns form? Biological organisms are characterized by their genomes. The letters of the genetic alphabet (the nucleotides), their precise arrangement in selected organisms (the gene sequence), and the molecular structure of a huge number of encoded proteins are public today. However, analysis of single gene and protein function is not sufficient to explain the complex pattern formation which results from the collective behavior of interacting molecules and cells. In the beginning of embryological development all cells are identical—equipped with basically the same set of genes. Accordingly, collective phenomena brought about by the interaction of cells with themselves and their surroundings are responsible for the differentiation and pattern formation characterizing subsequent developmental stages. It has become clear that mathematical modeling is strongly needed to discover the self-organization principles of interacting cell systems (Deutsch et al. 2004). But what are appropriate mathematical models, how can they be analyzed, and which specific biological problems can they address?

This chapter provides the motivation for the book. The basic problems are introduced and the connection between biological pattern formation and mathematical modeling is emphasized. An outline presents the book's structure and specific suggestions on how to read the book depending on the reader's background.

Principles of Biological Pattern Formation. The *morphogenesis* of multicellular organisms (fig. 1.1) as the development of characteristic tissue and organ arrangements, but also as the establishment and maintenance of life cycles distinguishing unicellular microorganisms (e.g., the slime mold *Dictyostelium discoideum* or myxobacteria; fig. 1.2) are manifestations of biological pattern formation.

Pattern formation is a *spatio-temporal* process: characterization of its principles therefore depends on the underlying space and time concept. In a *static* Platonian

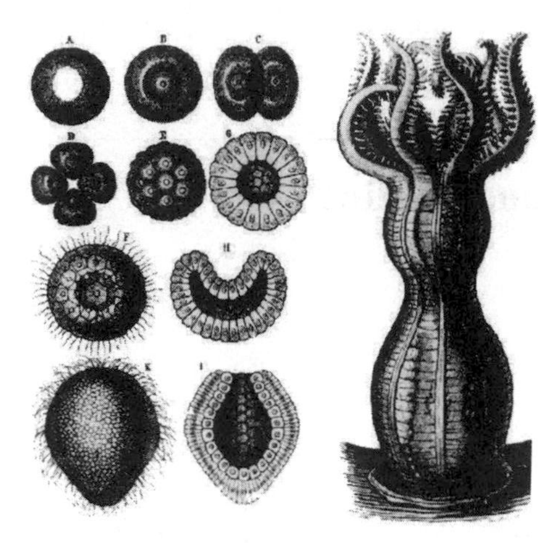

Figure 1.1. Morphogenesis of the coral *Monoxenia darwinii* (drawing by E. Haeckel). Left: from fertilized egg (top left) to gastrula stage (bottom right); right: adult stage. Courtesy of Ernst-Haeckel-Haus, Jena.

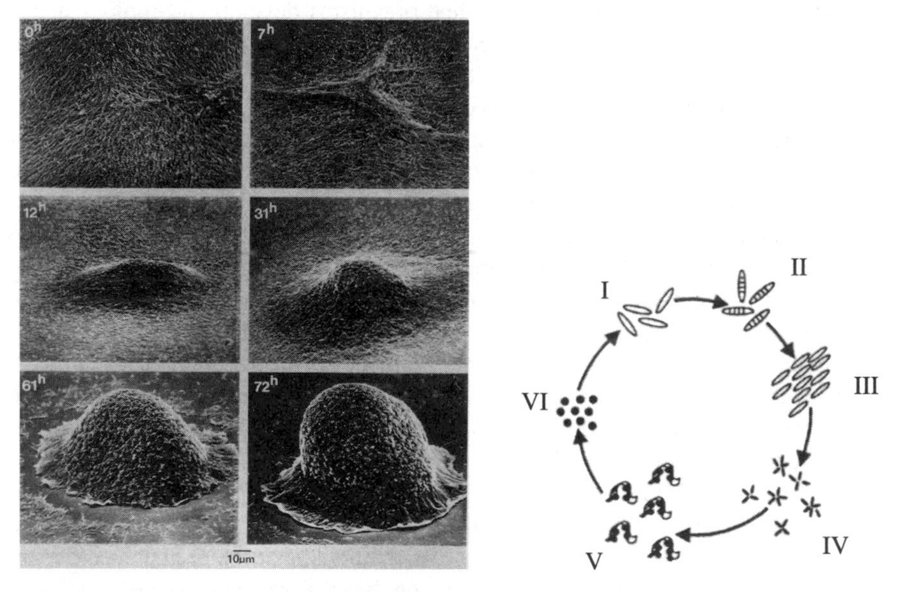

Figure 1.2. Pattern formation of unicellular organisms: myxobacteria. Left: fruiting body formation of *Myxococcus xanthus*. A developmental time series in submerged culture is shown. Initially asymmetric aggregates become round and develop intro three-dimensional fruiting bodies. Right: schematic sketch of myxobacterial life cycle. Rod-shaped vegetative cells (I) undergo cell division (II) and are able to migrate on suitable surfaces. Under certain conditions (e.g., starvation) cells cooperate, form streets (III), aggregate (IV), and develop fruiting bodies (V). Within these fruiting bodies cells differentiate into metabolically dormant myxospores (VI).

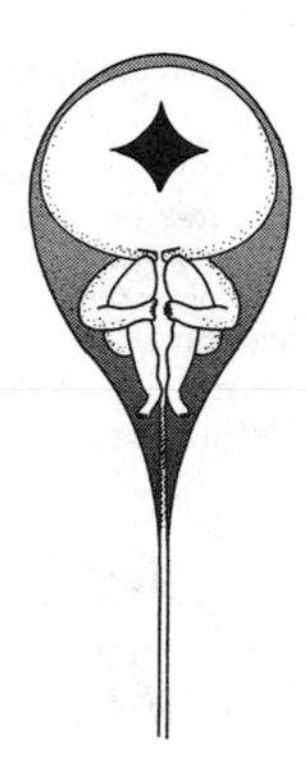

Figure 1.3. Preformed homunculus; inside a human spermatozoan a sitting "homunculus" was assumed which merely uncoils and grows during embryogenesis. Similar "homunculi" were expected to occupy the female egg (after Hartsoeker, 1694).

world view, any form (including biological forms) is regarded as *preformed* and static: the world is fixed (and optimal) without time, motion, or change. This concept only allows recycling of existing forms (see also fig. 1.3). In a *dynamic* Aristotelian world view, the need for epigenetic principles is emphasized to account for *de novo* pattern formation (fig. 1.4).

Life is characterized by an inherent small *ontogenetic* and a larger *phylogenetic* time scale, which define individual morphogenesis and evolutionary change, respectively (fig. 1.5). Darwin realized the importance of space (spatial niches) and time (temporal changes of varieties), which can result in different survival rates for organismic varieties. He contributed to an understanding of the *evolutionary change* of biological organisms with his theory of selection (Darwin 1859). But how do varieties develop during their individual lifetimes? At the end of the nineteenth century, developmental dynamics, i.e., *ontogenetic change,* became a target of experimentally oriented research. It was discovered that embryos do not contain the final adult form

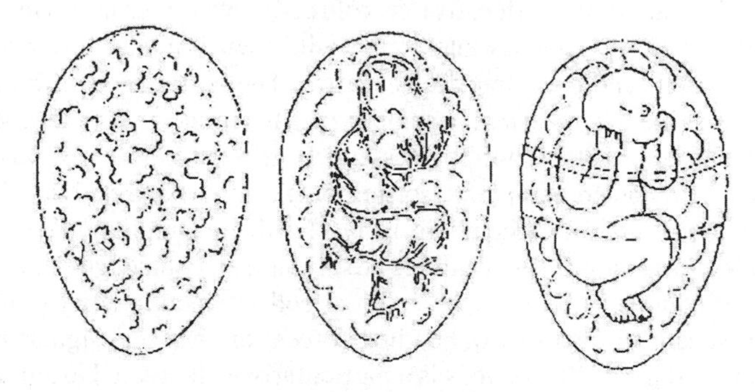

Figure 1.4. Aristotelian *epigenetic* view of development from a uniform distribution to a structured embryo (after Rueff 1554).

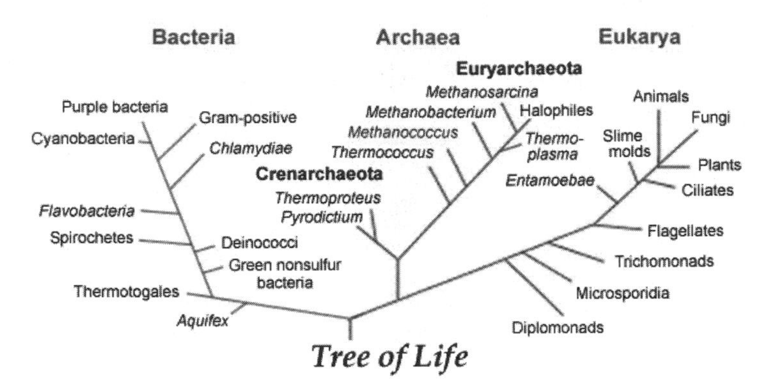

Figure 1.5. The tree of life or phylogenetic tree traces the pattern of descent of all life over millions of years into three major branches: bacteria, archaea, and eucarya. There is a controversy over the times at which archaea diverged from eubacteria and eukaryotes. One proposal is that eubacteria and archaea diverged around the time that eukaryotic cells developed, about 1.5 billion years ago.

in a mosaic prepattern and that all cells within an organism carry essentially the same genetical information. The need for *epigenetic principles* (e.g., regulation) became evident. Otherwise, cell differentiation and *de novo* formation of complex structures from a single cell in every new generation can not be explained.

The Problem. Even though a unified theory of morphogenesis comparable to Darwin's selection theory of evolution is still missing, one can address principles of biological pattern formation. Morphogenesis results from a limited repertoire of cellular activities: in particular, cells can change their shape, grow, divide, differentiate, undergo apoptosis, and migrate. It is the core of biological morphogenesis that cells do not behave independently of each other. To the contrary, cellular activities are intertwined and strongly rely on cooperative dynamics of *cell–cell interaction*, which may induce changes in cellular properties and activities.

Cells can interact directly (locally) or indirectly even over large distances. Local interaction of cells comprises interaction with their immediate environment, in particular other cells and the extracellular matrix. The importance of direct cell-cell interactions, in particular *adhesion*, became evident in tissue growth regeneration experiments (Holtfreter 1939, 1944; Townes and Holtfreter 1955). For example, phenomena of tissue reconstruction (e.g., sorting out) can be explained on a cell-to-cell basis by *differential adhesion* (Steinberg 1963). Further examples of direct cell interactions with their immediate surroundings are alignment, contact guidance or contact inhibition, and haptotaxis. In contrast, indirect cell interaction is mediated through long-range mechanical forces (e.g., bending forces) or chemical signals that propagate over large distances. Chemotaxis is a particularly well-studied example. Hereby, cells orient towards local maxima of a chemical signal gradient field.

Figure 1.6. Axolotl pigment pattern. The barred (or transverse band) pigment pattern of an *Ambystoma mexicanum* larva. An albino-black larva (stage 40; 9.5 mm long) lacking maternal pigment granules is shown. Lateral aspect, head to the left outside. Melanophores (dark transverse bands) and xanthophores (bright areas in between) alternate along the dorsal trunk. The periodic pattern has no resemblance at the individual cell level; it is a *collective phenomenon* brought about by interactions of the axolotl cells (courtesy of H. H. Epperlein, Dresden).

A *morphogenetic system* provides an example of self-organization. The system is composed of many individual components, the cells, that interact with each other implying qualitatively new features on macroscopic scales, i.e., scales that are far bigger than those of the individual cells (e.g., formation of a periodic pigment cell pattern, see fig. 1.6). The question is, What are essential cell interactions and how do corresponding cooperative phenomena influence organismic morphogenesis? Possible answers can be found by means of *mathematical modeling*, which allows one to abstract from specific component behavior and to analyze generic properties.

Mathematical Modeling. Starting with D'Arcy Thompson (1917), principles of morphogenesis have been studied with the help of mathematical models. In particular, Thompson considered the shapes of unicellular organisms and suggested the minimization of surface tension as a plausible hypothesis, which he analyzed by studying corresponding equations. Another pioneer was A. Turing, who contributed with the idea of *diffusive instabilities* in reaction-diffusion dynamics (Turing 1952). The *Turing instability* can explain pattern formation in disturbed spatially homogeneous systems if (diffusive) transport of *activator-inhibitor morphogens* is coupled to appropriate *chemical kinetics*. In order to develop, e.g., a (periodic) pattern it is necessary that the diffusion coefficients of activator and inhibitor species differ drastically. Reaction-diffusion systems have become paradigms of nonequilibrium pattern formation and biological self-organization (Britton 1986; Murray 2002).

Meanwhile, there exists an established arsenal of *macroscopic models* based on continuum equations to analyze cellular interactions in reaction-diffusion systems (Chaplain, Singh, and McLachlan 1999; Maini 1999; Meinhardt 1992; Othmer, Maini, and Murray 1993). While the continuum assumption is appropriate in systems dealing with large numbers of cells and chemical concentrations, it is not adequate in systems consisting of a small number of interacting *discrete cells*. The problem arises of how to design appropriate *microscopic models*, which allow the identification of individual cells.

Cellular Automaton Models. Interest in microscopic models, i.e., *spatial stochastic processes*, has grown lately due to the availability of "individual cell data" (genetic and proteomic) and has triggered the development of new cell-based mathematical models (for a recent review, see Drasdo 2003). Cell-based models are required if one is interested in understanding the organizational principles of interacting cell systems down to length scales of the order of a cell diameter in order to link the individual cell (microscopic) dynamics with a particular collective (macroscopic) phenomenon. Cell-based models, particularly cellular automata (CA), allow one to follow and analyze the spatio-temporal dynamics at the individual cell level.

Cellular automata are *discrete dynamical systems* and may be used as models of biological pattern formation based on cell–cell, cell–medium, and cell–medium–cell interactions. The roots of cellular automata can be traced back to the time when the origin of the genetic code was discovered (Watson and Crick 1953a, 1953b). Cellular automata were introduced by John von Neumann and Stanislaw Ulam as a computer model for self-reproduction, a necessary precondition for organismic inheritance (von Neumann 1966). Intensive research within the last few decades has demonstrated that successful model applications of cellular automata go far beyond self-reproduction. Since cellular automata have no central controller and are rule-based discrete dynamical systems, they can also be viewed as models of massively parallel, noncentralized computation. Cellular automata have become paradigms of self-organizing complex systems in which collective behavior arises from simple interaction rules of even more simple components. The automaton idea has been utilized in an enormous variety of biological and nonbiological systems. Accompanying the availability of more and more computing power, numerous automaton applications in physics, chemistry, biology, and even sociology have been studied extensively (Casti 2002; Hegselmann and Flache 1998; Mitchell 2002; Wolfram 2002).

Are there microscopic cellular automaton rules that can model the mechanisms of cell interaction? An important insight of complex system research is that macroscopic behavior is rather independent of the precise choice of the microscopic rule. For example, it was shown that simple collision rules can give rise to the intricate structures of hydrodynamic flow as long as the rules conserve mass and momentum (lattice-gas cellular automaton; Frisch, Hasslacher, and Pomeau 1986; Kadanoff 1986). Could it be that likewise in biological systems basic rules of cellular interaction are underlying complex developmental pattern formation? Contrary to differential equations representing a well-established concept, cellular automaton models of biological pattern formation are in a rather juvenile state. In particular, morphogenetic automaton classes have to be defined that allow for an analytic investigation. In this book, we introduce cellular automaton models in order to analyze the dynamics of interacting cell systems. We show how appropriately constructed stochastic automata allow for straightforward analysis of spatio-temporal pattern formation beyond mere simulation.

Outline of the Book. The book starts with a historical sketch of static and dynamic space-time concepts and shows how these concepts have influenced the understanding of pattern formation, particularly biological morphogenesis (ch. 2). Corresponding morphogenetic concepts are based on preformation, topology, optimization, and self-organization ideas. Furthermore, experimental approaches to the investigation of developmental principles are presented.

Ch. 3 introduces mathematical modeling concepts for analyzing principles of biological pattern formation. In particular, partial differential equations, coupled map lattices, many-particle systems, and cellular automata can be distinguished. In addition, *macroscopic* and *microscopic* modeling perspectives on biological pattern formation and their relations are discussed.

The cellular automaton idea is elaborated in ch. 4, starting from the biological roots of cellular automata as models of biological self-reproduction. We focus on the definition of deterministic, probabilistic, and lattice-gas cellular automata. Furthermore, routes to the linear stability (Boltzmann) analysis of automaton models are described. Stability analysis of the corresponding Boltzmann equation permits us to analyze the onset of pattern formation. It is demonstrated how particular cell interactions can be translated into corresponding automaton rules and how automaton simulations can be interpreted. It is also shown how to proceed from the (microscopic) automaton dynamics to a corresponding macroscopic equation.

An overview of cellular automaton models for different types of cellular interactions is presented in chs. 5–11. As a first example, the *interaction-free* (linear) automaton is introduced (ch. 5); this automaton can be viewed as a model of random cell dispersal. Stability analysis shows that all modes are stable and, accordingly, no spatial patterns can be expected. *Growth processes* are analyzed in ch. 6. In particular, probabilistic and lattice-gas automaton models for simple growth processes are proposed.

Adhesive interactions are the focus of ch. 7. Here, we consider adhesive interactions in systems consisting of a single cell type and two cell types, respectively. The underlying microdynamical equations of the proposed cellular automaton models are no longer linear. Stability analysis of the linearized Boltzmann equation indicates that the dominant (diffusive) mode can become unstable, implying spatial pattern formation visible as clustering and sorting out behavior (in the two-cell-type model). In particular, the two-cell-type system allows us to model and simulate a *differential adhesion* dynamics that is essential in key phases of embryonic development.

A cellular automaton based on *orientation-dependent (alignment) interaction* serves as a model of cellular alignment (swarming; see ch. 8). Cellular swarming is visible, e.g., in the formation of streets (rafts) of similarly oriented cells in certain microorganisms (e.g., myxobacteria). With the help of stability analysis it is possible to identify an "orientational mode" that indicates the swarming phase and that can destabilize. This behavior allows us to characterize the onset of swarming as a phase transition.

Ch. 9 takes up the problem of cellular pigment pattern formation. These patterns are easy to observe and evolve as the result of complex interactions between pigment cells and their structured tissue environment. Now, modeling has to incorporate both interactions between cells and interactions between cells and the extracellular matrix. An automaton model of pigment pattern formation is introduced that is based on solely local interactions (adhesive interaction as well as contact guidance) without including long-range signaling. Simulations are shown which exhibit the development of vertical stripes that are found in axolotl embryos (cp. fig. 1.6). However, the cellular automaton model can be modified to simulate other cellular pigment patterns (e.g., horizontal stripes) that arise in salamanders and fish.

Cellular automata can also be used to simulate the implications of long-range signaling, in particular chemotaxis. Chemotaxis is an example of long-range cellular interaction mediated by diffusible signal molecules. In ch. 10 it is demonstrated how *chemotaxis* may be modeled with cellular automata. Furthermore, we present a model for *contact inhibition*. By combining of model modules (e.g., chemotaxis, contact inhibition, adhesion, etc.) simple models of tissue and tumor growth can be constructed (see ch. 10).

The study of Turing systems and excitable media has contributed enormously to a better qualitative and quantitative understanding of pattern formation. In ch. 11 we introduce and analyze cellular automaton models for Turing systems and excitable media based on microscopic interactions. The analysis allows us to characterize heterogeneous spatial (Turing) and spiral patterns, respectively, and sheds light on the influence of fluctuations and initial conditions.

Ch. 11 summarizes the results presented throughout the book. Cellular automata can be viewed as a discrete dynamical system, discrete in space, time and state space. In the final chapter, we critically discuss possibilities and limitations of the automaton approach in modeling various cell-biological applications, especially various types of morphogenetic motion and malignant pattern formation (tumor growth). Furthermore, perspectives on the future of the cellular automaton approach are presented.

The chapters of the book can be studied independently of each other, depending on the reader's specific interest. Note that all models introduced in the book are based on the cellular automaton notation defined in ch. 4. The model of pigment pattern formation presented in ch. 9 and the model of tumor growth in ch. 10 use model modules explained in earlier chapters (chs. 7 and 8). Readers interested in the principles of the mathematical modeling of spatio-temporal pattern formation should consult chs. 2 and 3. In addition, it is highly recommended to study the "Further Research Projects" sections at the end of the individual chapters (for additional problems, see also Casti 1989; Wolfram 1985). Readers requiring an introduction to a specific modeling problem should concentrate on elementary interactions, especially those derived in chs. 5 and 6. Ch. 9 can serve as a good introduction on how to construct a complex model by combining individual model modules.

The book has greatly benefitted from various inputs by many friends and colleagues. Special thanks go to E. Ben-Jacob (Tel-Aviv), L. Berec (Ceske Budejovice), U. Börner (Dresden), D. Bredemann (Dresden), L. Brusch (Dresden), H. Bussemaker (Amsterdam), D. Dormann (Dundee), H. Dullin (Loughborough), H.-H. Epperlein (Dresden), A. Focks (Osnabrück), S. Gault (Montreal), A. Greven (Erlangen), M. Kolev (Warsaw), M. Lachowicz (Warsaw), A. Lawniczak (Toronto), H.

Malchow (Osnabrück), C. Meinert (Mosebeck), M. Meyer-Hermann (Dresden), S. Müller (Magdeburg), L. Olsson (Jena), M. Schatz (Atlanta), M. Spielberg (Bonn), A. Voss-Böhme (Dresden), and M. Wurzel (Dresden) for stimulating discussions, valuable comments, providing computer simulations and/or figures, and/or critically reading draft chapters of the manuscript. Many thanks to Oscar Reinecke (Dresden), who implemented the "Cellular Automaton Simulator" (see www.biomodeling.info). We wish to thank the publisher Birkhäuser, in particular Tom Grasso, Elizabeth Loew, and Seth Barnes for very competent, professional, and ongoing support of the project. Finally, we are grateful to our colleagues at the Center for High Performance Computing (ZHR) at Dresden University of Technology and its head, Wolfgang E. Nagel, for support and excellent working conditions.

2

On the Origin of Patterns

"In the beginning God created the heavens and the earth. And the earth was without form and empty. And darkness was on the face of the deep. And the Spirit of God moved on the face of the waters. And God said, Let there be light. And there was light. And God saw the light that it was good. And God divided between the light and the darkness...

(from the Book of Genesis, Old Testament)

This chapter dwells on the origins of patterns in nature. The driving question is, What are the principles of pattern formation in the living and nonliving world? Possible answers are intimately tied to the particular conception of the world and the underlying space-time concept. Mathematics allows us to formalize and analyze space-time concepts and has even triggered new concepts. In the first part of the chapter, a historical excursion highlights static and dynamic space-time and corresponding mathematical concepts. Pattern forming principles in biology are introduced in the second part.

2.1 Space, Time, and the Mathematics of Pattern Formation: A Brief Historical Excursion

2.1.1 Greek Antiquity: Static and Dynamic World Conceptions

Order Out of Chaos. Chaos is the formless, yawning (Gr.: *chainein*) void (emptiness) from which the ordered world, the cosmos, originated. How can the development from chaos to order be explained? This is one of the prominent questions of Greek philosophy. Philosophy (love of wisdom) starts as one begins to ask questions. The formulation of new (and critical) questions has definitely catalyzed the evolution of human culture from its very beginning. However, the majority of philosophical concepts can only be traced back to their origins in Greek antiquity; earlier evidence is rather limited.

What was the Greek conception of space and time? The particular Greek way of questioning and reasoning is captured in the word *logos*, which denotes *word*, *measure* but also *understanding* and *proof.* Each of these meanings represents an important Greek contribution to the *mode of scientific inquiry.* Thereby, *word* is a synonym for philosophical discourse and *measure* for scientific progress, while *understanding* and *proof* open the world to *ratio.* The specific combination of *discursive* and *measuring* interpretation characterizes the differences from Eastern philosophies, which predominantly focus on discursive (meditating) inquiry into *world processes* (Günther 1994).

Early mathematics arose out of the need for improvement of measurements and constructions. Although the Egyptians and Babylonians already knew complicated formulas for the solution of architectural problems (constructive mathematics), they didn't wonder about their *universal validity.* **Thales of Miletus** (approx. 624–544 B.C.) marks the transition from constructive to formal mathematics. Thales, the founder of Ionian natural philosophy and thereby antique philosophy, not only felt a need to prove universal validity but also showed that a proof is possible; the theorem of Thales[1] named after him is one example. Thales defended a *dynamic world conception*: according to him, all being evolved from water.

The Idea of the Unlimited. This (dynamic) conception was strongly criticized by **Anaximander** (610–546 B.C.), who asserted that the ancestor substance of all being should not consist of one of its forms (e.g., water). For Anaximander the primary cause of all being is the unlimited (eternal) undeterminable. Interestingly, he claimed that humans evolved from fish and tried to prove this by *reductio ad absurdum*: if humans had always been as at present, in particular with a long childhood period necessitating care and protection, they would not have been able to survive. Consequently, Anaximander argued, humans must have evolved from an animal that became self-sufficient earlier in life. From his observations of sharks feeding their juveniles, Anaximander concluded that the shark should be assumed to be this ancestor species.

Empedocles (490–430 B.C.) explained formation based on an *antagonistic principle*, namely mixing and demixing of just four *eternal substances*: fire, air, water, and earth. Thereby, union and separation are caused by love and hate, which were assumed to have a substantial (material) character (the idea of *field influences* was unknown at that time).

A World of Numbers. Clarifying the structure of numbers and their interactions has been an important driving force in the development of mathematics. **Pythagoras** (approx. 570/560–480 B.C.) is the founder of mystical number concepts (and the Pythagorean school): the non-substance-like *number* should be the basic principle of all being. Pythagoras believed that in order to understand the world, we just have to extract the underlying numbers. With regards to biological patterns, there is an interesting relationship between a well-known number series, the Fibonacci series,

[1] The diameter divides a circle into two equally large parts.

Figure 2.1. Fibonacci spirals in the sunflower. One can count left- and right-turning spirals. Their numbers (here: 34 and 55) are subsequent elements of the Fibonacci sequence.

and phyllotaxis (the sequence of flowers, leaves, or petals along an axis[2]). Certain pairs of adjacent Fibonacci numbers can be identified by appropriate description of phyllotactic plant patterns (Strassburger 1978; see fig. 2.1). Meanwhile, the origin of the phyllotactic Fibonacci series can be explained by means of mathematical models (Richter and Dullin 1994).

According to the Pythagoreans there is a certain relationship between space and numbers since numbers are entities that can be viewed as points with a spatial dimension. In particular, *rational numbers* may always be expressed as integer multiples of an appropriately chosen entity that is a rational number itself. Pythagoreans claimed that all numbers should be rational. But the Pythagorean theorem

$$a^2 + b^2 = c^2,$$

where a and b are the short sides and c the long side of a rectangular triangle, implies the existence of the (irrational) number $\sqrt{2}$ since there exists a corresponding rectangular triangle with $a = b = 1$, i.e., $c^2 = 1^2 + 1^2 = 2$, yielding $c = \sqrt{2}$. This was the first time that a mathematical abstraction, finally, implied that a prejudice was abandoned. It is reported that the discoverer of the *irrational number* $\sqrt{2}$ was thrown into the sea because he had seriously threatened the belief in harmony based on ratios of integer numbers.

Pythagoras' theory of ideas is based on the *dichotomy* of the infinite *apeiron* ideas that are spiritually perceivable and the transient *peras* ideas that are sensually tangible. This conception not only marks a division of the world but also induces a nucleus of asymmetry, since only the spiritually perceivable is of infinite validity. On one hand, Pythagoras introduced a unifying concept (in his opinion), the numbers,

[2] In the Fibonacci series, each number is the sum of its two predecessors, i.e., 1, 1, 2, 3, 5, 8, 13, 21, 34, 55 ... The series' name goes back to Leonardo di Pisa, known as Fibonacci, who tried to describe the population dynamics of rabbits. Under certain simplifying assumptions the numbers of the Fibonacci series appear as the numbers of rabbit pairs per generation.

but on the other hand, he is also the father of dichotomous (biased) world conceptions that try to split up *the one world* (Russell 1993). Pythagoras's theory was further developed by **Plato** (428–349 B.C.) and **Socrates** (469–399 B.C.).

Periodic Dynamics. Heraclitus (approx. 540–480 B.C.) postulated an eternal circle of natural phenomena (e.g., day, night, summer, winter), which is behind the idea of a periodic dynamic world (*panta rhei*: everything is in flow). According to Heraclitus, all contrasts will finally be resolved at a higher level, the "world *logos*" (final understanding).

Static Cosmos and Optimality. It is noteworthy that static world conceptions appeared rather late in Greek antiquity. One of the early protagonists of a static cosmos was **Parmenides of Elea** (approx. 510–440 B.C.): The unit of all being is an *eternal substance*, of which there is neither creation nor disappearance. Only thinking can uncover this eternal substance, while sensual perceptions just communicate opinions, a view very much reminiscent of Pythagoras's dichotomous ideas. Parmenides is the founder of material monism: for him the world is a fixed, materially uniform sphere without *time, motion*, or *change*. *Formative ideas* are impossible within this static, materialistic concept. One can speculate why Parmenides' world has spherical shape, most probably because the sphere is regarded as optimal, i.e., of perfect form. This optimality can be specified in mathematical terms: out of all three-dimensional shapes of a given volume, the sphere has the smallest surface area. This is an early indication of *optimality principles* which were later used to explain e.g., the shapes of single cells (Thompson 1917).

Finite and Infinite Space. Zeno (490–430 B.C.) was a student of Parmenides and is known for the paradox of Achilles and the turtle: According to Zeno's reasoning, in a competition with the turtle Achilles will never reach the turtle as long as the turtle starts with a little advantage. Then, the turtle will have moved a tiny distance while Achilles is running to her starting point. And while Achilles is running to the turtle's new position she will again have a small advantage, and so on. Every time Achilles reaches the next position the turtle will have moved on a little, implying that Achilles will approach the turtle but will never reach her. This dilemma arises if a *Pythagorean* finite space (a line consisting of discrete points) is assumed—an example of unjustified parametrization. Indeed, the turtle is ahead at the infinite sequence of times $t_1, t_2, \ldots$ at which Achilles has made it to the point where the turtle has been a moment ago. So what? There is no reason to restrict discussion to this (nevertheless converging) temporal sequence. Fig. 2.2 shows how the solution can be determined as the limit of an infinite sequence (the time steps) (Davis and Hersh 1981).

Achilles' dilemma can be resolved by allowing for continuity of space and time. The tension between finiteness and infinity creates one of the driving forces behind mathematical development. Note that with respect to the mathematical modeling of populations (of biological cells) it is essential to choose an appropriate mathematical structure. While for large populations a density description (implicitly assuming

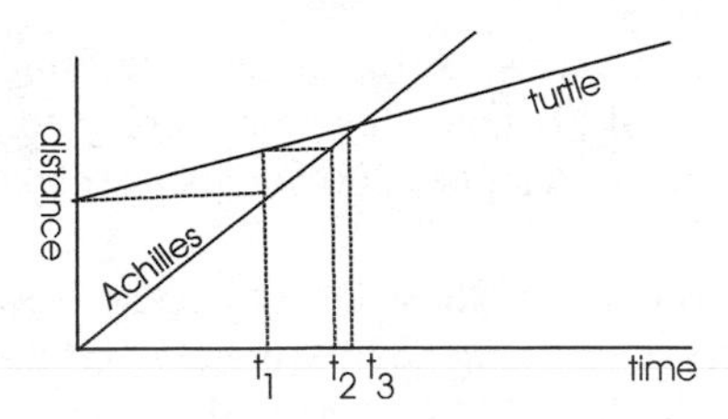

Figure 2.2. The paradox of Achilles and the turtle as an example of unjustified parametrization (for explanation, see text).

infinite populations or infinite space) is often appropriate, for small populations a discrete model has to be chosen. Cellular automaton approaches to be presented in this book are inherently discrete models, discrete in space and time. However, an approximative macroscopic continuum description can be derived from the discrete microscopic dynamics (cp. ch. 4, especially sec. 4.5).

Breakthrough of Geometry. Plato was a student of Socrates. According to Plato, the world is grounded on geometry. The harmony expressed in the Platonian bodies is manifested in all elements since they are composed of these bodies. This *geometric view* strongly influenced **Descartes** (1596–1650) and **Einstein** (1879–1955). *Time* is completely disregarded in the Platonian world since creation (formation) is only considered real insofar as it takes place in the ideas, i.e., eternal laws. Numbers lose their spatial correspondence: they are merely abstract ideas, implying that an addition of numbers is impossible. This is a rather modern concept reminiscent of ideas formulated by **Frege** (1848–1925), **Russell** (1872–1970), or **Whitehead** (1861–1974) (Davis and Hersh 1981).

Plato's theory of ideas introduced an additional dichotomy. Again, there is the (Pythagorean) dichotomy of finite *peras* (the Platonian *one*) and the infinite *apeiron*. But the latter undeterminable (Plato's *dyad*) is further subdivided into the good male principle *logos* and the evil female principle *eros*. Particularly, the Platonian good male/bad female dichotomy had longlasting and devastating effects on the evolution of human culture.

Ptolemy (approx. 170–100 B.C.) and **Apollonius of Alexandria** (262–190 B.C.) were strongly influenced by Plato. Ptolemy was an astronomer in Alexandria and is known for his geocentric representation of planetary orbits by means of epicycles. The Ptolemaic world conception survived until **Nicolaus Copernicus** (1473–1543). Cycles (as spheres) can be viewed as optimal, perfect forms, since among all planar figures with given area, cycles have the shortest perimeter. Thus, an *optimality principle* is applied to explain planetary orbits.

Apollonius reinforced geometry and demonstrated that rotation surfaces can be viewed as specific conic sections. Thereby, he could deduce geometrical structures (straight line, parabola, ellipse, hyperbola, and circle) as special cases of just one common structure, the conic section. This is an example of how an appropriately chosen mathematical structure allows one to unravel relationships between apparently unrelated phenomena (e.g., patterns) and to suggest unifying explanations. Apollonius also introduced so-called trochoids (cycloids, epicycloids, and hypocycloids), which were used as cell shape analogues 2000 years later by Thompson (Thompson 1917) (cp. fig. 2.12).

Euclid (approx. 300 B.C.) completed Greek geometry with his *Elements* comprising 13 books (chapters) containing approximately 500 theorems and proofs. For the first time, the success of the axiomatic method was convincingly demonstrated. Euclid implicitly assumed the "parallel axiom." Euclid's geometrical conception of the three-dimensional space would dominate until the end of the 19th century, when new spatial concepts (in particular those neglecting the parallel axiom) were introduced.

Logos and Logic. Aristotle (384–322 B.C.) substantially contributed to logic, the mathematics of proof. The importance of Aristotle for the development of logic is comparable to Euclid's influence on geometry. The basic motivation for Aristotle was that the *why* is more important than the *how*. Aristotle developed a "theory of causality" distinguishing between material and formal aspects of cause. In particular, the formal aspect comprises effective and finalistic causes, the latter denoted by motivation, teleology or *entelechy* (Mayr 1982). One can interpret scientific progress as resulting from the attempt to replace finalistic with effective causes. With respect to morphogenesis this implies that phenomena first explained by concepts of preformation can be deduced to regulation, i.e., *de-novo* synthetical (epigenetic) concepts, marking a shift from finalistic to effective arguments (cp. Aristotle's parabola of the house described in sec. 3.2). Aristotle had a very modern process-oriented concept of time: as much as a thing demands some space, it requires corresponding time. In addition, Aristotle postulated an intimate correspondence between space and time (Russell 1993).

What Are Principles of Pattern Formation in Greek antiquity? The transition from constructive to formal mathematics indicated the curiosity of Greek thinkers about questions of *general importance*. Early mathematical developments included concepts such as abstraction, numbers, axiomatic organization, logic, infinity, approximation, and proof. Many of these ideas were really new; they simply had no predecessors (at least none that we know of). Static and dynamic conceptions of the world were present but mathematics still focused on the description of a static world (Euclidean geometry).

Time is generally considered to be of little importance since dichotomous conceptions (Pythagoras, Plato) attributed a higher value to phenomena of timeless, eternal validity. If there are *forming principles*, no clear distinction is possible between evolutionary and ontogenetic formation: all views that can be put in the context of the history of evolutionary thought may be placed equally well in the context of the history of ontogenetic concepts. World and beings are generally believed to exist and to change in harmony due to God's influence and his invisible hands. Formation in the Greek sense predominantly meant transformation or redistribution along the lines spanned by *optimal (perfect) forms* as Platonian bodies, circles, or spheres.

2.1.2 The Prescientific Age

The influence of Greek antiquity accompanied the rise of "revelation religions" such as Judaism, Christianity, and Islam. Plato's influence is, for example, apparent in the work of **Augustine** (354–430). For Augustine God is the creator of a world comprising substance and (universal) order. God has an architect role outside the world. This view clearly differs from a pantheistic identification of God and the world. For Augustine God is subject to neither causality nor historic development. In creating the world, God created time. Questions refering to pre-creational events are therefore meaningless (Küng 1996). According to Augustine, the present is the only reality; the past just survives as a present memory, while the future can be regarded as present expectations. Therewith, Augustine attributed to time a subjective character. He tried to comprehend the subjective character of time as part of the spiritual experience of the created human (Russell 1993). This is in some sense an anticipation of Cartesian doctrines according to which the only reality that cannot be doubted is the reality of thinking (*cogito, ergo sum*). Fifteen hundred years later **Immanuel Kant** (1724–1804) would propose the subjective perception of time as a particular type of knowledge.

It was **Thomas Aquinas** (1225–1274) who built on the Aristotelian (realist) foundation by demanding a duality in the sphere of knowledge. Rational knowledge stems from sensual perceptions while revealed knowledge is based on belief (in God), e.g., in the triune God, resurrection, and the Christian eschatology. **Albertus Magnus** (approx. 1193–1280) was probably Aquinas' most influential teacher. Albertus had written an encyclopedia on Aristotelian thinking (Weisheipl 1980). This was revolutionary at that time, since Aristotle did not accept creation but was convinced of an eternal world, which strongly opposed the leading opinion of the church (in 1264 Pope Urban IV had forbidden the study of Aristotle). Albertus proposed a theory of plant form based on celestial influences (Magnus 1867). In particular, the semispherical shape of tree tops should be due to the effects of solar heat, which acts from all directions, inducing the boiling and extension of a postulated "plant fluid." This would lead to a subsequent spread of branches into all directions (Balss 1947). While plant shape is due to God (celestial agents), in Magnus's thinking, the actual emergence is delegated to physico-chemical processes (e.g., heating). This can well be called an early "theory of plant morphogenesis" motivated by the search for effective causes of pattern formation. Albertus Magnus also contributed with a description of leaf forms and tried to explain their differences physically by different consistencies of the "plant fluid" (Magnus 1867). This is a clear indication of a *physical modeling approach* to a morphological problem.

2.1.3 The Deterministic World of Classical Mechanics

The Analytical Method. Wilhelm von Ockham (1295–1349) marked the transition to a new *scientific thinking*. As a radical, antimetaphysical empiricist Ockham's name is tied to the following principle (Ockham's razor): it is useless to do something with more that could be done with less. This introduces an optimization principle by referring to a category of maximum (practical) expediency. The Franciscan friar Ockham also claimed that God should not necessarily be the primary cause of all being. God's unity and infiniteness are not provable.

René Descartes (1596–1650) is not only the founder of modern philosophy, but also the founder of the analytical method, which initiated and strongly influenced the Western mode of scientific inquiry. In his *Discourse on Method* (1637) Descartes described this (empirical) method: Each problem should be divided into precisely as many parts as are necessary for its solution. Thoughts should follow a bottom-up order from the simplest to the most intricate. If there is no order, it should be created. In addition, every conclusion should be tested thoroughly. Descartes himself provided examples for the success of the analytical method. He could, e.g., describe features of a large number of different curves by means of simple (algebraic) equations. This marked the birth of analytical geometry, which can be interpreted as a scientific crossover, namely the application of algebraic methods in geometry.

According to Descartes's natural philosophy physical phenomena may be explained as a consequence of conservation and collision laws (not force!). His concepts of space and time are explained in his *Meditations on First Philosophy* (1641). Matter can only be perceived if it possesses some spatial dimension and therefore geometrical shape (Plato) together with a possibility of motion. Matter may be divided infinitely often, a view that leaves no space for (indivisible) atoms. Since there is no distinction between space and matter, there is no vacuum. Consequently, because motion cannot leave empty space (vacuum), all motion in the universe should proceed along circular lines (optimal form!). Descartes claimed that his "vortex theory" would also explain the formation of the universe and planetary motion. The universe, conceived as a mechanical system, received its first impetus from God, who shaped the world like a watchmaker (Russell 1993).

Once established, the analytical method triggered new key questions and helped to deduce common origins of apparently unrelated phenomena. A further important crossover is tied to **Nicolaus Copernicus** (1473–1543). Copernicus realized the identity of terrestrial and celestial matter. Furthermore, the Copernican heliocentric system would replace the geocentric (Ptolemaic) system. In *De revolutionibus orbium coelestium* (*On the Revolutions of the Heavenly Spheres*, 1543) Copernicus described a world in which planets move along circular orbits around the sun. **Johannes Kepler** (1571–1630) helped the Copernican system to a breakthrough. The *Kepler laws* formulate empirical observations. In particular, Kepler discovered that planetary orbits do not have optimal, circular form, but exhibit ellipses, with the sun in one of their focal points.

Reversibility, Determinism, and Transformation. Issac Newton (1643–1727) could explain Kepler's empirical laws with the help of the laws of gravitation and inertia, thereby implicitly applying a *force concept* that was vaguely present already in Galileo's and Kepler's work. In his *Philosophiae naturalis principia mathematica* (*Mathematical Principles of the Philosophy of Nature*, 1687) Newton formulated an axiomatically grounded *dynamic geometry* as a deterministic model of motion based on differential and integral calculus. The axiomatic method does not try to explain the origin of gravity, its existence is just postulated as an axiom, Newton's universal law of gravitation: Two bodies attract each other with a force that is proportional to the bodies' masses and reciprocally quadratically proportional to the bodies' distance. After Euclid had created an axiomatic "geometric foundation" of a static world (see above), Newton's work demonstrated the success of the axiomatic method in a dynamic conception of the world.

Newton's model is fully reversible: his concept of space assumes an infinitely large box provided with an absolute coordinate system. Time is also absolute: Newton postulated a global reference time, a clock. Specification of initial conditions allows us to predict not only any future configuration but also historical states of the world. It has to be mentioned that *de novo* pattern formation is not possible in a Newtonian world since any pattern reflects the prepattern imposed by the initial conditions; initial and boundary conditions completely determine the system's fate. Newtonian dynamics merely describes a *transformation* of states without the potential to create any new quality. Later, **D'Arcy Thompson** (1860–1948) would contribute to the theory of biological form by describing various organismic forms as transformations by means of affine mappings (cp. sec. 2.2.6).

Newton's work triggered a competition between force and nonforce concepts in physics (cp. Descartes' "vortex theory"). A final decision had to wait until the 18th century, when precise measurements of polar and equator regions were performed (1736/1737). These measurements demonstrated the flattening of the poles, which supported Newtonian physics, in particular the effect of gravity forces.

Metaphysical Optimization Principles. Differential and integral calculus were independently introduced by **Gottfried Wilhelm Leibniz** (1646–1716) (*Acta eruditorum*, Engl. *Philosophical transactions,* 1684). Leibniz's symbols are, thereby, closer to the forms used nowadays. About the same time that Newton and Leibniz discovered the *infinitesimal calculus*, the light microscope was invented by **Antoni van Leeuwenhoek** (1632–1723) which permitted the investigation of "infinitesimally small" organisms (fig. 2.3).

For Leibniz the world is the best of all possible worlds. This (metaphysical) optimization principle assumes *prestabilized harmony*. God has created all being in a way that appears as a mirror of the harmonic whole. The validity of the concept is expressed in the purposefulness of the world (Russell 1993).

Leibniz is the creator of *monad theory* (1714), according to which space and time are merely sensual (subjective) perceptions, not realities. The universum is a composition of monads that are spiritual, spaceless force entities whose essence is perception and endeavor. God is the primary monad; all other monads are just *emissions*. Every monad circles around itself, but mirrors the whole universe (an example of self-similarity).

Leibniz is also the founder of the dyadic, the 0-1 system. For Leibniz the 0-1 transition is a model of *creatio ex nihilo* ($0 \rightarrow 1$) with the help of God, demonstrating that Leibniz was aware of the problem of *formation*. Leibniz started his genealogy of the Welfen family by focusing on the evolution of earth, lands, mountains, and minerals through the forces of fire and water. It is one of Leibniz's merits to realize the tempo-

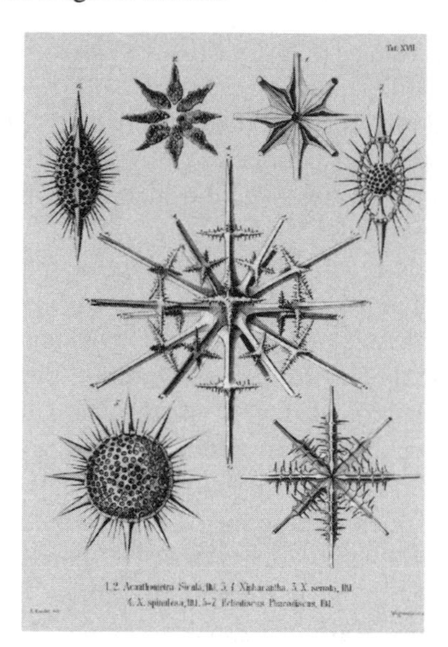

Figure 2.3. Various species of *Radiolaria* (a type of marine unicellular protozoa, approximate size in the order of μm; drawing by E. Haeckel). Courtesy of Ernst-Haeckel-Haus, Jena.

ral development of geological processes (cp. C. Lyell, 1797–1875). Immanuel Kant would later assign a temporal dimension to astronomical systems, while Darwin's theory of evolution can be characterized as the appreciation of a temporal dimension in biological systems.

Furthermore, Leibniz proposed a dynamic organization principle, the "principle of the least action" to account for temporal changes: if there occurs any change in nature, the amount of action necessary for this change must be as small as possible (Hildebrandt and Tromba 1996). Regarding the principle of the least action, there was a priority dispute with **Louis Moreau de Maupertuis** (1698–1759), a participant of the great Lapland expedition whose scientific results undoubtedly supported Newton's ideas with respect to the effect of gravitational forces (see above on the competition between Cartesian "vortex theory" and Newtonian force-based physics). The principle of the least action marks another example of crossover discoveries, namely application of a moral concept, parsimony, born of the Calvinist tradition, to physics.

Leonhard Euler (1707–1783) is the known author of more than a thousand publications. In *Theoria motus* (*Theory of the Motions of Rigid Bodies*, 1765) Euler focused on dynamics, the relationship between force and motion. Without knowing Maupertuis' work, Euler could prove that the principle of the least action is sufficient to explain the motion of a mass point in a conservative force field,[3] planetary motion around the sun being an example. Note that the principle of the least action is an optimality principle. Euler argued that all phenomenona in the universe were due to a maximum or minimum rule and based his argument on the metaphysical plausi-

[3] In conservative systems, there exists a potential energy U, such that the force F can be expressed as $F = -\text{grad}\, U$.

bility of God-given harmony: because the shape of the universe is perfect, created by the wisest God, only optimality principles should account for his creations. Euler laid the foundations of variational calculus, and the theory of deducing dynamics from optimality principles, by publishing the first textbook on the matter, *Methodus inveniendi lineas curvas maximi minimive proprietate gaudentes* (*On a Method to Find Curved Lines Fulfilling a Maximum Condition*, 1743). Euler replaced effective causes, inertia, and gravitation by a finalistic (metaphysical) cause, the optimality principle. This indicates a reverse direction of inquiry, which is rather atypical in the history of science. Typically, scientific research advances by replacing finalistic with effective causes (in the Aristotelian sense).

Equilibrium Systems and Minimal Surfaces. Johann Bernoulli (1661–1748), a teacher of Euler, characterized the *equilibrium state* by means of an extremum principle, namely his *principle of virtual work*: if a mechanical system is in equilibrium, no work is necessary to, infinitesimally, displace it. By this, a stationary state is traced back to a dynamic principle. The principle of the least action and the principle of virtual work are both optimization principles; one governs nonstationary processes, the other one rules the states at rest (Hildebrandt and Tromba 1996).

Carl Friedrich Gauss (1777–1855) focused on Bernoulli's principle of virtual work in his *Principia generalia theoriae fluidorum in statu equilibri,* Engl. *General Principles of Fluids in a State of Equilibrium,* (1830). In particular, Gauss showed that this principle allows one to interpret thin soap films spanned in a wire frame (in equilibrium) as surfaces which assume a minimum area among all virtual positions or are at least stationary states of the area (Hildebrandt and Tromba 1996). The *theory of minimal surfaces* inspired Thompson's work on biological shapes (cp. sec. 2.2.6). **Joseph Louis Lagrange** (1736–1812) proposed a minimal surface equation and made important contributions to variational calculus by freeing time from its particularity, for Lagrange time is just an additional dimension. Seemingly different phenomena, space and time are traced back to a common origin. Lagrange's space-time concept is one of the roots of relativity theory.

Relativistic World Concepts. Further developments were necessary to allow for relativistic interpretations of the world. According to the *world postulate* of **Hermann Minkowski** (1864–1909), substance in any *world point* can always be interpreted as resting after appropriate definition and scaling of space and time. Therewith, a highly dynamic situation is traced back to a static situation in which Einstein's electrodynamic equations replace Newton's axioms. It was **David Hilbert** (1862–1943) who showed that these equations also follow an *effective optimality principle* that can be interpreted as a geometric variational principle in Minkowski's four-dimensional world. Newton's force concept is substituted by local curvatures in the relativistic space-time world. Historically, it is interesting that this conception appears as a renaissance of Descartes's natural philosophy, trying to explain all natural phenomena by means of conservation and collision laws.

Furthermore, time has lost its total and global character: in his special theory of relativity (1905) **Albert Einstein** introduced a local, still reversible time. According to the special theory of relativity there is no physical process that allows one to determine if a system is in absolute rest or proceeding linearly and homogeneously. The space-time continuum is a field (of curvatures and matter/energy) consisting of heavy and inert masses. Curvature of space depends on mass density. The conclusions are that energy (E) and mass (m) are equivalent ($E = m \cdot c^2$), and acceleration and the effect of the gravity field are equivalent. The velocity of light, $c = 300{,}000$ km/sec, is the maximum possible. Euclidean geometry is still a good approximation for short distances (spatial and temporal), in particular for the space-time scales effective in biological morphogenesis.

Pattern Formation in a Deterministic World. The world conception of classical mechanics, acknowledging the existence of temporal dynamics, is strictly deterministic. Even in its most sophisticated relativistic interpretations no randomness or stochasticity is encountered. Pattern formation is envisaged as *transformation* and proceeds according to *optimality principles*, partly based on metaphysical reasoning, finding their mathematical formulation in variational calculus. Time plays no particular role, at least there is no distinguished direction of time that is assigned a reversible character. On the other hand, the historic development of life and earth was apparent by the end of the 19th century. It is the merit of thermodynamics to acknowledge the importance and the direction of time.

2.1.4 Discovering the History of Time

Diffusion and Irreversibility. Pierre Simon de Laplace (1749–1827), on the one hand, completed Newton's deterministic celestial mechanics and, on the other hand, demonstrated the need for probabilistic concepts. Laplace convincingly showed that probability concepts are mathematically tractable. Another route to dynamic phenomena initiated by **Jean-Baptiste Joseph Fourier** (1768–1830) is to abstract from the exact position and velocity information of particles or molecules to a macroscopic description, e.g., by means of pressure or temperature. In 1811, Fourier won the prize of the French Academy of Sciences for his discovery that a linear is able to describe heat flow in such different media as solid matter, fluids, and gases. Fourier proposed an analytical theory of heat flow in his *Théorie analytique de la chaleur* (*Analytical Theory of Heat*, 1822) and became one of the founders of theoretical physics. In the *heat equation*[4]

$$\frac{\partial}{\partial t} u(x,t) = D \frac{\partial}{\partial x^2} u(x,t),$$

with temperature u, heat flow J is assumed to be proportional to the gradient of temperature, i.e., $J = -D\frac{\partial u}{\partial x}$. Describing a dynamic phenomenon through such an equation is very different from the classical mechanical approach in several aspects. First, heat energy is a property of a large number of particles that are not followed

[4] The equation is formulated for a one-dimensional system with space coordinate x and time t.

individually. Instead, a macroscopic characterization, temperature, is used. Furthermore, Fourier realized the importance of initial and boundary conditions. Starting from a heterogeneous initial distribution and zero-flux boundary conditions, diffusion dynamics always leads to a homogeneous distribution. Accordingly, the heat (diffusion) equation is a manifestation of irreversibility.

Conservation and Conversion. The diffusion equation is a conservation equation, stating the conservation of mass. **James Prescott Joule** (1818–1889) introduced the concept of *conversion*, thereby realizing the close relationships among chemistry, the science of heat, electricity, and magnetism. "Something" can be quantitatively conserved and, simultaneously, qualitatively transformed. This "something" was later called "the equivalent." Joule introduced the concept of energy conservation. It is a generalization of (friction-free) mechanical motion whereby total energy is conserved while constantly transformed between kinetic and potential energy. Joule appeals to religion by relegating the conservation concept to the sovereign will of God.[5]

Entropy as a Measure of Irreversibility. In order to distinguish conservation and reversibility, **Rudolf Clausius** (1822–1888) introduced the entropy concept. While mechanical transformation always implies reversibility and conservation, in physicochemical transformations (with typically many more degrees of freedom) it is possible that conservation coincides with irreversibility. For example, heat conduction (described as a diffusion process by Fourier) leads to an (irreversible) heat balance (while conserving particle number). Entropy can be regarded as energy that is irreversibly converted into thermal fluctuations. Clausius formulated the *second law of thermodynamics*: the energy of the world is lost while the entropy approaches a maximum. In other words: the most likely states have largest entropy for isolated systems.[6]

Probability Explains Irreversibility. By the middle of the 19th century, the existence of irreversible processes could no longer be ignored. Consequently, the conception of a static world had to be abandoned. In geology, **Charles Lyell** (1797–1875) described an earth in flux; i.e., he gave the earth a historical dimension, while Charles Darwin's selection theory provided an explanation for the (irreversible) history of life.

It was **Ludwig Boltzmann** (1844–1906) who was able to give a probabilistic explanation for the irreversible entropy increase in isolated systems. Boltzmann in-

[5] "Indeed the phenomena of nature, whether mechanical, chemical, or vital, consist almost entirely in a continual conversion of attraction through space, living force (N.B. kinetic energy) and heat into one another. Thus it is that order is maintained in the universe—nothing is deranged, nothing ever lost, but the entire machinery, complicated as it is, works smoothly and harmoniously. . . —the whole being governed by the sovereign will of God" (Joule 1884, cited in Prigogine and Stengers 1984).

[6] Thermodynamical systems are classified relative to their environment as open (exchange of matter and energy), closed (only energy exchange), adiabatic (energy exchange but not as heat), and isolated (no exchange).

troduced probability into physics as an explanatory principle and could show that the attractor state is the macroscopic state corresponding to the microscopic configuration occuring with the largest probability. This was the first time that a physical concept was explained probabilistically.

The essential bridge between microscopic configuration and macroscopic state is Boltzmann's definition of *entropy*:

$$S = k \cdot \ln P,$$

where S is entropy, P is the number of complexions, i.e., the number of "possible microscopic realizations," and k is Boltzmann's constant.[7] Hence, entropy is not a property of a microscopic state—it is a property of an **ensemble.**[8] To understand Boltzmann's idea let us consider the following thought experiment: n (distinguishable) balls shall be distributed within two boxes B_1 and B_2. How many possibilities P (complexions) are there, if one additionally demands that the number of balls in box B_1 be m $(\leq n)$? Obviously, the (combinatoric) answer is $P = \binom{n}{m}$. A number of mathematical observations can be made with regard to the value of $\binom{n}{m}$. First, for fixed n the maximum of P is achieved for $m \approx \frac{n}{2}$. Furthermore, for large n, differences between $\binom{n}{m_1}$, $\binom{n}{m_2}$, and $m_1 \neq m_2$, increase. Accordingly, the probability

$$P(m \text{ particles in } B_1, n - m \text{ particles in } B_2) = \frac{\binom{n}{m}}{\sum_{j=0}^{n} \binom{n}{j}}$$

is maximal for $m \approx n/2$. This led Boltzmann to the following conclusion: in systems consisting of a huge number of particles (e.g., $n \approx 10^{23}$) all states differing from the homogeneous distribution (i.e., $m = n - m \approx n/2$), corresponding to maximum disorder, are very improbable. This is the probabilistic explanation of the maximum entropy attractor studied in equilibrium thermodynamics.

For closed or isolated systems described by a Hamiltonian (energy function), the equilibrium phase space distribution is a Gibbs distribution. In particular, for closed systems (in contact with an energy source to maintain a fixed temperature T), the equilibrium distribution over all microstates is the Gibbs distribution for a canonical ensemble and is given by

$$\rho(q, p) = \frac{1}{Z(q, p)} e^{-\beta H(q,p)},$$

where

$$Z(q, p) = \int_q \int_p e^{-\beta H(q,p)} dp\, dq, \quad \beta = \frac{1}{kT}.$$

$q = (q_1, \ldots, q_n)$ and $p = (p_1, \ldots, p_n)$ are the space and the momentum coordinates of the particles, respectively; $H(q, p)$ is the Hamiltonian.

[7] Boltzmann's constant $k = 1.3807 \cdot 10^{-23}$ Joules/Kelvin.

[8] A large number of identically prepared systems is called an ensemble. Imagine a very large collection of systems evolving in time. A snapshot of the state of each of these systems at some instant in time forms the ensemble.

2.1.5 From Equilibrium to Self-Organizing Systems

Linear Nonequilibrium Thermodynamics. The idea of (thermodynamic) equilibrium was soon transferred to chemical systems leading to the *law of mass action*. However, chemical processes are generally irreducible. This was just one reason that interest in nonequilibrium systems increased. In 1931, **Lars Onsager** (1903–1976) presented a theory of linear thermodynamics. Onsager's reciprocity relations define the range of validity in which the linear approximation is reasonable (Onsager 1931). He could show that in this region the system evolves into a stationary state that corresponds to least entropy production for given boundary conditions (e.g., two points of the system with a fixed temperature difference). Note that here the stationary state is a nonequilibrium state in which dissipative processes take place. The entropy production can be viewed as a potential of the system. Existence of a potential again implies independence of initial conditions. Thus, the situation in linear thermodynamics is similar to equilibrium thermodynamics: any *de novo* pattern formation cannot be explained.

A theory of *de novo* pattern formation would first arise in the theory of nonlinear thermodynamics which postulates a new ordering principle: *order through fluctuations*. But for its breakthrough a number of prerequisites were necessary.

Modern Space-Time Concepts: Process Philosophies. While for Aristotle the *one whole* was dominant and his logical analysis of causes does not distinguish between present, past, and future, the Cartesian reductionist approach assumed a splitting of space and time. In particular, any cause of organization should merely be found in the past. Nevertheless, the end of the 19th century saw a renaissance of holistic approaches contrasting the evolution of a (technical) Cartesian world: **Immanuel Kant** (1727–1804) tried to unify rationalistic and empirical concepts. For him the starting point are the categories (Aristotle) that transform experience into insight. Kant distinguished between analytical and synthetical insight. *Noumena* are open to analytic insight that can be determined to be true by purely logical analysis. This truth is timeless and independent of experience. Thus, analytic insight is always *a priori* insight. In contrast, *phaenomena* (including the sciences) are perceivable by sensual impressions and can be (empirically) disproved by experience. Note that such synthetic insight can be *a priori*; space and time are important examples. Consequently, for Kant Euclidean geometry is a mere empirical expression of a prevailing (static) space concept, while arithmetics represents an intuitive (sequential) time concept. It is to Kant's credit that his work allowed consideration of alternative space and time concepts, e.g., Riemann geometry, one of the roots for Einstein's relativity theory (see above).

Georg Wilhelm Friedrich Hegel (1779–1831), the father of dialectics, denied dualism, the possibility of splitting the world. The whole is the only reality. There is *self-development* of the absolute spirit: every developmental stage is preliminary and removes its opposite to join with it on a higher level. Hegel was idealist and doubted that one could know anything about an object without knowing everything.

Figure 2.4. Symbol of interaction: mandala. Mandala is a Sanskrit term for circle, polygon, community, or connection. In particular, the mandala represents the "idea of interaction" visualized by a multitude of gates connecting the different regions of the pattern.

A further step into deobjectivating the world was taken by the anti-Hegelian **Søren Kierkegård** (1813–1855), one of the founders of existentialism. He distinguished between the realms of ratio and belief (cp. Thomas Aquinas). In accordance with romantics, there is no objective truth—all truth is subjective. Prior to ratio is the individual will. Will was evil for **Arthur Schopenhauer** (1788–1860). According to Schopenhauer insight into the void (*nirvana*) and therewith *the one world* is only possible by killing the will. This mystic understanding is reminiscent of Buddhist process-oriented concepts. **Siddharta Buddha** (approx. sixth century B.C.) described his enlightenment thus: "The reality of all beings is self-less, void, free and relative." (Thurman 1996) (see figs. 2.4 and 2.5).

Here, *self-less* means that beings do not exist independently of each other. In particular, they are free (*void*) of any essence or isolated substance. Relativity implies that there is no origin in the universe that is not relative. While for Schopenhauer the evil will was responsible for all suffering, according to Buddhism suffering is caused by misunderstanding the *selflessness of the self*. It is essential to free oneself from this misunderstanding in order to overcome suffering and to be able to real sympathy (interaction!).

Process thoughts were introduced into Western philosophy by **Alfred North Whitehead** (1861–1947) and **Martin Heidegger** (1889–1976). Heidegger was influenced by Kierkegård, focused on the relationship between "being" and "becom-

Figure 2.5. Letter of interaction: the chinese letter "JIAN" is symbolized by a sun enclosed by two gates and expresses any "between" (interaction), in space as well as in time. Drawing by C. Meinert, Mosebeck; used by permission.

ing" and hoped to go beyond the identification of being with *timelessness* (Prigogine and Stengers 1984). Interestingly, for Heidegger the *void* has a positive connotation. For Whitehead, space and time are derived concepts. The ultimate compounds of reality are events. In *Process and Reality* Whitehead described his metaphysics of organisms (Whitehead 1969). Whitehead claimed a *creative* evolution of nature. While elements are not permanent, processes are the underlying reality. The identity of the beings stems from their relations with other beings. This is a philosophy of *innovative becoming* that overcomes the subject–object (organizer–the organized) dichotomy. The nature of the related things must derive from these relations, while at the same time the relations must derive from the nature of the things. This is clearly an early indication of *self-organization* concepts. Note that the cellular automaton models introduced in this book focus on interactions. The precise (material) character of the interacting entities is of far less importance, if of any. The systems studied are defined by their interactions.

There is a similar understanding of God in process philosophies and process religions. Whitehead viewed God as an extensive continuum; for Spinoza God is the energy stream in evolution, while Buddhism knows the *shunyata*, imprecisely translated as the *emptiness*.

Henri Bergson (1859–1941) distinguished between physical and existential time. Contrary to physical time, existential time has a duration. Bergson broke with positivistic thinking to argue that scientific rationality is incapable of understanding duration since it reduces time to a sequence of instantaneous states linked by a deterministic law. In contrast, reality is creative development. In its duration (*durée*) the desire to live (*élan vital*) creates more and more *creatures*.[9]

While in the 19th century, the importance of time (history) had been generally accepted (particularly, in geology, thermodynamics, biology), the 20th century (re)introduces the unity of space and time, namely *process thinking*. This is just one of the prerequisites for the appreciation of new ordering principles of *de novo* pattern formation.

A *De Novo* Ordering Principle: Creative Instability. Boltzmann had proposed a probabilistic argument to explain equilibrium states in thermodynamic systems. However, Hamiltonian systems are not the only dynamical systems that are of interest with respect to spatio-temporal pattern formation. By the middle of the 20th century, a number of nonbiological nonequilibrium systems became known, for which the Boltzmann order principle is not applicable, e.g., turbulence, Bénard rolls (fig. 2.6), later the Belousov-Zhabotinskii reaction (Ross, Müller, and Vidal 1988; fig. 2.7). Those structures correspond to neither a maximum of complexions nor a minimum of free energy. The systems turned out as paradigmatic systems of *self-organization*.

It is a benefit of nonlinear nonequilibrium thermodynamics, synergetics, and the study of autopoietic systems to appreciate instability, randomness, fluctuations as important preconditions for self-organization (Haken 1977; Jantsch 1980; Nicolis and

[9] "Time is invention or it is nothing at all" (H. Bergson in *L'evolution créatrice*, Engl. *Creative Evolution*).

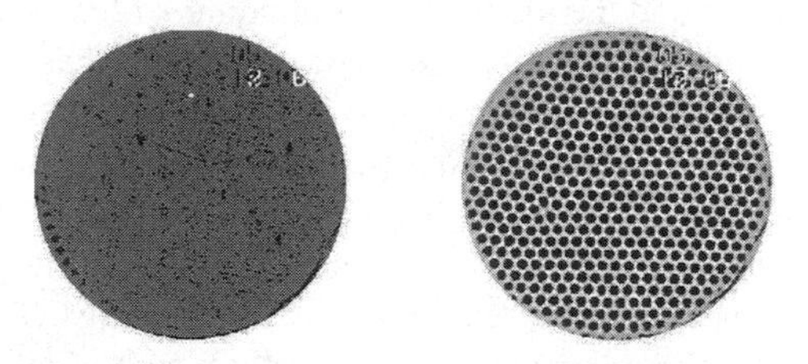

Figure 2.6. Bénard rolls (Raleigh-Bénard pattern formation) as an example of self-organization. A fluid container (filled with silicone oil) layer is heated from below. The hexagonal pattern (right) arises when a critical temperature difference between the top and the bottom is reached (Haken 1978a). Until recently, large discrepancies existed between theory and experiment regarding the onset of convection. The picture illustrates the abrupt onset for slowly increasing control parameter, which differs by less than 1 part in 10^3 between the two images (from Schatz et al. 1995, used with permission).

Prigogine 1977). In self-organized systems there is no dichotomy between the organizer and the organized. Such systems can be characterized by antagonistic competition between interaction and instability. No pattern forms if there is just interaction or mere instability in the system. Creativity results from interplay. Note that whenever there exists a potential, this can be viewed as an organizer since the system will eventually reach the corresponding attractor state. It is not clear if there exists a potential in self-organizing systems. Most probably the variation principle undergoes an evolution itself.

In particular, self-organization is typical of biological systems, pattern formation as ontogenetic morphogenesis and phylogenetic evolution being their most prominent manifestations. One of the morphogenetic biological model systems for studying self-organized pattern formation is the slime mold *Dictyostelium discoideum* (for a review, see Dallon et al. 1997). Meanwhile, many further model systems have been

Figure 2.7. Belousov-Zhabotinskii reaction. Concentric waves (target pattern) and a counter-rotating pair of spirals propagate through an excitable medium: An organic substrate, malonic acid, reacts with bromate in the presence of a redox-catalyst (e.g., ferroin) in sulphuric acid solution (an experimental recipe can be found in Deutsch 1994). Courtesy of S. C. Müller, Magdeburg.

analyzed and different instabilities (e.g., diffusive, chemotactic, Turing, and orientational) have been discovered (see sec. 2.2 and Table 2.1).

> *"THE METAMORPHOSIS OF PLANTS.*
> *...None resembleth another, yet all their forms have a likeness;*
> *Therefore, a mystical law is by the chorus proclaim'd; ... Closely observe how the plant, by little and little progressing,*
> *Step by step guided on, changeth to blossom and fruit! First from the seed it unravels itself, as soon as the silent*
> *Fruit-bearing womb of the earth kindly allows Its escape, And to the charms of the light, the holy, the ever-in-motion,*
> *Trusteth the delicate leaves, feebly beginning to shoot. Simply slumber'd the force in the seed; a germ of the future,*
> *Peacefully lock'd in itself, 'neath the integument lay, Leaf and root, and bud, still void of color, and shapeless; ..."*
>
> (Johann Wolfgang von Goethe)

2.2 Principles of Biological Pattern Formation

Depending on the underlying space-time concept, various organization principles can be distinguished, in particular preformation, optimization, and self-organization principles (cp. sec. 2.1). How can these principles contribute to an understanding of biological morphogenesis? In this section we will try to link organization principles to specific aspects of biological pattern formation.

Many developmental hypotheses appear rather absurd today, even if they survived for many centuries. One of these dates back to the Roman naturalist **Caius Plinius Secundus** (Pliny the Elder, 23–79). His "licking theory" is an attempt to explain the development of bears. According to Plinius, bears are born as meatballs that are subsequently licked into the appropriate form by the meatball's mother. Nevertheless, this licking can be viewed as a means of regulation of an initially homogeneous structure, the meatball, and therewith anticipates rather modern theories of embryonic regulation (fig. 2.8).

2.2.1 Preformation and Epigenesis

Development means "de-veiling," un-coiling, an indication that development, originally, was a preformistic concept since in the process of uncoiling nothing new can be created; structure is simply uncovered. Note that also the word *evolution* (Latin: *evolutio*: unrolling, from *evolvere*) bears a preformistic notion, namely expression of the preformed germ (Bowler 1975). *Preformed development* only allows for differential growth or uncoiling. Until Darwin's acknowledgment of historical evolution, i.e., a subsequent change of biological organisms in the course of evolution, the preformation hypothesis was the best conceivable rational conception of morphogenesis in accordance with the Christian understanding of a perfect static world.

Organization principle	**Morphogenetical competence**
Preformation	*Uncoiling* → Homunculus *Self-assembly: key-lock principle* → macromolecular structure, virus form (Waddington 1962) *Substitution: L-systems* → branching (Lindenmayer 1982) *Prepattern: positional information, gradient model* → differentiation (Wolpert 1981)
Optimization	*Minimal surfaces* → cell form (Thompson 1917) *Differential adhesion* (minimization of free energy) → sorting out, aggregation (Glazier and Graner 1993)
Topology	*Allometric transformation* → fish shapes (Thompson 1917) *Catastrophe theory* → organ forms etc. (Thom 1972)
Self-organization	*Turing inst.* (reaction-diffusive inst.) (Meinhardt 1982; Turing 1952) → phyllotaxis (Yotsumoto 1993) *Diffusive inst.* → diffusion-limited aggregation (Witten and Sander 1981) *Mechanical inst.* → plant patterns, e.g., whorls (Green 1996), blastulation, gastrulation (Drasdo and Forgacs 2000) *Mechano-geometrical inst.* → cell shape (Pelce and Sun 1993) *Mechano-chemical inst.* → segmentation (Oster, Murray, and Harris 1983; Oster, Murray, and Maini 1985) *Hydrodynamic-chemical inst.* → cell deformation, cell division, active motion (He and Dembo 1997; Lendowski 1997) *Orientational inst.* → swarming, collective motion (Börner, Deutsch, Reichenbach, and Bär 2002; Bussemaker, Deutsch, and Geigant 1997; Mogilner, Deutsch, and Cook 1997; Mogilner and Edelstein-Keshet 1995), cytoskeleton (Civelekoglu and Edelstein-Keshet 1994) *Excitable media* → *Dictyostelium discoideum* taxis, aggregation (Dallon et al. 1997)

Table 2.1. A classification of organization principles and selected examples of their morphogenetical competences.

Preformation implies that each generation contains the complete information to form all subsequent generations, since no information is created *de novo*. Consequently, if organisms are allowed to change to new varieties, preformation is only possible if the number of generations is limited. Ultimately, preformation is a theological concept since it must assume an arbitrary beginning, a creational act, introducing an obvious dichotomy of *creator* and the created (preformed) beings: Within each *animalcule* is a smaller animalcule, and within that a smaller one, and so on (emboitement principle) (Bard 1990). Thus, in the ovaries of Eve (or the testicles of Adam) should be the forerunner of every successive human. In a naive interpretation human sperm consists of a coiled *homunculus* that simply needs to uncoil in its mother like plant seeds in a flower bed (fig. 1.3).

For a long time experimental results seemed to "prove" preformistic ideas while, simultaneously, the idea of *spontaneous generation* was not supported. **Marcello Malpighi** (1628–1694) noticed that the outlines of chicken embryonic form were already present at the earliest stages of development that he could observe (after the egg had moved down the oviduct) (Bard 1990). **Jan Swammerdam** (1637–1680) observed that after hardening a chrysalis with alcohol, a perfectly formed butterfly formed within. Swammerdam deduced that the butterfly structure was present but masked within the caterpillar and hence within the egg.

Caspar Friedrich Wolff (1734–1794) experimentally disproved preformation and suggested a vitalistic *theory of epigenesis* based on a germ-inherent *vis essentialis*. Experimentally, Wolff demonstrated at the resolution of his microscope that blood vessels of the chick blastoderm were not present from the very beginning but emerged from islands of material surrounded by liquid (Wolff 1966). Furthermore, he discovered that the chick gut was initially not a tube, but formed by a folding of the ventral sheet of the embryo that was mere reminiscent of coiling than of uncoiling. Wolff applied his epigenetic theory to plant development and obtained similar results as J. W. von Goethe regarding the function of plant stalk and leaves (cp. the quotation at the beginning of the section). Note that Aristotle and **William Harvey** (1578–1657) had already regarded epigenetic concepts as important for embryogenesis (fig. 1.4).

Figure 2.8. Plinius' "licking theory," according to which a mother licks a cub into adult shape (Lat.: *Lambendo paulatim figurant*).

2.2.2 Ontogeny and Phylogeny

The 19th century was the century of Darwin and his selection theory proposing an explanation of biological evolution by means of natural selection: *The Origin of Species* appeared in 1859. What now needed to be explained was morphological flexibility and the ability for progressive change, including its twin aspects, ancestry and novelty. Darwin designed his *pangenesis theory* of heredity to connect the levels of ontology and phylogeny in order to explain the transmission of evolutionarily acquired changes into morphologies.

> *Through a series of generations, the results of adaptive variations acquired at one stage are transferred directly into the egg, which is the sole vehicle for transmission to the next generation: the mechanism is a Lamarckian inheritance of the effects of the environment (e.g., growth of a chicken neck in response to an acquired food source). The onus of explanation of pattern change was on the hereditary mechanism, while the embryologic events are non-contributory and, implicitly, preformative.* (quoted from Horder 1993)

Ernst Haeckel (1834–1919) aimed to relate ontogenesis, i.e., embryology, to phylogenesis. According to Haeckel's *biogenetic law*, developmental stages through which an embryo passes as it approaches the mature form are a reflection of adult evolution; in short, "ontogeny recapitulates phylogeny." For Haeckel, not an embryologist himself, the sole purpose of embryology lies in the verification of this (purely phenomenological) law. **Karl Ernst von Baer** (1792–1876) modified the law: developmental stages, through which the embryo of a higher animal passes as it matures, are a reflection of the embryos, but not the adult forms of lower animals (fig. 2.9). Thus, "ontogeny recapitulates embryonic phylogeny". Von Baer proposed the *theory of germ layers*:[10] The shape of all metazoa would result from the same two or three blastema regions pointing to a common phylogenetic origin.

The biogenetic law cannot explain any pattern formation; it is a mere (questionable) description. The embryologist **Edmund B. Wilson** (1856–1939) acknowledged as early as 1898 that

> *Development more often shows, not a definite record of the ancestral history, but a more or less vague and disconnected series of reminiscences. Today, it has been realized that embryos pass through a phylotypic stage when traits typical of a particular phylum are determined. For example, all vertebrate embryos at a certain stage—about four weeks in humans—have the same body plan, including a dorsal rod of cells called a notochord and a set of paired muscles. Yet, before this stage, embryos may look very different, and afterwards, their development takes them down a variety of paths to finned, feathered, or footed adults.* (quoted from Pennisi and Roush 1997)

August Weismann (1834–1914) proposed a separation of pathways for transmission of hereditary factors between generations (phylogeny) from the actual manifestations of the hereditary factors (ontogeny). The hereditary mechanism is identified

[10] Germs are examples of blastemas, i.e., embryonic, not yet differentiated tissues.

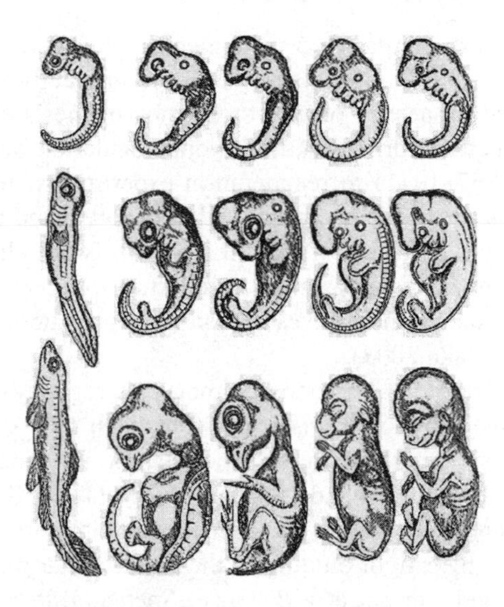

Figure 2.9. Biogenetic law. The embryonic development of five vertebrate species (from left to right: fish, turtle, chicken, rabbit, human) shows remarkable similarities in the early stages.

as nuclear and chromosomal. New morphologies are viewed as the result of new assortments (due to crossing over and biparental mixing) of independent hereditary factors acquired directly from the previous generation. Pattern expression depends entirely on the way the factors are transferred from the chromosomes: Weismann hypothesized patterned localization within the egg cytoplasm (*mosaic embryo*). Later embryonic events as such were ignored, implying a preformationist view of embryogenesis. By focusing on the biparental assortment of hereditary factors and removing the Lamarckian element (from Darwin's pangenesis theory), Weismann and the rapidly ensuing rediscovery of Mendel's laws opened the gateway directly into modern genetics.

2.2.3 On Organizers and Embryonic Regulation

Today, developmental theories comprise preforming as well as *epigenetic* concepts, the latter focusing on *de novo* synthesis of patterns. The crucial question is whether an embryo should be interpreted as *mosaic* or *regulative*. According to the mosaic conception embryonic (and adult) structure is directly determined by DNA-coded information laid down in the eggs. The information appears as a mosaic (spatial) distribution of form-giving substances, determinants of form, arranged along the egg's periphery according to the adult's shape.[11] *Genetic program* and the notion of locus-responsive control genes (pattern genes) are modern transcriptions of an old concept,

[11] *Preformation* should not be confused with *predetermination*, indicating a maternal influence before fertilization which definitely cannot be neglected.

preformation. In contrast, regulative development implies that structure is not fully specified in the DNA code; it arises later and more indirectly from changes in the properties of cells and tissues; in other words embryogenesis *per se* is assumed to play a crucial role in pattern formation. Embryonic regulation was first demonstrated by **Hans Driesch** (1867–1941) in regeneration experiments: removal of a part of the early embryo does not result in a partial adult (as predicted by Weismann) but a complete final morphology. It follows that the pattern is not already established in the egg—parts of the egg and early embryo are in fact *totipotent*—and that the later events of embryogenesis are themselves responsible for pattern formation (see also the recent review in Tanaka 2003).

Hans Spemann (1869–1941) proposed embryonic regulation in order to find an explanation for the phenomenon of *induction* (Spemann 1938). The term *induction* refers to any mechanism whereby one cell (population) influences the development of neighboring cells. Inducing signals fall into three classes. Some are attached to the cell surface and are available only to immediate neighboring cells. Others may be highly localized by their tight binding to the extracellular matrix, and yet others are freely diffusible and can act at a distance. Spemann discovered an *organizer* region in the blastula/early gastrula stage with long-lasting organizing effects in the sea urchin *Xenopus laevis*. In particular, this region organizes the formation of the main body axis. *Spemann's organizer* is the oldest and most famous example of an embryonic *signaling center*.

Note that, originally, epigenesis was regarded an "irrational concept" since it was associated with "spontaneous generation," while preformation seemed to be the only rational explanation in the framework of a static world (Horder 1993). With the discovery of embryonic regulation the situation changed dramatically and preformation was now considered the more irrational point of view. While preformation assumes an arbitrary beginning, epigenesis cannot explain any beginning. The question, "Which came first, the chicken or the egg?" cannot be answered. A hen could be viewed as the egg's way of making another egg (Horder 1993). To evolutionary theorists, embryos are "just a way to carry genes from one generation to the next." Nevertheless, there are subtle relationships between ontogenetic development and evolution, and development comprises a *whole* on individual and historic time scales. All aspects of organismic life cycles are subject to selection. Thus, it is obvious that the set of causal factors stored in any given, existing genome gives us only a fragmentary glimpse of the totality of causes actually underlying the development of an organism: most causes (e.g., the chain of intervening ancestors and the selective forces applied to them) no longer exist. It seems reasonable to assume a *developmental cascade* (Horder 1993).

Experimental Approach. *Embryological data* are produced by descriptive experimental embryology whose main tools are microscopy (e.g., bright field, phase optics, time-lapse cine, scanning and transmission, Nomarski optics, confocal) and staining techniques primarily provided by histochemistry and immunofluorescence. There are, in principle, two strategies: one can describe the *normal* embryonic situation, or one can manipulate an embryo by means of dissecting or mutational techniques.

Furthermore, one can focus on a cellular approach, i.e., description of single cells (bottom-up) or, alternatively, follow a *whole organ* (top-down) approach. Early cellular approaches focused on individual cell movement *in vitro* (Harrison 1907) and reassembling abilities of sponge cells (Wilson 1907). The whole organ approach was initiated by **Wilhelm His** (1831–1904) and Thompson's influential work *On Growth and Form* (His 1874; Thompson 1917) (cp. sec. 2.2.6). The beginning of the twentieth century marked the great days of experimental embryology represented by Driesch, Spemann, and many others (Hilbert 1991). Today the whole spectrum of molecular genetic methods is applied, which particularly allows us to analyze problems of developmental biology from an evolutionary perspective (Pearson 2001).

2.2.4 Molecular and Genetic Analysis

Conrad H. Waddington (1905–1975) was among the first to study the "developmental action of genes" (Waddington 1940). In recent years, efforts have been concentrated on identifying the *subcellular*, i.e., molecular basis of morphogenesis. On one hand, descriptive methods help us understand the role of intracellular components as the cytoskeleton for cellular pattern formation, particularly shape change and motion. On the other hand, the importance of extracellular structures, in particular the extracellular matrix and basal lamina (laid down by mesenchymal cells and epithelia), can now be investigated. Nowadays, efforts are focusing on the role of genomic structure and regulation as well as cell–cell interactions mediating complex changes in embryonic development.

Recently, the importance of the close relationship between ontogeny and phylogeny has been rediscovered. It has been realized that development is for the most part modular (Pennisi and Roush 1997). For example, whether a crustacean's thorax sprouts a feeding leg that grabs food and shoves it towards the mouth, or a leg used for swimming or walking may depend on whether two very similar genes called *Ubx* and *Abd-A* are turned on or off in budding limbs. This is a prime example of how small changes may have dramatic effects on body plan and adaptation. Similar genes generating quite different body plans have been identified in very different organisms.

Among other candidates, one superfamily of genes presents itself as capable of regulating developmental decisions: the homeobox genes. As a general principle, homeobox genes encode transcription factors that play key roles in the determination and maintenance of cell fate and cell identity (Manak and Scott 1994). Homeobox genes share a common nucleotide sequence motif (the homeobox) encoding the roughly 61-amino-acid homeodomain. The homeodomain, in turn, is a helix-turn-helix DNA binding domain of the functional transcription factor. Evolutionary relationships and family classifications are determined based upon the degree of identity and similarity among homeodomains followed by comparative analysis of amino acid sequences both amino-terminal and carboxyl-terminal to the homeodomain (Scott, Tamkun, and Hartzell 1989). These terminal sequences vary considerably from protein to protein and, indeed, may demonstrate no evidence of evolutionary or functional relationship whatsoever. Homeobox genes are found in animals

Figure 2.10. *Drosophila* segmentation (sketch). Distribution of various proteins at early developmental stages; from left to right: maternal coordinate proteins (e.g., *bicoid*), gap proteins (e.g., *Krüppel*), pair-rule proteins (e.g., *fushi tarazu*), and segment polarity proteins (e.g., *engrailed*). Anterior is top, posterior is bottom.

ranging from hydra to humans (as well as in fungi and plants). Over evolutionary time, the number of homeobox genes has increased and their functions have been reengineered to meet the demands of increasingly diverse developmental processes. To date, there has been well over 100 homeobox genes identified in the human, with a comparable number of homologs characterized in the mouse (Stein et al. 1996). In mammals, homeobox genes reign over the specification of the overall body plan and are known to play key roles in a variety of developmental processes, including central nervous system and skeletal development, limb and digit specification, and organogenesis. Mutations in homeobox genes can cause dramatic developmental defects, including loss of specific structures as well as "homeotic transformations," in which one body part or segment is converted to the likeness (identity) of another. Some homeobox genes appear to serve cell autonomous functions in differentiation and cell cycle control; others serve noncell autonomous functions such as pattern formation and mediation of reciprocal tissue interactions. In the fruit fly *Drosophila melanogaster*, for example, a *homeobox* gene (*Abd-B*) helps to define the posterior end of the embryo (various protein distributions accompanying the development are well known (fig. 2.10)), while a similar family of genes in chicks helps to partition a developing wing into three segments. Such molecular discoveries suggest that evolution was not primarily ignited by new genes but by the utilization and regulation of existing genes in complex ways.

2.2.5 Self-Assembly

At the level of preformation, no principle of pattern formation can be envisaged since preformation merely recycles preformed patterns in a reversible Newtonian world. The suggestion of **Conrad H. Waddington** (1905–1975) to focus on generation of forms by *self-assembly* reduces pattern generation to a geometrical problem induced by preformed and disjunct elements—in accordance with a geometry-based Platonian conception of the world (Waddington 1962). It is questionable whether self-assembly can account for much more than the shape of a virus. Nevertheless, there still exist mathematical models with strong preformistic/static elements, L-systems,

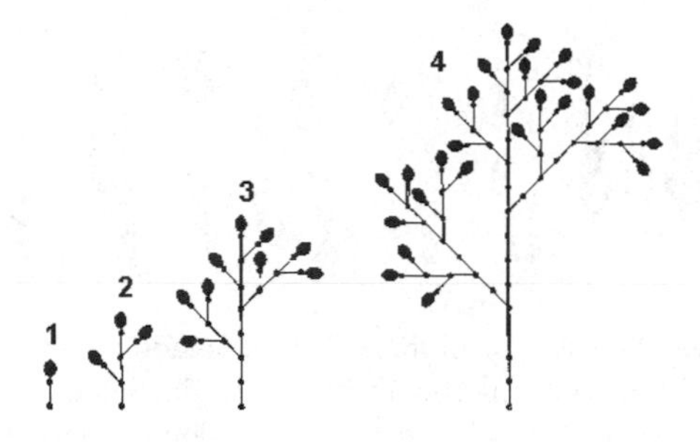

Figure 2.11. Branching pattern formation in L-system simulation. Patterns are generated by substitution of "pattern elements;" there is just one pattern element in the example, namely the "stalk-leaf structure" (generation 1). The rewriting rule for this element is defined pictorially by the transition from the first to the second generation. Subsequent generations are generated by successive application of this rewriting rule.

in which an initial structure develops by subsequent replacement of certain pattern elements with the help of appropriately chosen rewriting rules (Lindenmayer 1982; see fig. 2.11).

The problem first addressed by August Weismann, namely how randomly arranged discrete particles in the nucleus can be selected out in different cells in order to produce the patterned differentiation in adult morphology, has been reinforced by the increasing certainty that, not only does the egg cytoplasm provide minimal preformed organization, but, throughout development, all cells have the same (complete and often multipotential) array of genes. In order to explain the adjustments that could underlie regulation and the way in which integrated pattern generation might be achieved through embryogenesis, a variety of concepts have been introduced.

2.2.6 Physical Analogues

Only at the end of the 19th century did serious research on efficient causes of biological pattern formation begin. Wilhelm His, whom we have already met as an initiator of the "whole organ approach," proposed the first physical explanation of form formation (His 1874; cp. sec. 2.2.3). He convincingly demonstrated that a developing gut can be modeled as a rubber tube under the influence of complex tensions. His' model not only served as a mechanical analogue assuming a field of forces, but also prepared the field for Haeckel's student, **Wilhelm Roux** (1850–1924), and his intention to deduce biological phenomena from the laws of physics. Roux' work on "developmental mechanics"[12] marks the transition from teleological interpretations of embryology to the search for efficient (mechanical) causes in the Aristotelian sense.

It was the Scottish naturalist and mathematician **D'Arcy Thompson** (1860–1948) who claimed that an optimization principle, namely minimization of curvature,

[12] Roux founded the journal *Archiv für Entwicklungsmechanik*, Engl.: *Roux's Archives of Developmental Biology*. Recently, the name was changed to *Development, Genes, and Evolution*.

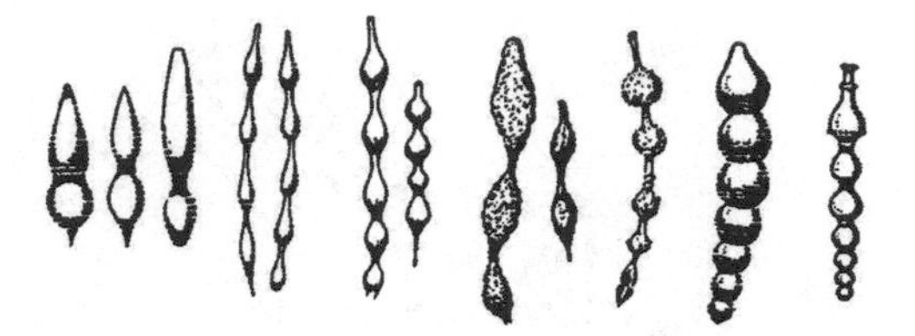

Figure 2.12. "Minimal surface as optimal shape:" unduloid form of some Foraminifera species (*Nodosaria*, *Rheopax*, *Sagrina*) (from Thompson 1917). The unduloid is a rotationally symmetric surface of constant curvature (H-surface of revolution). Plateau experimentally showed that there are exactly six H-surfaces of revolution: the plane and the catenoid with mean curvature zero, the cylinder, the sphere, the unduloid and the nodoid with mean curvature nonzero. These surfaces can be constructed by appropriate rotation of conic sections (see e.g., Hildebrandt and Tromba 1996).

should account for single cell shapes (Thompson 1917; see also fig. 2.12). Thompson tried to explain form and evolutionary change of form as the result of the immediate, primarily mechanical forces operating on the developing embryo and developed a theory of allometric transformations (fig. 2.13). Changing morphologies are explained solely as the result of coordinated differential growth during development (preformation concept).

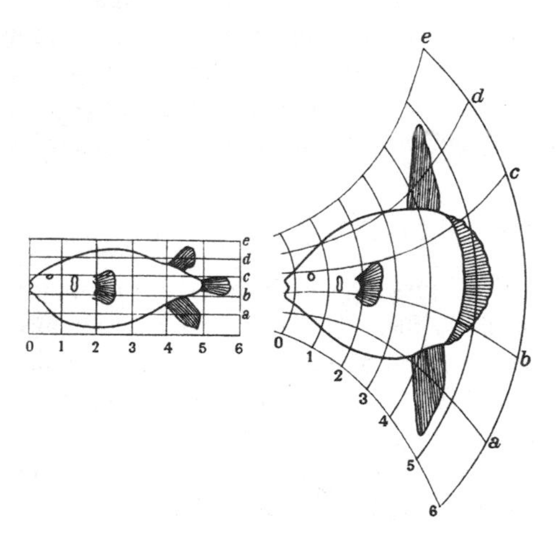

Figure 2.13. Transformation: Thompson (1917) suggested allometric transformations (affine mappings) to (phenomenologically) describe the body form of several fish. The figure shows an example. Left: *Diodon* (porcupinefish); by transforming the vertical coordinates to a system of concentric circles and the horizontal coordinates to curves approximating a system of hyperbolas, the *Orthagoriscus* (sunfish) appears.

This approach has been continued, essentially unchanged, by "neo-Darwinist" evolutionary biologists (of the 1930s and 1940s) in their attempts to integrate evolution theory with genetics (Jepsen, Mayr, and Simpson 1949; Mayr 1982). Changes in adult morphologies were interpreted as results of gradualistic growth changes, using concepts like allometry, in turn linked to genetics by concepts such as "rate genes" or "gene balance." But these approaches are rather limited and fail to explain, for example, changes of somite number or differentiation.

2.2.7 On Gradients and Chemical Morphogens

The notion of gradients and "morphogenetic fields" dates back particularly to the work of H. Driesch (cp. sec. 2.2.3) and **Theodor Boveri** (1862–1915) (Boveri 1910). In particular, the concept that a gradient was built into the egg and subdivided within daughter cells of the early embryo was introduced by Boveri. The essential assumption is thereby that concentrations of substances, *morphogens*, can be precisely measured by cells inducing corresponding cell behavior changes, e.g., differentiation (fig. 2.14). In principle, two gradients can provide an exact *positional information* within a cell, e.g., a fertilized egg (Wolpert 1981). Meanwhile, the existence of such gradients has been proven. For example, in the *bicoid* mutant of the famous fruitfly *Drosophila melanogaster* it results from (passive) diffusion of proteins activated by certain developmental genes (Nüsslein-Volhard 1991). Also, it has been shown that when cells from the animal pole of an early *Xenopus laevis* embryo are exposed to the signaling molecule activin, they will develop as epidermis if the activin concentration is low, as muscle if it is a little higher, and as notochord if it is a little higher still. The normal role of activin in the intact *Xenopus* embryo, however, is uncertain.

Some problems go ahead with gradient theories. First, they assume a rather precise measurement of signal concentrations by the cells. It is far from clear how these measurements can be achieved. Also, even in the case of one of the best studied biological systems, *Drosophila melanogaster*, development of polarity, i.e., the primary prepattern responsible for creation of the initial gradient and all subsequent gradients is not fully understood. Furthermore, while within a single cell proteins and

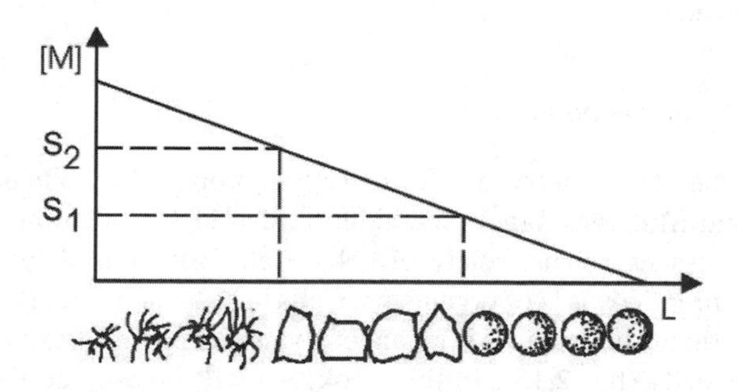

Figure 2.14. Pattern formation viewed as resulting from a spatial morphogen gradient along an axis L. Differentiation into various cell types is induced by different levels of the morphogen concentration $[M]$.

other substances can more or less freely diffuse, diffusional transport is much less dominant in conglomerates of cells, e.g., tissues. Therefore, search for alternative pattern forming principles is essential.

2.2.8 Self-Organization and Morphogenesis

Turing (Diffusive) Instability. The key to understanding *de novo* pattern formation in biological morphogenesis is to view it as a *self-organization system* (Jantsch 1980). In a pioneering paper **Alan M. Turing** (1912–1954) analyzed a system of diffusively coupled reacting cells (Turing 1952). Counter-intuitively, this system can produce periodic patterns. The *Turing instability* is the first demonstration of *emergence* with respect to biological pattern formation: Starting from a slightly perturbed homogeneous situation spatial patterns arise. It is assumed that cells can read concentrations and react, correspondingly; i.e., it is a *theory of prepattern formation.* Note that Turing analyzed a *macroscopic system* in terms of concentration changes of appropriately chosen chemicals (morphogens).

Such indirect (macroscopic) approaches have been further investigated in particular by Gierer and Meinhardt (1972; see also Meinhardt 1982). They introduced a dichotomy that poses conceptual difficulties since a distinction is made between "the organization" (of the prepattern), which admittedly is a self-organization process, and "the organized": the final pattern, which is viewed as the "hand of the prepattern." It is noteworthy that only recently, experimental manifestations of the Turing instability, the CIMA and the PAMBO reactions, were discovered (Castets, Dulos, Boissonade, and de Kepper 1990; Ouyang and Swinney 1991; Watzl and Münster 1995). It is today questionable if the Turing instability is important in biological morphogenesis. Nevertheless, the *instability principle* is essential and a number of *morphogenetic instabilities* have been identified, particularly chemotactical, mechano-chemical, and hydrodynamical instabilities (He and Dembo 1997; Höfer, Sherratt, and Maini 1995; Murray and Oster 1984). It has been demonstrated that self-organization can account for differentiation, aggregation, taxis, cell division, or shape formation. Overviews of corresponding morphogenetic models can be found in Chaplain, Singh, and McLachlan 1999; Deutsch 1994; Othmer, Maini, and Murray 1993; Rensing 1993.

2.2.9 Cell–Cell Interactions

As early as in the 19th century, a *cell theory* was proposed by **Theodor Schwann** (1810–1882) and **Matthias Jacob Schleiden** (1804–1881), according to which animal and plant tissues consist solely of cells. Cell formation should be the common *developmental principle* of organisms. Cells have an ambiguous character: they are closed functional units and, simultaneously, are capable to exchange *information* with other cells (fig. 2.15). In this book, we will investigate the influence of *cell–cell interactions* on pattern formation. **Johannes Holtfreter** (1901–1992) was among the first to realize the importance of cell–cell interactions, particularly differential adhesivity for cellular pattern formation (Holtfreter 1943; Steinberg 1963).

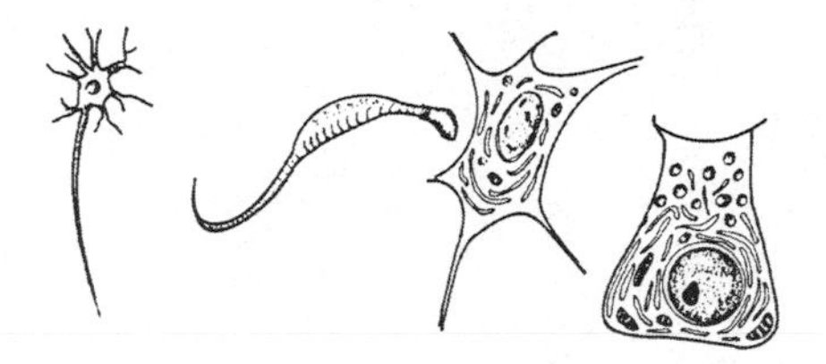

Figure 2.15. Examples of human cell types. From left: nerve cell (neuron), spermatocyte, bone cell (osteocyte), pancreatic acinar cell.

In particular, Holtfreter's pioneering study combining observations on dissociated and reassociated embryonic tissue with analysis of possible mechanisms underlying tissue movements and specificity apparently led directly to Steinberg's differential adhesion hypothesis of cell sorting (Mostow 1975; cp. also ch. 7 in this book).

Many developmental signals are local, based on cell–cell inductive interactions. This self-organization view does not assume a prepattern at the genomic level. According to the *cell-interaction*-based self-organization hypothesis, patterns are built up by local interactions between cells with discrete states, such that it is unnecessary to postulate genes responsive to position as such or control genes whose specific purpose is to control pattern layout. In this book, we will develop and analyze cellular automaton modeling of self-organized cell interaction.

Summary. In this chapter spatio-temporal organization principles have been introduced and their application to biological pattern formation has been described. Emphasis was placed on demonstrating how a particular spatio-temporal conception directs possible principles of pattern formation, particularly preformation, optimization, and self-organization. This allows one to classify morphogenetic processes according to the underlying organization principle (see Table 2.1). Note that the table is not intended to present a complete overview of morphogenetic competences but to name characteristic representatives. Typically, in biology, various organization principles are combined to explain the formation of specific biological patterns.

Today, morphogenetical research is focusing on the characterization of paradigmatic systems, experimentally and theoretically. Such systems are in particular microorganisms such as the slime mold *Dictyostelium discoideum* or myxobacteria (cp. ch. 8); but higher organisms may also serve as illustrations of important morphogenetic concepts (e.g., pigment pattern formation in salamander larvae; cp. ch. 9). More recently, the focus of morphogenetic modeling has even been directed to tumor growth (cp. ch. 10).

3

Mathematical Modeling of Biological Pattern Formation

We shall assume…
…that all cows are spherical.
(old joke)

The application of mathematical models to explain the dynamics of biological pattern formation started with the work of D'Arcy Thompson who showed that mathematics can describe not only static form but also the change of form (Thompson 1917; cp. also sec. 2.2.6). In this chapter, an overview of mathematical models of biological pattern formation is presented.

Models of (biological) pattern formation combine concepts of *space, time, and interaction.* Pattern formation arises from the interplay of (active or passive) cell motion and/or short- or long-range mechano-chemical interaction of cells and/or signaling molecules. What are appropriate mathematical structures for the analysis of such systems? One can distinguish *microscopic* and *macroscopic* modeling perspectives on biological pattern formation, focusing on the individual component (e.g., molecules or cells) or the population level, respectively. Macroscopic (typically deterministic) modeling ideas have traditionally been employed and predominantly formulated as partial differential equations (e.g., Meinhardt 1982; Murray and Oster 1984). Interest in microscopic approaches, particularly *spatial stochastic processes,* has lately grown due to the recent availability of "individual cell data" (genetic and proteomic) and has triggered the development of new mathematical models, e.g., *cellular automata.* Such models allow us to follow and analyze spatio-temporal dynamics at the individual cell level.

3.1 The Art of Modeling

Models are tools for dealing with reality (Bossel 1994). They are "caricatures" of the real system[13] and built to answer questions related to the real system. By capturing

[13] Loosely stated, a "system" is defined as a set of interrelated elements that are related to a precisely defined environment (Bossel 1994).

a small number of key processes and leaving out many details, simple models might be designed to gain a better understanding of reality (Gurney and Nisbet 1989). The investigation of the "final model" is important, but so is the process of modeling, since this forces the modeler to conduct a detailed system analysis. Other objectives are to test different scenarios and assumptions,[14] to demonstrate that certain ideas should not or cannot be realized, or to give predictions for the future. One can distinguish between qualitative and quantitative models. Models that are not designed to quantitatively reproduce or to predict concrete field situations, but to obtain an understanding of the mechanisms, are called *strategic* (Gurney and Nisbet 1989). Formal models[15] for dynamical systems, in which the set of assumptions about reality is expressed in mathematical (*mathematical model*) or a computer (*simulation model*) language, have turned out to be especially useful.

Note that not all mathematical models are accessible to mathematical analysis but that all of them can be simulated on a computer. If an analytic solution is available, this may provide a complete characterization of the system dynamics. Many simulation models cannot be described in a coherent mathematical framework in such a way that they become accessible to an analytical mathematical analysis.[16] Those models have to be investigated by means of statistical analysis of large numbers of simulation runs. The choice of a model approach depends on the characteristics of the dynamical system itself and on the aspects of the dynamical system that are emphasized according to the model's purpose (Britton 2003; Hastings 1994). Therefore, interdisciplinary approaches are essential because those who are experts on the structure of the particular application have to work together with those who are experts on the structure of the mathematical modeling approaches. This is particularly true for the design of models for biological pattern formation, which requires both experimental and mathematical knowledge.

3.2 How to Choose the Appropriate Model

Faced with the problem of constructing a mathematical model of biological pattern formation, the first modeling step is to clarify the level of organization in which one is primarily interested with regard to *space, time, states,* and *interactions* (Deutsch and Dormann 2002).

One way to classify approaches to modeling spatially extended dynamical systems is to distinguish between continuous and discrete state, time and space variables. A classification of different approaches is shown in Table 3.1.

The answer to the question of whether a process is viewed as state-, time-, space-discrete, or continuous is essential. If a real system does not exactly fit into any

[14] Especially in order to find out which assumptions are essential.

[15] In contrast to physical models, and pure semantic models, which are mainly used in psychology and social sciences.

[16] For example, agent-based models and models in the framework of artificial life research (Langton 1989).

Model approach	*Space variable*	*Time variable*	*State variable*
I. PDEs, integro-differential eqs.	Continuous	Continuous	Continuous
II. Spatial point process, set of rules	Continuous	Continuous	Discrete
III. Integro-difference eqn.	Continuous	Discrete	Continuous
IV. Set of rules	Continuous	Discrete	Discrete
V. Coupled ODEs	Discrete	Continuous	Continuous
VI. Interacting particle systems	Discrete	Continuous	Discrete
VII. Coupled map lattices, system of difference eqns., lattice-Boltzmann models	Discrete	Discrete	Continuous
VIII. Cellular automata, lattice-gas cellular automata	Discrete	Discrete	Discrete

Table 3.1. Mathematical modeling approaches to spatio-temporal pattern formation (ODE: ordinary differential equation, PDE: partial differential equation; see Berec 2002 for details).

of these categories, the particular choice should not influence the results too much within ranges of relevant parameter values. In other words, the model results should be relatively robust with respect to the chosen space-time-state framework. However, the choice of a particular framework may influence the results significantly. For example, while the time-continuous logistic ordinary differential equation describes a simple growth process, the time-discrete logistic map leads to "complex" dynamical behavior, including chaotic motion (Kaplan and Glass 1995). In the modeling process the scales of all involved processes and their relation to each other have to be specified. For instance, a variable (e.g., temperature) could be regarded as constant if the chosen time scale is short. In contrast, if the time scale is large, the detailed dynamics of the variable might become important.

In particular, the appropriate choice of the model together with the appropriate space and time scales is not a mathematical problem, but a *preliminary decision of causes,* since one has to determine and distinguish the phenomena to be considered and those to be ignored. The modeling strategy reminds one of what Thompson called *Aristotle's parabola* (taken from Gould 1976): "We must consider the various factors, in the absence of which a particular house could not have been built: the stones that compose it (material cause), the mason who laid them (efficient cause), the blueprint that he followed (formal cause), and the purpose for which the house was built (final cause)." Applied to biological morphogenesis, the house corresponds to the adult organism and the stones to the molecular components, the material cause of pattern formation. While it is clear that genetic information (blueprint), natural selection, and physico-chemical constraints all contribute to efficient, formal, and final causes of morphogenesis, the precise role and interactions of external and internal factors cannot easily be (if at all) distinguished. For example, when trying to explain cellular shape, shall we focus on cytoskeleton architecture as arising from reaction-diffusion intracellular macromolecular dynamics? Or, alternatively, shall we investigate curvature dynamics of the cell cortex, e.g., appearing as a minimum property of an appropriately defined potential energy?

In modeling the temporal evolution of spatially distributed systems, describing, e.g., the interaction of cytoskeleton molecules or the formation of cellular tissue, we can think of the system as a *game of formation,* which is defined by specifying the players (cells and/or molecules) together with the rules of interaction. In particular, the players' *internal state space* representation has to be defined. *Internal state* may refer to position, velocity, acceleration, orientation, or age of cells or molecules.

In specifying the players' interactions, namely the rules of the game, one can choose several levels of description involving different resolutions of spatial detail, ranging from *microscopic* to *macroscopic* perspectives. Transitions from one level to the other involve *approximations* with regards to the spatio-temporal nature of the underlying interactions (fig. 3.1).

Model I: Coupled differential equations

An example for this approach is Turing's cellular model (Turing 1952), the purpose of which is to explore how spatial structures (forms, patterns) can emerge from a homogeneous situation (see also ch. 11). Space is divided into discrete compartments $r = 1, \ldots, L$ (spatially homogeneous cells) in which different components $\sigma = 1, \ldots, \varsigma$ are governed by ordinary differential equations. The variables $a_\sigma(r, t) \in \mathbb{R}$ usually represent macroscopic quantities, e.g., densities (of cells or molecules). Spatial interactions are modeled by discrete coupling of components of different cells. In particular, Turing suggested the following nearest neighbor coupling of a ring of cells for a two-component one-dimensional system:

$$\frac{\partial a_1(r,t)}{\partial t} = F_1\left(a_1(r,t), a_2(r,t)\right) + D_1[(a_1(r-1,t) - 2a_1(r,t) + a_1(r+1,t))],$$
$$\frac{\partial a_2(r,t)}{\partial t} = F_2\left(a_1(r,t), a_2(r,t)\right) + D_2[(a_2(r-1,t) - 2a_2(r,t) + a_2(r+1,t))],$$

with continuous time $t \in \mathbb{R}$ and "diffusivity" $D_\sigma \in \mathbb{R}$, which expresses the rate of exchange of species σ at cell r and the neighboring cells ("diffusive coupling," discrete diffusion).

An advantage of coupled differential equations is the possible adaptation of the transport scheme (coupling) to particular system demands, e.g., to extended local neighborhood relations.

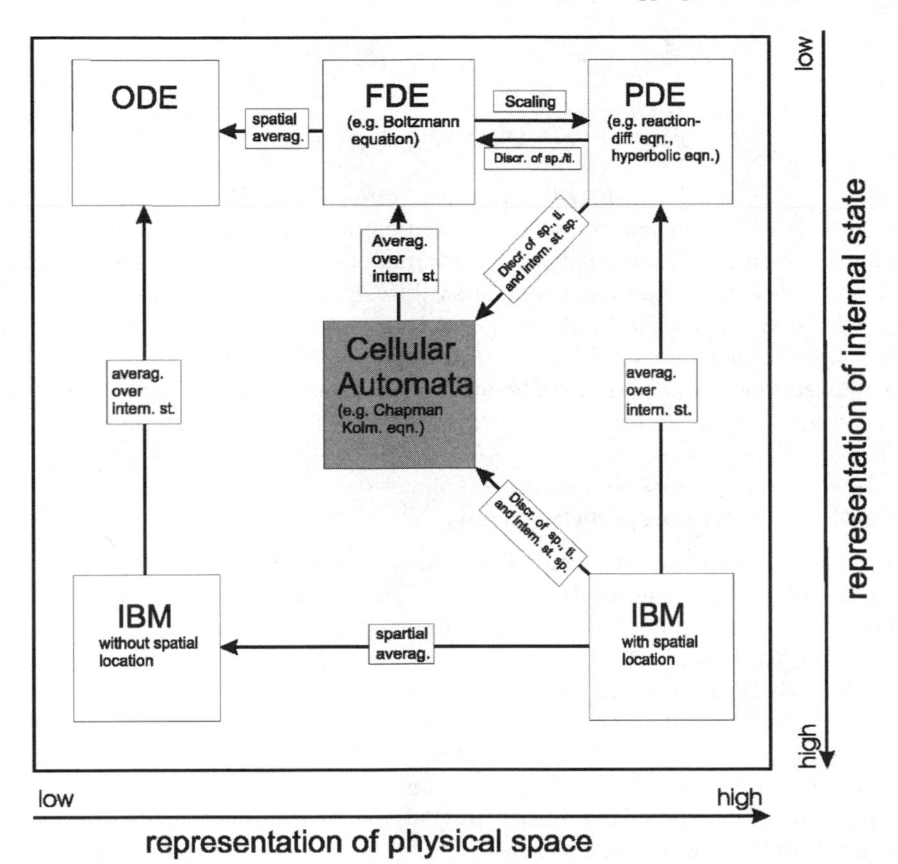

Figure 3.1. Model relationships with respect to physical space and internal state representation. The sketch shows the relations between various model levels. Averaging and limiting procedures allow transitions between ordinary differential equations (ODEs), partial differential equations (PDEs), finite-difference equations (FDEs), and individual-based models (IBMs) or interacting particle systems. Cellular automata can be alternatively viewed as a separate model class or as discretizations of partial differential equations or individual-based models.

Model II: Partial differential equations

The classical approach to modeling spatially extended dynamical systems is based on (deterministic) partial differential equations, which model space as a continuum,[17] $x \in \mathbb{R}^d$, where d is the space dimension. In the simplest version particle transport is assumed to be passive diffusion. Then, a two-component one-dimensional system is described by

[17] Note that this implies an infinite amount of information about the state values in any arbitrarily small space–time volume.

$$\frac{\partial a_1(x,t)}{\partial t} = F_1\left(a_1(x,t), a_2(x,t)\right) + D_1 \frac{\partial^2 a_1(x,t)}{\partial x^2},$$
$$\frac{\partial a_2(x,t)}{\partial t} = F_2\left(a_1(x,t), a_2(x,t)\right) + D_2 \frac{\partial^2 a_2(x,t)}{\partial x^2},$$

where $t \in \mathbb{R}$ and $D_\sigma \in \mathbb{R}$ ($\sigma = 1, 2$) are "diffusion coefficients." A vast literature deals with a framework based on "reaction-diffusion" models. They describe interaction processes and demographic dynamics, commonly called reaction, which are combined with various (not necessarily diffusive) transport processes (reviewed in Okubo and Levin 2002). Pattern-generating mechanisms in partial differential equations include diffusive instabilities in reaction-diffusion equations (see ch. 11, and Okubo and Levin 2002), density-dependent diffusion (Mimura 1981), and models with aggregation terms (Levin 1992; see Murray 2002 for further examples).

Model III: Interacting particle systems

Interacting particle systems model interactions between finitely or infinitely many particles (e.g., cells or molecules). They are stochastic models consisting of a collection of spatial locations called sites and a finite set of states. Each site can be in one particular state at each time $t \in \mathbb{R}$. The temporal evolution is described by specifying a rate at which each site changes its state. The rate depends upon the states of a finite number of neighboring sites. In the absence of interaction, each site would evolve according to independent finite or countable state Markov chains (Liggett 1985). When an event occurs at a constant rate α, then the time intervals between successive events are exponentially distributed (Poisson distribution) with expectation α (Durrett 1993). The ensemble of states of all lattice sites defines a configuration or a microstate of the system.[18]

Model IV: Coupled map lattices

If both space (r) and time (k) are subdivided into discrete units, one refers to coupled map lattices or *time- and space-dependent difference equations*.[19] An example for a one-dimensional two-component system is

$$\begin{aligned} a_1(r, k+1) &= F_1\left(a_1(r,k), a_2(r,k)\right) \\ &\quad + D_1\left(a_1(r-1,k) - 2a_1(r,k) + a_1(r+1,k)\right), \\ a_2(r, k+1) &= F_2\left(a_1(r,k), a_2(r,k)\right) \\ &\quad + D_2\left(a_2(r-1,k) - 2a_2(r,k) + a_2(r+1,k)\right). \end{aligned}$$

This model approach has been introduced in order to study spatio-temporal chaos, which is important in the study of turbulence (Kaneko 1993) and has found biological application (Hendry, McGlade, and Weiner 1996). Coupled map lattices are

[18] A good introduction into this modeling field can be found in Durrett 1999 and Liggett 1985.
[19] Or finite-difference equations.

also used as a tool for numerical studies of partial differential equations. *Lattice-Boltzmann models* are a particular case of coupled map lattices. They can be derived from a microscopic description (lattice-gas cellular automaton) of (physical) systems,[20] which are composed of many "particles" (cp., e.g., fluid dynamics, McNamara and Zanetti 1988). Then, the state variables $a_\sigma(r, k) \in [0, 1]$ are defined by averaging over an ensemble of independent copies of the lattice-gas cellular automaton; i.e., they represent the probability of the presence of a particle at a cell r at time k. When spontaneous fluctuations and many-particle correlations can be ignored, this approach offers an effective simulation tool in order to obtain the correct macroscopic behavior of the system and provides a "natural interpretation" of the numerical scheme (Chopard and Droz 1998).

3.2.1 Model Perspectives

The View from the Top: Macroscopic Level of Description

The model perspective obviously depends on the central reference level (e.g., cells or molecules). The simplest (macroscopic) approach is to assume that the (spatially extended) cellular (molecular) system is homogeneously mixing over its entire extent and to model the dynamics by a system of *ordinary differential equations* (continuous time) or *difference equations* (discrete time steps). This assumption is of minor value in systems in which the precise spatio-temporal patterns are of interest. In such systems, individuals can move about in (continuous or discrete) geometrical space; the appropriate mathematical description is a system of coupled differential equations (see Model I) or a reaction-diffusion or advection system (see Model II).

Regarding the modeling perspective, the partial differential equation approach (e.g., reaction-diffusion or advection) can lead to a satisfactory model if a sufficiently large number of cells or molecules allows evaluation of a local density. Quantification of densities implies a spatial average, which is only meaningful if the effects of local fluctuations in the number of the considered components are sufficiently small. Accordingly, this macroscopic modeling approach has some shortcomings: the determination of a temporal derivative implicitly involves a limit which is only justified under the assumption that the population size gets infinitely large or, equivalently, that individuals become infinitesimally small. The following question immediately arises: How can one treat effects of local stochasticity and the fact that individuals (cells or molecules) are obviously discrete units?

Two strategies, macroscopic and microscopic, respectively, are possible: one can start from the top (macroscopic perspective), abandon the continuum of spatial scales and subsequently subdivide available space into patches. This procedure leads to deterministic *finite-difference* (with respect to space) *patch models* or *coupled map lattice models* (see Model IV). Note that numerical solution methods of partial differential equations typically also replace spatial and/or temporal derivatives by finite differences, which is equivalent to a discretization of space and/or time.

[20] For further details see sec. 4.4.2.

Stochastic fluctuations can be studied after introduction of noise terms into the reaction-diffusion equation. This so-called *Landau approach* (Landau and Lifshitz 1979) offers possibilities for analyzing the effect of fluctuations in corresponding stochastic partial differential equations. The approach is somewhat paradoxical, since one has averaged over the microscopic fluctuations to obtain the macroscopic (mean-field) equation (Weimar 1995). Alternatively, one can start from the bottom, i.e., a microscopic stochastic description (Lagrangian approach).

Viewing from the Bottom: Microscopic Level of Description

Interacting Particle Systems. *Interacting particle systems* consist of a finite population of particles (cells or molecules) moving about in discrete space and continuous time (see Model III). Particles are characterized by their *position* and *internal state*. There is no restriction of possible interactions to be studied within the modeling framework of interacting particle systems: particle configurations may change due to direct or indirect, local or nonlocal interactions (some examples that have been treated mathematically can be found in Liggett 1985). Particle traces governed by Newtonian dynamics can in principle (in nonquantum systems) be followed individually—an example of an *individual-based description.*

Simulations based on principles of *molecular dynamics* (continuous space and time) are most effective for high particle densities, since particle traces are individually evaluated by integration. In molecular dynamical algorithms integrations are performed even if the situation remains physically almost unchanged, which is typical for the low-density situation. Under such circumstances *dynamical Monte-Carlo methods* generating random sequences of events (e.g., Metropolis algorithm (Metropolis et al. 1953) or simulated annealing (van Laarhoven and Aarts 1987)) are preferable (Baldi and Brunak 1998). For further (statistical) analysis a large number of simulations have to be performed (Haberlandt et al. 1995). A "morphogenetic application" of Monte-Carlo methods is provided by simulations of tissue pattern formation based on direct cell–cell interactions (Drasdo 1993).

Cellular Automata. Cellular automata are introduced in the next chapter (ch. 4). *Cellular automata* can be interpreted as discrete dynamical systems: discrete in space, time, and state. Spatial and temporal discreteness are also inherent in the numerical analysis of approximate solutions to, e.g., partial differential equations. As long as a stable *discretization scheme* is applied, the exact continuum results can be approximated more and more closely as the number of sites and the number of time steps is increased—the numerical scheme is convergent. The discreteness of cellular automata with respect to the limited (discrete) number of possible states is not typical of numerical analysis, where a small number of states would correspond to an extreme *round-off* error. The problem of state space limits has not yet been addressed in a rigorous manner. It is possible to devise cellular automaton rules that provide approximations to partial differential equations and vice versa (Omohundro 1984; Toffoli 1984). For cellular automaton models certain strategies have been developed to analyze continuous approximations (Schönfisch 1993, 1996, Stevens

1992; cp. also ch. 4). Cellular automata differ from coupled map lattices, which are characterized by a continuous state space.

3.2.2 From Individual Behavior to Population Dynamics

Again it is essential to specify the central reference, which could be the cellular or molecular scale in particular. In the following, we start from cellular "individuals," but similar arguments hold for other "individuals," especially molecules. The aim of determining how the (macroscopic) population dynamics evolves with time, in terms of the (microscopic) individual behavior that governs the motion of its constituents, is similar to the main goal of nonequilibrium statistical mechanics. The *master equation*, which completely specifies the change of the probability distribution over a given interacting particle system's phase space, is far too complicated in its generality. For system analysis certain assumptions are necessary. Historically, two different directions have developed: the *Brownian motion* theory (initiated by Einstein, Smoluchowski, and Langevin) and the *kinetic* theory started by Clausius, Maxwell and Boltzmann (Résibois and de Leener 1977).

With regards to interacting particle systems as models of biological pattern formation, both theories can be rediscovered as modeling strategies: The first strategy uses a (stochastic) assumption on the dynamics of individual cell motion; this is the *active walker* or *Langevin approach*. The second approach is based on a *kinetic* or *Boltzmann interpretation* and focuses on the dynamics of particle distribution functions, which are observables depending on statistical properties, not on the details of individual cell motion. Interestingly, in many relevant situations the well-known Boltzmann equation in terms of the single particle distribution function already captures essential system characteristics (Pulvirenti and Bellomo 2000).

Further strategies for deriving macroscopic descriptions for individual-based models have been developed. For example, alternative approximations of Eulerian and Lagrangian approaches were proposed in a model of swarming and grouping based on density-dependent individual behavior (Grünbaum 1994). Another possibility is the method of *adiabatic approximation* (quasi-steady state assumption) that has been applied to an individual-based ecological model (Fahse, Wissel, and Grimm 1998). A further strategy for analyzing stochastic spatial models is to derive approximations for the time evolution of the moments (mean and spatial covariance) of ensembles of particle distributions. Analysis is then made possible by "moment closure," i.e., neglecting higher order structure in the population. Potential applications are predator–prey and host–parasite models (for an ecological application of moment equations see, e.g., Bolker and Pacala 1997).

Pair approximation is a particular moment closure method in which the mean-field description is supplemented by approximate ordinary differential equations for the frequency of each type of neighboring site pairs, e.g., the fraction of neighboring sites in which both sites are empty. Higher order frequencies (e.g., for triplets) are approximated by pair frequencies in order to obtain a closed system of equations (Rand 1999). The equations are nonlinear but typically fairly low dimensional. Pair approximation techniques have been applied widely to various biological phenomena, e.g., to host–pathogen dynamics, reproduction modes in plants, forest gap dynamics, and bacterial allelopathy (see Iwasa 2000 for a review). Note that pair approximation has also been used for the analysis of a stochastic model of adhesively interacting cells producing cell sorting (Mochizuki et al. 1996, 1998). The pair approximation method has been recently extended to account for multiple interaction scales, which provides a useful intermediate between the standard pair approximation for a single interaction neighborhood and a complete set of moment equations for more spatially detailed models (Ellner 2001).

Active Walker or Langevin Approach

We want to illuminate the Langevin approach by means of an example that has been studied particularly well in recent years. This is chemotactic aggregation as a model of bacterial pattern formation (Ben-Jacob, Shochet, Tenenbaum, Cohen, Czirók, and Vicsek 1994; Othmer and Stevens 1997; Stevens 1992). Here, it is assumed that a chemical (diffusible or nondiffusible) substance (produced by the bacterial cells) determines motion of the cells insofar as these search for local maxima of the *chemoattractant* substance. If cells communicate by means of an external field, e.g., a chemical concentration field $s(x, t)$ which they actively produce (or destroy), one can view them as active (or communicative) *Brownian walkers* (Schimansky-Geier, Schweitzer, and Mieth 1997). Typically, the individual walker's motion is described as linear superposition of (passive) diffusive (Brownian) and reactive (active) parts. The underlying dynamics are governed by Newton's law: force = mass · acceleration (Haken 1977). The equation of motion for the ith walker ($i = 1, \ldots, N$) can be formulated as[21]

$$v_i(t) = \frac{dq_i(t)}{dt}, \qquad \frac{dv_i(t)}{dt} = -\gamma \cdot v_i(t) + \nabla s(q_i(t), t) + \sqrt{2\epsilon\gamma}\, \xi_i(t), \tag{3.1}$$

where the mass of the walker has been normalized to 1, t is time, $q_i(t)$ is the position of the walker, $v_i(t)$ is its velocity, $\gamma \cdot v_i(t)$ is the viscous force that slows down the particle's motion (γ is a friction coefficient), and $\xi_i(t)$ is Gaussian white noise with intensity ϵ. The gradient $\nabla s(x, t)$ introduces the taxis behavior, since this form assures that particles search for local maxima of the chemoattractant $s(x, t)$, which follows an appropriate reaction-diffusion equation. Eqn. (3.1) is known as the *Langevin equation*, which is an inhomogeneous linear stochastic differential equation that can be regarded as a phenomenological *ad hoc* description of individual behavior.

In some cases, there are averaging strategies to proceed from a fully stochastic Langevin description of single walker motion to density equations (Haken 1977), especially for spatially homogeneous situations in the mean-field limit. Particularly, in the chemotaxis model (eqn. (3.1)) it can be shown that under certain conditions

[21] We follow the notation of Stevens and Schweitzer 1997.

(particularly regarding the interaction, which has to be moderate[22]) the continuous chemotaxis system provides a good approximation of the interacting particle model (Stevens 1992).

The relation of reaction-diffusion and interacting particle systems has been systematically studied with regards to limiting behavior; in particular, hydrodynamic, McKean-Vlasov, and moderate limits can be distinguished. Recently, an alternative model of aggregation (without an external diffusive medium), characterized solely by density-dependent long-range weak (or McKean) and short-range moderate interactions has been analyzed by a Langevin strategy (Morale 2000).

Kinetic or Boltzmann Perspective

While the active walker approach starts with a (stochastic) description of individual particle dynamics, the *kinetic interpretation* neglects details of individual motion which had to be assumed in an *ad hoc* manner anyway. The kinetic approach focuses on *bulk behavior*, i.e., analysis of statistical properties of cellular interactions fully contained in the *single particle distribution function*, which represents the probability of finding a particle with a given velocity at a given position. It turns out that in both continuous and discrete interacting cell systems, a particularly simple description at a kinetic level can be gained by describing the state of the system in terms of the single particle distribution functions and to discard the effect of correlations (*Boltzmann approximation*). The corresponding equation is known as the *Boltzmann equation*, which arises as an approximate description of the temporal development in terms of the single particle distribution functions if all pair, triplet, and higher order correlations between particles are completely neglected. Boltzmann showed that this assumption is approximately true in dilute gases. Nevertheless, it has turned out that the Boltzmann (or mean-field) approximation is a reasonable assumption for many other, particularly cellular interactions (see the examples throughout this book).

With regards to biological pattern formation the Boltzmann strategy has not been utilized in a fully continuous interacting cell system so far, but in cellular automata, i.e., models discrete in space, time, and state space (cp. Bussemaker, Deutsch, and Geigant 1997). However, the *kinetic approach* is not restricted to discrete models. A systematic comparison of the Langevin and the Boltzmann approach with respect to biological pattern formation applications is still lacking. Note that Boltzmann-like models have also been suggested and analyzed in the context of traffic dynamics (see Helbing 2001 for a review).

---

[22] Consider an interacting particle system with N particles in the limit $N \to \infty$ and suppose that the particles are located in R^d. Therefore, in the macroscopic space-time coordinates the typical distance between neighboring particles is $O(N^{-1/d})$. Then, the system is studied in the (a) *McKean-Vlasov limit*, if any fixed particle interacts with $O(N)$ other particles in the whole system; (b) *moderate limit*, if any fixed particle interacts with many ($\approx N/\alpha(N)$) other particles in a small neighborhood with volume $\approx 1/\alpha(N)$, where both $\alpha(N)$ and $N/\alpha(N)$ tend to infinity; (c) *hydrodynamic limit*, if any fixed particle interacts with $O(1)$ other particles in a very small neighborhood with volume $\approx 1/N$ (Oelschläger 1989).

Summary. In this chapter we have presented mathematical modeling approaches to biological pattern formation. While reaction-diffusion models are appropriate to describe the spatio-temporal dynamics of (morphogenetic) signaling molecules or large cell populations, microscopic models at the cellular or subcellular level have to be chosen if one is interested in the dynamics of small populations. Interest in such "individual-based" approaches has recently grown significantly as more and more (genetic and proteomic) data are available. Important questions arise with respect to the mathematical analysis of microscopic individual-based models and the precise links to macroscopic approaches. While in physical processes typically the macroscopic equation is already known, the master equations in biological pattern formation are far from clear. In the remainder of the book we focus on cellular automata, which can be interpreted as an individual-based modeling approach.

Part II

Cellular Automaton Modeling

4

Cellular Automata

"Biological systems appear to perform computational tasks using methods that avoid both arithmetical operations and discrete approximations to continuous systems."

(Hasslacher 1987)

In this chapter the biological roots of cellular automata (CA) are described and formal definitions of CA are provided. CA are characterized by a regular lattice, an interaction neighborhood template, a set of elementary states, and a local space- and time-independent transition rule which is applied to each cell in the lattice. In particular, we introduce deterministic, probabilistic and lattice-gas cellular automata (LGCA). Furthermore, we present strategies to analyze spatio-temporal pattern formation in CA models. In sec. 4.4.2, the so-called mean-field theory is presented as an approximative method to study dynamic properties of CA. After assigning a probability to each of the elementary states of the automaton, the probabilities of the next generation are calculated on the basis of the usual combinatorial rules of probability, assuming that the probabilities for each of the sites in a neighborhood are independent. It turns out that mean-field equations for deterministic and probabilistic CA are time-dependent difference equations, whereas mean-field equations for LGCA are space- and time-dependent difference equations, called lattice-Boltzmann equations. The space dependence in the latter case results from the propagation step that supplements the interaction dynamics in LGCA. Since the state values of CA mean-field approximations are continuous, standard analytical tools for studying the system dynamics can be applied. We focus on a linear stability analysis. In sec. 4.4.3, we outline the linear stability analysis for systems of time-dependent difference equations. We also show how to deal with systems of space- and time-dependent difference equations. The stability of spatially homogeneous steady states is investigated by analyzing the spectrum of the linearized and Fourier-transformed system of space- and time-dependent difference equations, called Boltzmann propagator. We characterize situations in which Fourier modes can indicate the formation of spatial patterns. In subsequent chapters specific examples of LGCA and their analysis are provided.

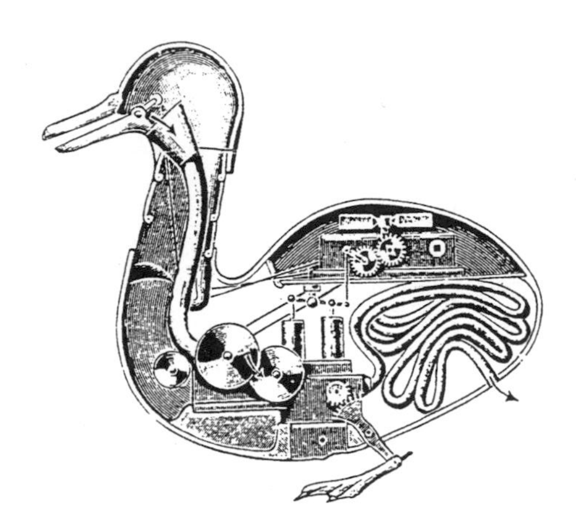

Figure 4.1. Vaucanson's mechanical duck from 1735 is an early "automaton" which can move its head, tail, and wings (more details about the duck can be found in Chapuis and Droz, 1958).

4.1 Biological Roots

The notion of a CA originated in the works of **John von Neumann** (1903–1957) and **Stanislaw Ulam** (1909–1984). CA as discrete, local dynamical systems (to be formalized later in this chapter) can be equally well viewed as a mathematical idealization of natural systems, a discrete caricature of microscopic dynamics, a parallel algorithm, or a discretization of partial differential equations. According to these interpretations distinct roots of CA may be traced back in biological modeling, computer science, and numerical mathematics that are well documented in numerous and excellent sources (Baer and Martinez 1974; Bagnoli 1998; Casti 1989; Emmeche 1994; Farmer, Toffoli, and Wolfram 1984; Ganguly, Sikdar, Deutsch, Canright, and Chaudhuri 2004; Gerhardt and Schuster 1995; Lindenmayer and Rozenberg 1976; Sigmund 1993; Vollmar 1979; Wolfram 1984; Wolfram 1986b; Wolfram 2002). Here, we focus on a sketch of important roots and implications of *biologically motivated cellular automata*.

Self-Reproduction and Self-Reference. The basic idea and trigger for the development of CA as biological models was a need for *noncontinuum* concepts. There are biological problems in which continuous (e.g., differential equation) models do not capture the essentials. A striking example is provided by self-reproduction of discrete units, the cells. In the 1940s John von Neumann tried to solve the following problem: Which kind of logical organization makes it possible that an automaton (viewed as an "artificial device") reproduces itself? An early example of a *biological automaton* is Jacques de Vaucanson's mechanical duck (fig. 4.1) from 1735, which can not only move its head, tail, and wings but is even able "to feed". However, the duck automaton is unable to reproduce.

John von Neumann's lectures at the end of the 1940s clearly indicated that his work was motivated by the self-reproduction ability of biological organisms. Additionally, there was an impact of achievements in automaton theory (Turing machines) and Gödel's work on the foundations of mathematics, in particular the incompleteness theorem ("There are arithmetical truths which can, in principle, never be proven"). A central role in the proof of the incompleteness theorem is played by self-referential statements. Sentences such as "This sentence is false" refer to themselves and may trigger a closed loop of contradictions. Note that biological self-reproduction is a particularly clever manifestation of self-reference (Sigmund 1993). A genetic instruction such as "Make a copy of myself" would merely reproduce it*self* (self-reference), implying an endless doubling of the blueprint, but not a construction of the organism. How can one get out of this dilemma between self-reference and self-reproduction?

The first model of self-reproduction proposed by von Neumann in a *thought experiment* (1948) is not bound to a fixed lattice; instead the system components are fully floating. The clue of the model is the twofold use of the (genetic) information as uninterpreted and interpreted data, respectively, corresponding to a syntactic and a semantic data interpretation. The automaton actually consists of two parts: a flexible construction and an instruction unit referring to the duality between computer and program or, alternatively, the cell and the genome (Sigmund 1993). Thereby, von Neumann anticipated the uncoding of the genetic code following Watson and Crick's discovery of the DNA double helix structure (1953), since interpreted and uninterpreted data interpretation directly correspond to molecular translation and transcription processes in the cell (Watson and Crick 1953a, 1953b). Arthur Burks, one of von Neumann's students, called von Neumann's first model the *kinematic model* since it focuses on a kinetic system description. It was Stanislaw Ulam who suggested a "cellular perspective" and contributed with the idea of restricting the components to discrete spatial cells (distributed on a regular lattice). In a manuscript of 1952/1953, von Neumann proposed a model of self-reproduction with 29 states. The processes related to physical motion in the kinematic model are substituted by information exchange of neighboring cells in this pioneering *cellular automaton model*. Edgar F. Codd and Chris Langton, one of the pioneers of *artificial life* research, reduced this self-reproducing automaton model drastically (Codd 1968; Langton 1984). Meanwhile, even "sexually reproducing" CA have been proposed (Vitanyi 1973). Furthermore, an autopoietic self-reproducing micellar system has been modeled by a LGCA (Coveney, Emerton, and Boghosian 1996).

Game of Life. While the fundaments of CA were laid out in the 1950s, they became known to a larger audience only at the end of the 1960s when von Neumann's collected work on CA appeared under the auspices of A. Burks (Burks 1970; von Neumann 1966). Shortly after, John Conway proposed the *Game of Life* automaton through Martin Gardner's column in *Scientific American* as a simple metaphor of birth and death. This game is typical of the cellular automaton approach: it is simple, and purely local rules can give rise to rather complex behavior, a self-structuring phenomenon (Boerlijst 1994; cp. fig. 4.2). The development of CA is governed by a

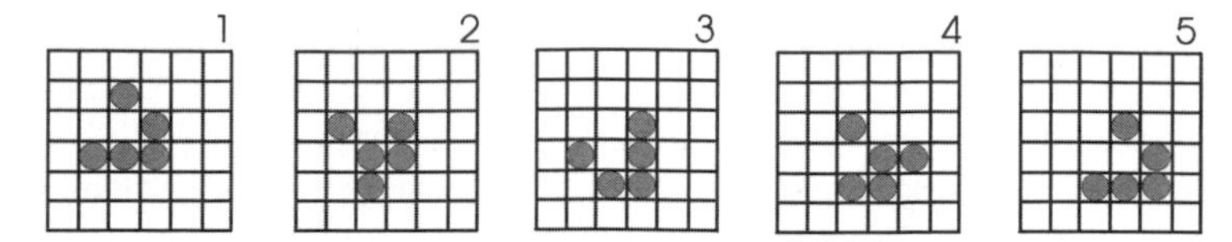

Figure 4.2. Game of Life: Five subsequent time steps are shown. Gray-filled squares represent "living" cells, while empty squares are "dead" cells. In each time step, the following rule is applied to each square simultaneously: A "living" cell surrounded by less than three or more than three "living" cells amongst its eight nearest neighbors dies of isolation or overcrowdedness. On the other hand, a "dead" cell will come to "life" if there are exactly three "living" cells amongst its eight nearest neighbors.

kind of *microcausality*: "States of spatially distant cells, or of the distant past exert no influence on what is happening here and now; the next generation is entirely determined by the present one; the future depends on the past, but only via the present" (quoted from Sigmund, 1993).

The Game of Life led to an unprecedented popularization of the CA idea insofar as the rules are easy to program and even the lay person can enjoy simulations on the home computer.[23] The upcoming digital computers made it easy to investigate the effects of rule alterations and to "simulate biology on computers" (cp. Feynman 1982). Thus, a game and the computer were triggering a burst of *new rules* motivated by various types of biological interactions, among them the first suggestions for CA models of biological pattern formation (Gocho, Pérez-Pascual, and Rius 1987; Meakin 1986; Young 1984). Simultaneously, interest in the principles of self-organizing systems arose and CA seemed to offer a paradigm of universal computation and self-organization (Mitchell 2002; Varela, Maturana, and Uribe 1974; Wolfram 1984; Wolfram 2002).

Lattice Models and Pattern Formation. CA are examples of *lattice models* since spatial configurations are described with respect to an underlying (regular) lattice. Various other lattice models with biological applications have been proposed. "Lattice proteins" (Bornberger 1996) are, for instance, a valuable tool to address basic questions of the sequence–structure relation and foldability of biopolymers and may be viewed as abstractions of biopolymers: residues are presented at a unified size by placing each, but at most one at a time, on one node of a regular lattice (Dill et al. 1995). It is commonly assumed that only the sequence determines the unique *native* structure which corresponds to the equilibrium minimum free energy state. Furthermore, cell–gene interactions and heterogeneity of tissue arrays have been studied in coupled map lattice models (Bignone 1993; Klevecz 1998). *Lattice swarms* of interacting agents can be viewed as a model of complex architectures of social insects (Theraulaz and Bonabeau 1995a; Theraulaz and Bonabeau 1995b).

[23] Meanwhile, the Game of Life has been extended to the "larger than life version" by considering larger interaction neighborhoods (Evans 2001).

4.2 Cellular Automaton Models of Pattern Formation in Interacting Cell Systems

CA models have been proposed for a large number of biological applications, including ecological (Bagnoli and Bezzi 1998, 2000; Cannas, Paez, and Marco 1999; de Roos, McCauley, and Wilson 1991; Phipps 1992), epidemiological (Schönfisch 1993, 1995), ethological (game theoretical) (Solé, Miramontes, and Goodwin 1993; Herz 1994), evolutionary (Bagnoli 1998; Boerlijst 1994), and immunobiological aspects (Agur 1991; Ahmed 1996; de Boer, van der Laan, and Hogeweg 1993; Hogeweg 1989; Meyer-Hermann 2002). Here, we are interested in CA models of biological pattern formation in interacting cell systems. While von Neumann did not consider the spatial aspect of cellular automaton patterns per se—he focused on the pattern as a unit of self-reproduction—we are particularly concerned with the precise form of the patterns as well as the spatio-temporal dynamics of pattern formation.

Various automaton rules mimicking general pattern-forming principles (as described in the first chapter) have been suggested and may lead to models of (intracellular) cytoskeleton and membrane dynamics, tissue formation, tumor growth, and life cycles of microorganisms or animal coat markings. Automaton models of cellular pattern formation can be roughly classified according to the prevalent type of interaction. *Cell–medium interactions* dominate (nutrient-dependent) growth models, while one can further distinguish *direct cell–cell* from *indirect cell–medium–cell interactions*. In the latter, communication is established by means of an extracellular field. Such (mechanical or chemical) fields may be generated by tensions or chemoattractant produced and perceived by the cells themselves.

Cell–Medium or Growth Models

Growth models typically assume the following scenario: A center of nucleation is growing by consumption of a diffusible or nondiffusible substrate. Growth patterns mirror the availability of the substrate since the primary interaction is restricted to the cell-substrate level. Bacterial colonies may serve as a prototype expressing various growth morphologies, in particular dendritic patterns (Ben-Jacob et al. 1992, 1994). Various extensions of a simple *diffusion-limited aggregation* (DLA) rule can explain dendritic or fractal patterns (Ben-Jacob, Shmueli, Shochet, and Tenenbaum 1992; Boerlijst 1994; Witten and Sander 1981) (fig. 4.3; cp. also sec. 5.5). A CA model for the development of fungal mycelium branching patterns based on geometrical considerations was suggested in Deutsch 1991, 1993 and Deutsch, Dress, and Rensing 1993. Recently, various CA have been proposed as models of tumor growth (cp. sec. 10.4; Moreira and Deutsch 2002). Note that CA have also been introduced as pattern recognition tools, in particular for the detection of genetic disorders of cancerous cells (Moore and Hahn 2000, 2001).

Cell–Medium–Cell Interaction Models

Excitable Media and Chemotaxis. Spiral waves can be observed in a variety of physical, chemical, and biological systems. Typically, spirals indicate the *excitabil-*

Figure 4.3. Fractal growth pattern of a bacterial colony (*Paenibacillus dendritiformis*). Bacteria are grown in a Petri dish containing growth medium. Different nutrient concentrations may imply various growth morphologies. Courtesy of E. Ben-Jacob, University of Tel-Aviv.

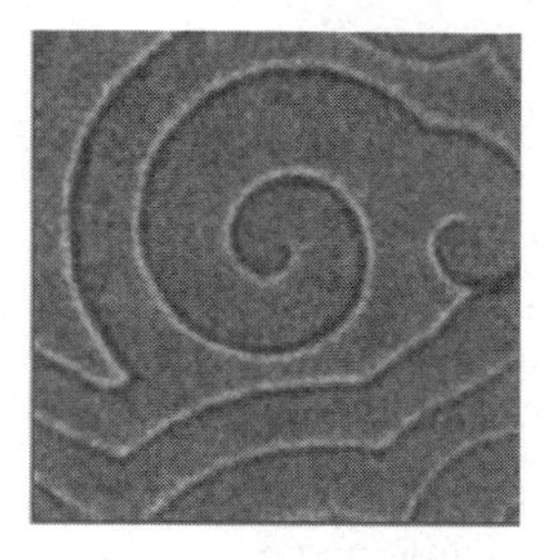

Figure 4.4. The slime mold *Dictyostelium discoideum* as an excitable medium. (Courtesy of D. Dormann, University of Dundee.) Excitable systems are characterized by wave-like signal propagation and typical cycles of excitability which particularly contain a refractory period. Periodic aggregation signals (cAMP pulses) spread in concentric or spiral form and lead to corresponding changes in *Dictyostelium* cell shape and motility, which are visualized using dark field optics. The mechanism driving migration in *Dictyostelium* is chemotaxis towards the cAMP signal. The picture shows a temporal sequence in the wave pattern of the cAR3/R19-mutant (period length appr. 11.6 min, wave speed appr. 260 μm/min).

ity of the system. Excitable media are characterized by resting, excitable, and excited states (see also sec. 11.2). After excitation the system undergoes a recovery (refractory) period during which it is not excitable (Tyson and Keener 1988). Prototypes of *excitable media* are the Belousov-Zhabotinskii reaction (fig. 2.7) and aggregation of the slime mold *Dictyostelium discoideum* (fig. 4.4; see also Dallon, Othmer, v. Oss, Panfilov, Hogeweg, Höfer, and Maini 1997; Ross, Müller, and Vidal 1988).

A number of CA models of excitable media have been proposed that differ in state space design, actual implementation of diffusion, and the consideration of random effects (Gerhardt and Schuster 1989; Gerhardt, Schuster, and Tyson 1990a; Gerhardt, Schuster, and Tyson 1990b; Markus and Hess 1990). A stochastic cellular automaton was constructed as a model of chemotactic aggregation of myxobacteria (Stevens 1992). In particular, a nondiffusive chemical, the slime, and a diffusive chemoattractant are assumed in order to produce *realistic* aggregation patterns.

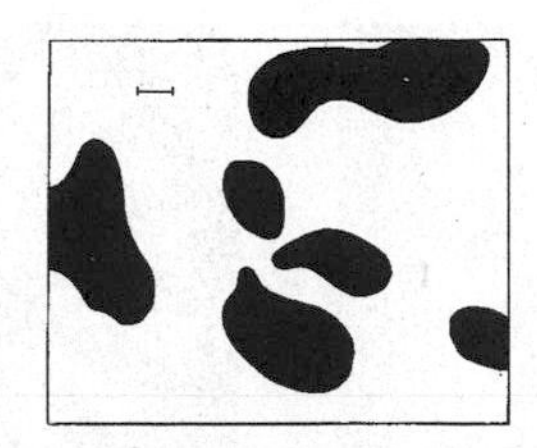

Figure 4.5. Historic Turing simulation (Turing 1952): an example of a "dappled pattern" resulting from a reaction-diffusion morphogen system. The simulation was performed "by hand" by Turing himself (a marker of unit length is shown).

Turing systems. Spatially stationary Turing patterns are brought about by a diffusive instability, the *Turing instability* (Turing 1952). The first (two-dimensional) CA of Turing pattern formation based on a simple activator-inhibitor interaction was suggested by Young 1984. Simulations produce spots and stripes (claimed to mimic animal coat markings) depending on the range and strength of the inhibition (cp. also fig. 4.5).

An extension is the work by Markus and Schepers 1993 on Turing patterns. Turing patterns can also be simulated with appropriately defined reactive LGCA (Dormann, Deutsch, and Lawniczak 2001; Hasslacher, Kapral, and Lawniczak 1993; cp. also ch. 11).

Activator–inhibitor automaton models might help to explain the development of ocular dominance stripes (Swindale 1980). Ermentrout et al. introduced a model of molluscan pattern formation based on activator–inhibitor ideas (Ermentrout, Campbell, and Oster 1986). Further CA models of shell patterns have been proposed (Gunji 1990; Kusch and Markus 1996; Markus and Kusch 1995; Plath and Schwietering 1992; Vincent 1986). An activator–inhibitor automaton also proved useful as a model of fungal differentiation patterns (Deutsch 1993; cp. also fig. 4.6).

Cell–Cell Interaction Models

Differential Adhesion. In practice, it is rather difficult to identify the precise pattern-forming mechanism, since different mechanisms (rules) may imply phenomenologically indistinguishable patterns. It is particularly difficult to decide between effects of cell–cell interactions and indirect interactions via the medium. For example, one-dimensional rules based on cell–cell interactions have been suggested as an alternative model for animal coat markings (Gocho, Pérez-Pascual, and Rius 1987). Such patterns have traditionally been explained with the help of reaction-diffusion systems based on indirect cell interaction (Murray 2002). Meanwhile, several CA are available as models of cell rearrangement and sorting out due to differential adhesion (Bodenstein 1986; Goel and Thompson 1988; Mostow 1975). A remarkable three-dimensional automaton model based on cell–cell interaction by differential adhesion and chemotactic communication via a diffusive signal molecule is able to model aggregation, sorting out, fruiting body formation, and motion of the slug in the slime

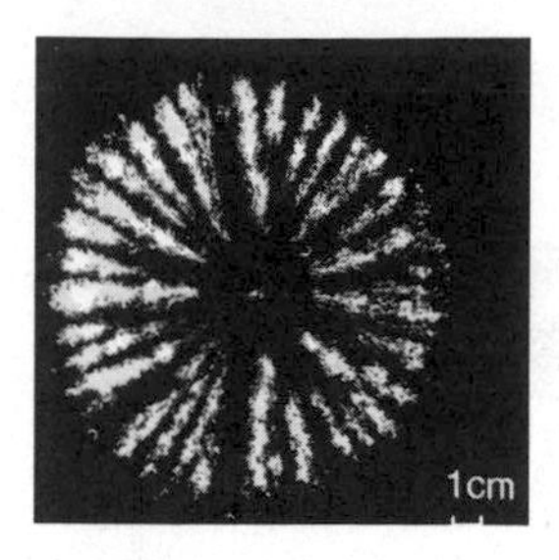

Figure 4.6. Radial pattern in the spore distribution of the ascomycete *Neurospora crassa* (left) and in CA simulation. The fungus is inoculated in the middle of a Petri dish filled with nutrient. The cellular automaton models an activator–inhibitor system in a growing network (see Deutsch 1991, 1993 for explanations).

mold *Dictyostelium discoideum* (Savill and Hogeweg 1997). This model has been extended to include the effects of phototaxis (Marée and Hogeweg 2001).

Alignment, Swarming. While differential adhesion may be interpreted as a *density-dependent* interaction one can further distinguish *orientation-dependent* cell–cell interactions. An automaton modeling alignment of oriented cells has been introduced in order to describe the formation of fibroblast filament bundles (Edelstein-Keshet and Ermentrout 1990). An alternative model of orientation-induced pattern formation based on the lattice-gas automaton idea has been suggested (Deutsch 1995; cp. ch. 8). Within this model the initiation of swarming can be associated with a phase transition (Bussemaker, Deutsch, and Geigant 1997). A possible application is street formation of social bacteria (e.g., myxobacteria; cp. fig. 8.1). Parallel bacterial motion is the precondition for the rippling phenomenon observed in myxobacteria cultures, which occurs prior to fruiting body formation and can be characterized as a standing wave pattern. Recently, an automaton model for myxobacterial rippling pattern formation based on "collision-induced" cell reversals has been suggested (Börner et al. 2002).

Cytoskeleton Organization, Differentiation

Beside the spatial pattern aspect, a number of further problems of developmental dynamics have been tackled with the help of CA models. The organization of DNA can be formalized on the basis of a one-dimensional CA (Burks and Farmer 1984). Microtubule array formation along the cell membrane is in the focus of models suggested by Smith et al. (Hameroff, Smith, and Watt 1986; Smith, Watt, and Hameroff 1984). Understanding microtubule pattern formation forms an essential precondition for investigations of interactions between intra- and extracellular morphogenetic dynamics (Kirschner and Mitchison 1986). In addition, alternative rules for microtubule dynamics—based on electrical excitation of elementary states—have been investigated (e.g., Hameroff, Smith, and Watt 1986; Rasmussen, Karampurwala, Vaidyanath, Jensen, and Hameroff 1990). Furthermore, automaton models of membrane dynamics exist, in particular focusing on the formation of channels by means of ion desorption (Kubica, Fogt, and Kuczera 1994).

Nijhout et al. 1986 proposed a rather complicated CA model for differentiation and mitosis based on rules incorporating morphogens and mutations. Another automaton model addresses blood cell differentiation as a result of spatial organization (Mehr and Agur 1992). It is assumed in this model that spatial structure of the bone marrow plays a role in the control process of hemapoiesis (Zipori 1992). The problem of differentiation is also the primary concern in a stochastic CA model of the intestinal crypt (Potten and Löffler 1987).

It is typical of many of the automaton approaches sketched in this short overview that they lack detailed analysis: the argument is often based solely on the beauty of simulations—for a long period of time people were just satisfied with the simulation pictures. This *simulation phase* in the history of CA characterized by an overwhelming output of a variety of cellular automaton rules was important since it produced a lot of challenging questions, particularly related to the quantitative analysis of automaton models.

4.3 Definition of Deterministic, Probabilistic and Lattice-Gas Cellular Automata

In the 1950s John von Neumann and Stanislaw Ulam proposed the concept of cellular automata. In recent years, the originally very restrictive definition has been extended to many different applications. In general, a **cellular automaton** (CA) is specified by the following definition:

- a regular discrete lattice ($\mathcal{L}$) of cells (nodes, sites) and boundary conditions,
- a finite—typically small—set of states ($\mathcal{E}$) that characterize the cells,
- a finite set of cells that defines the interaction neighborhood ($\mathcal{N}^I$) of each cell, and
- a rule ($\mathcal{R}$) that determines the dynamics of the states of the cells.

In this section we define CA, especially the classical (von Neumann) *deterministic* CA, the *probabilistic* CA, and the *lattice-gas* CA (LGCA). To illustrate the idea of deterministic and probabilistic CA we introduce a simple model for the *spread of plants*.

4.3.1 Lattice Geometry and Boundary Conditions

We start with the definition of a "cellular space" (*regular lattice*) in which the automaton is defined. A regular lattice $\mathcal{L} \subset \mathbb{R}^d$ consists of a set of cells,[24] which homogeneously cover (a portion of) a d-dimensional Euclidean space. Each cell is

[24] These are regular polygons, i.e., polygons whose sides all have the same length and whose angles are all the same.

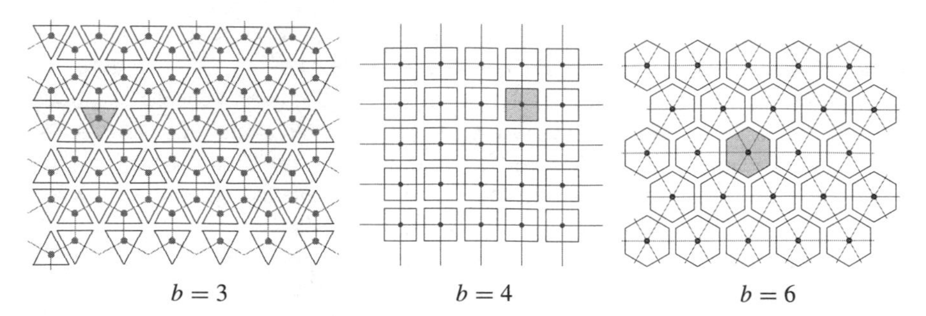

Figure 4.7. Cells and nodes (dots) in two-dimensional lattices; gray areas indicate cell shape, which is a triangle, a square, or a hexagon, respectively.

labeled by its position $r \in \mathcal{L}$. If LGCA are concerned, the cells are commonly called nodes, while they are referred to as sites in stochastic process models. The spatial arrangement of the cells is specified by nearest neighbor connections (links), which are obtained by joining pairs of cells, usually in a regular arrangement. For any spatial coordinate $r \in \mathcal{L}$, the nearest lattice neighborhood $\mathcal{N}_b(r)$ is a finite list of neighboring cells and is defined as

$$\mathcal{N}_b(r) := \{r + c_i : c_i \in \mathcal{N}_b,\ i = 1, \dots, b\},$$

where b is the *coordination number*, i.e., the number of nearest neighbors on the lattice. By $\mathcal{N}_b$ we denote the *nearest-neighborhood template* with elements $c_i \in \mathbb{R}^d$, $i = 1, \dots, b$.

A *one-dimensional* ($d = 1$) regular lattice consists of an array of cells, in which each cell is connected to its right and left neighbor ($b = 2$). Then, for example,

$$\begin{aligned} \mathcal{L} &\subseteq \mathbb{Z} = \{r : r \in \mathbb{Z}\}, \\ \mathcal{N}_2 &= \{1, -1\}. \end{aligned}$$

In *two dimensions* ($d = 2$), the only regular polygons that form a regular tessellation of the plane are triangles ($b = 3$), rectangles ($b = 4$), and hexagons ($b = 6$), such as are shown in fig. 4.7 (Ames 1977). The nearest neighborhood templates are defined as

$$\mathcal{N}_b = \left\{ c_i,\ i = 1, \dots, b : c_i = \left(\cos\left(\frac{2\pi(i-1)}{b}\right), \sin\left(\frac{2\pi(i-1)}{b}\right) \right) \right\}. \tag{4.1}$$

For example, for the square lattice ($b = 4$),

$$\begin{aligned} \mathcal{L} &\subseteq \mathbb{Z}^2 = \left\{ r : r = (r_1, r_2),\ r_j \in \mathbb{Z},\ j = 1, 2 \right\}, \\ \mathcal{N}_4 &= \{(1,0), (0,1), (-1,0), (0,-1)\}. \end{aligned}$$

The number of cells in each space direction i is denoted by $L_i, i = 1, \dots, d$, and the total number of cells by

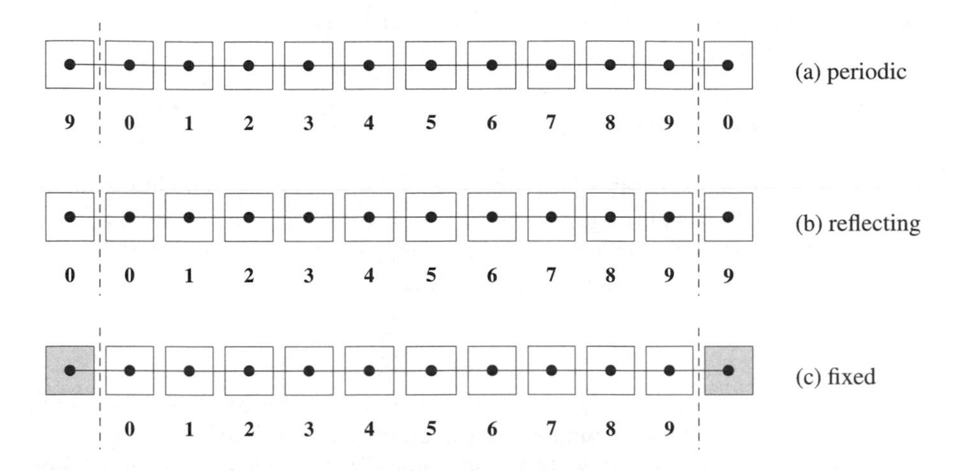

Figure 4.8. Boundaries in a one-dimensional lattice; cells $r \in \mathcal{L} = \{0, \ldots, 9\}$, $|\mathcal{L}| = 10$, represent the lattice of a CA. The single cells at the left and right of the dashed lines define the corresponding left and right nearest neighbor cells for the boundary cells $r = 0$ and $r = 9$, respectively. Gray cells are cells with a prescribed fixed value.

$$|\mathcal{L}| = L_1 \cdot \cdots \cdot L_d.$$

CA are usually explored by computer experiments in which the lattice has to be considered finite. In finite lattices, i.e., $|\mathcal{L}| < \infty$, it is necessary to impose *boundary conditions* that define the set of nearest neighbors for cells at the boundary of the lattice.

A lattice can be periodically extended; i.e., opposite bounds of the lattice are glued together (fig. 4.8a). In one dimension this leads to a ring, and in two dimensions to a torus. These so-called *periodic* boundary conditions are often used for the approximation of a simulation on an infinite lattice. Note that they can introduce artificial spatial periodicities into the system. In order to define a boundary condition that is comparable to a zero-flux (Neumann) boundary condition used in continuous diffusion models, the lattice is *reflecting* at the boundary (see fig. 4.8b). To model boundary conditions corresponding to Dirichlet boundary conditions in partial differential equations, cells at the boundary have nearest neighbors with a prescribed *fixed* value (see fig. 4.8c; Schönfisch 1993). Also, *absorbing* boundary conditions can be defined: if an individual steps off the lattice, it is lost forever. Of course, all types of boundary conditions can be combined with each other.

If interaction processes of more than one component (species) have to be examined, it is useful, especially for LGCA, to consider separate lattices for each component. From now on, in the first place we give the formal description of one-component systems. Subsequently, for LGCA we note very briefly the extension to multicomponent systems. The component dependence of any variable will be expressed by adding the subscript σ, $\sigma = 1, \ldots, \varsigma$; i.e., the lattice for component σ is $\mathcal{L}_\sigma$. We will assume that all $\mathcal{L}_\sigma$ have the same topology, are labelled identically, and consist of the same number of cells. Then, solely for simplicity of notation, we identify all lattices and denote them by $\mathcal{L}$.

4.3.2 Neighborhood of Interaction

An interaction neighborhood $\mathcal{N}_b^I(r)$ specifies a set of lattice cells that influence the state at cell r. Following the CA definition of John von Neumann, its size and topology do not depend on the lattice cell, and it does not change in time. Therefore, the *interaction neighborhood* is defined as an ordered set

$$\mathcal{N}_b^I(r) = \{r + c_i \, : \, c_i \in \mathcal{N}_b^I\} \subseteq \mathcal{L},$$

where $\mathcal{N}_b^I$ is an *interaction neighborhood template* that can be chosen in several ways. Note that this definition implies translation invariance of the interaction neighborhood. Furthermore, for cells at the lattice boundaries the interaction neighborhood has to be additionally specified if nonperiodic boundary conditions are imposed. In the case of periodic boundary conditions, we always assume that the sum $r + c_i$ is taken modulo some appropriate value such that $r + c_i \subset \mathcal{L}$ for all $c_i \in \mathcal{N}_b^I$. Famous examples in a two-dimensional square lattice are the *von Neumann neighborhood* (Burks 1970; cp. fig. 4.9(a))

$$\mathcal{N}_4^I = \{(0,0), (1,0), (0,1), (-1,0), (0,-1)\} = \mathcal{N}_4 \cup \{(0,0)\},$$

and the *Moore neighborhood* (Moore 1962; cp. fig. 4.9(b))

$$\mathcal{N}_4^I = \{(0,0), (1,0), (1,1), (0,1), (-1,1), (-1,0), (-1,-1), (0,-1), (1,-1)\}.$$

In some applications (i.e., Gerhardt and Schuster 1989; Wolfram 1984; Young 1984), extended neighborhoods are considered. They are defined as either *R-radial* (fig. 4.9(c)):

$$\mathcal{N}_4^I = \left\{c = (c_x, c_y) : c_x, c_y \in \{-R, \dots, R\} \wedge \sqrt{c_x^2 + c_y^2} \leq R\right\},$$

or *R-axial* (fig. 4.9(d)):

$$\mathcal{N}_4^I = \left\{c = (c_x, c_y) : \; c_x, c_y \in \{-R, \dots, R\}\right\}.$$

In most applications the interaction neighborhood template is chosen to be symmetrical, but asymmetrical schemes are also possible. Furthermore, the cell r itself does not necessarily have to be included in the neighborhood of interaction, i.e., $r \notin \mathcal{N}_b^I(r)$. If this is important to stress, we will refer to an *outer interaction neighborhood* $\mathcal{N}_{bo}^I(r)$. Moreover, in an extreme case, the interaction neighborhood contains only one element: the cell r itself, i.e., $\mathcal{N}_b^I = \{r\}$. Then, each cell is independent of other cells. The *number of nodes* in an interaction neighborhood is denoted by

$$\nu := |\mathcal{N}_b^I(r)|, \qquad \text{accordingly} \quad \nu_o := |\mathcal{N}_{bo}^I(r)|.$$

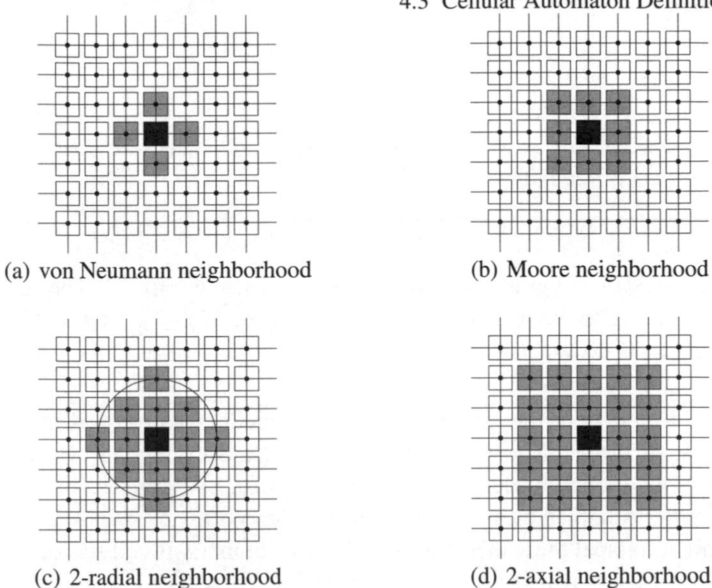

Figure 4.9. Examples of interaction neighborhoods (gray and black cells) for the black cell in a two-dimensional square lattice (see also Durrett and Levin 1994b).

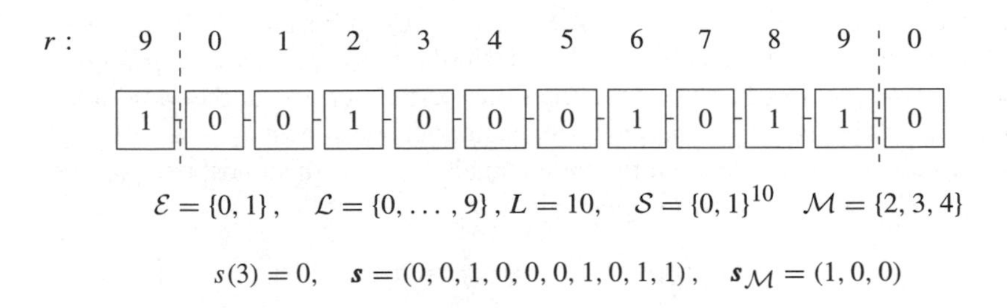

Figure 4.10. Example of a one-dimensional global ($\boldsymbol{s}$) and local ($\boldsymbol{s}_{\mathcal{M}}$) lattice configuration (periodic boundary conditions).

4.3.3 States

To each cell $r \in \mathcal{L}$ we assign a *state value* $s(r) \in \mathcal{E}$ which is chosen from the (usually "small") finite set of elementary states $\mathcal{E}$, i.e.,

$$s : \mathcal{L} \to \mathcal{E}.$$

The elements of $\mathcal{E}$ can be numbers, symbols, or other objects (e.g., biological cells). A simple example is shown in fig. 4.10.

A *global configuration* $\boldsymbol{s} \in \mathcal{E}^{|\mathcal{L}|}$ of the automaton is determined by the state values of all cells on the lattice (fig. 4.10), i.e.,

$$\mathbf{s} := \big(s(r_1), \ldots, s(r_{|\mathcal{L}|})\big) = (s(r_i))_{r_i \in \mathcal{L}} \,.$$

This vector assumes values in the *state space* $\mathcal{S} = \mathcal{E}^{|\mathcal{L}|}$.

A *local configuration* is a vector $\mathbf{s}_{\mathcal{M}}$ consisting of state values of cells in an ordered subset $\mathcal{M}$ of the lattice $\mathcal{L}$ (fig. 4.10), i.e.,

$$\mathbf{s}_{\mathcal{M}} := \big(s(r_1), \ldots, s(r_{|\mathcal{M}|})\big) = (s(r_i))_{r_i \in \mathcal{M}} \,, \quad \mathcal{M} \subset \mathcal{L}.$$

$\mathbf{s}|_{\mathcal{M}} = \mathbf{s}_{\mathcal{M}} \in \mathcal{S}|_{\mathcal{M}}$ is the global configuration $\mathbf{s}$ restricted to $\mathcal{M}$. The local configuration according to the interaction neighborhood $\mathcal{N}_b^I(r) \subseteq \mathcal{L}$ of a focal cell r is denoted by $\mathbf{s}_{\mathcal{N}(r)} \in \mathcal{S}_{\mathcal{N}(r)} := \mathcal{S}|_{\mathcal{N}_b^I(r)}$.

Example: In a model for the "spread of plants" a field is partitioned into equally sized squares, such that each square may contain at most one plant. Hence, in the language of CA, each cell (e.g., square of a field) of a one-dimensional (periodic) lattice $\mathcal{L}$ is either occupied by a plant or is vacant, i.e., $s(r) \in \mathcal{E} = \{plant,\ vacant\}$. It is convenient to label these elementary states by numbers: if cell r is occupied by a plant: $s(r) = 1$ and if it is vacant: $s(r) = 0$, i.e., $\mathcal{E} = \{0, 1\}$. A possible lattice configuration is illustrated in fig. 4.10.

States in Lattice-Gas Cellular Automata. In LGCA (velocity) *channels*, (r, c_i), $c_i \in \mathcal{N}_b$, $i = 1, \ldots, b$, are associated with each node r of the lattice. In addition, a variable number $\beta \in \mathbb{N}_0$ of rest channels (zero-velocity channels), (r, c_i), $b < i \leq b + \beta$, with $c_i = \{0\}^\beta$ may be introduced. Furthermore, an *exclusion principle* is imposed. This requires that not more than one particle can be at the same node within the same channel. As a consequence, each node r can host up to $\tilde{b} = b + \beta$ particles, which are distributed in different channels (r, c_i) with at most one particle per channel. Therefore, state $s(r)$ is given by

$$s(r) = \big(\eta_1(r), \ldots, \eta_{\tilde{b}}(r)\big) =: \boldsymbol{\eta}(r),$$

where $\boldsymbol{\eta}(r)$ is called *node configuration* and $\eta_i(r), i = 1, \ldots, \tilde{b}$, *occupation number*. Occupation numbers are Boolean variables that indicate the presence ($\eta_i(r) = 1$) or absence ($\eta_i(r) = 0$) of a particle in the respective channel (r, c_i). Therefore, for LGCA the set of elementary states $\mathcal{E}$ of a single node is given by

$$\mathcal{E} = \{0, 1\}^{\tilde{b}} \ni s(r).$$

The *total number of particles* present at a node r is denoted by

$$\mathrm{n}(r) := \sum_{i=1}^{\tilde{b}} \eta_i(r).$$

The *local configuration* $\mathbf{s}_{\mathcal{N}(r)}$ is given by

$$\mathbf{s}_{\mathcal{N}(r)} = (\eta(r_i))_{r_i \in \mathcal{N}_b^I(r)} =: \boldsymbol{\eta}_{\mathcal{N}(r)}.$$

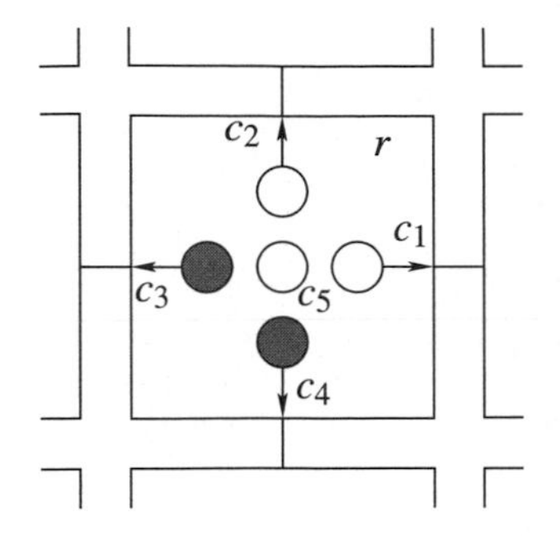

$$\tilde{b} = 5$$

velocity channels: $(r, c_1), (r, c_2), (r, c_3), (r, c_4)$

rest channel: (r, c_5)

$$\begin{aligned}\boldsymbol{\eta}(r) &= (\eta_1(r), \eta_2(r), \eta_3(r), \eta_4(r), \eta_5(r)) \\ &= (0, 0, 1, 1, 0)\end{aligned}$$

$$\mathrm{n}(r) = 2$$

Figure 4.11. Node configuration: channels of node r in a two-dimensional square lattice ($b = 4$) with one rest channel ($\beta = 1$). Filled dots denote the presence of a particle in the respective channel.

Fig. 4.11 gives an example of the representation of a node in a two-dimensional lattice with $b = 4$ and $\beta = 1$, i.e., $\tilde{b} = 5$.

In multicomponent LGCA, ς different types (σ) of particles reside on separate lattices ($\mathcal{L}_\sigma$) and the exclusion principle is applied independently to each lattice. The state variable is given by

$$s(r) = \boldsymbol{\eta}(r) = (\boldsymbol{\eta}_\sigma(r))_{\sigma=1}^{\varsigma} = \left(\eta_{\sigma,1}(r), \ldots, \eta_{\sigma,\tilde{b}}(r)\right)_{\sigma=1}^{\varsigma} \in \mathcal{E} = \{0, 1\}^{\tilde{b}\varsigma}.$$

Since we can define an interaction neighborhood $\mathcal{N}_{b,\sigma}^I(r) \subseteq \mathcal{L}_\sigma$ for each component σ, the local configuration $\boldsymbol{s}_{\mathcal{N}(r)}$ is given by

$$\boldsymbol{s}_{\mathcal{N}(r)} = \left((\boldsymbol{\eta}_\sigma(r_i))_{r_i \in \mathcal{N}_{b,\sigma}^I(r)} \right)_{\sigma=1}^{\varsigma} = \boldsymbol{\eta}_{\mathcal{N}(r)}.$$

4.3.4 System Dynamics

The dynamics of the automaton are determined by a *local transition rule* $\mathcal{R}$, which specifies the new state of a cell as a function of its interaction neighborhood configuration, i.e.,

$$\mathcal{R} : \mathcal{E}^\nu \to \mathcal{E}, \qquad \text{where } \nu = |\mathcal{N}_b^I|.$$

The rule is *spatially homogeneous;*, i.e., it does not depend explicitly on the cell position r. However, the rule can be extended to include spatial and/or temporal inhomogeneities. A typical example of spatial inhomogeneity is a CA with fixed boundary conditions (cp. p. 69). Similarly, time-dependent rules can be introduced, for example, by alternating two rules at even and odd time steps, as used for CA models for directed percolation or the model for Ising spin dynamics (Chopard and Droz 1998). In sec. 5.3.1 we present a time- and space-dependent local transition rule that allows us to model a simultaneous random walk of many particles in the

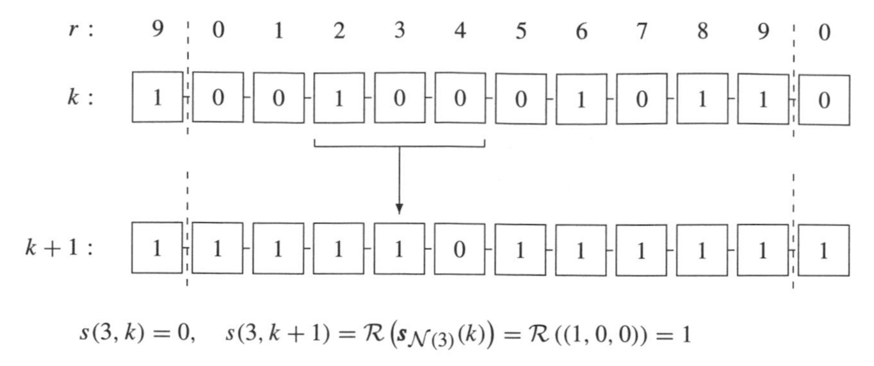

Figure 4.12. Example for the action of a deterministic local rule ("spread of plants") in a one-dimensional lattice with 10 nodes and periodic boundary conditions. The interaction neighborhood template is given by $\mathcal{N}_2^I = \{-1, 0, 1\}$ and the set of elementary states is $\mathcal{E} = \{0, 1\}$.

framework of probabilistic CA. However, in general, the transition rule is assumed to be spatially homogeneous.

In deterministic CA the local rule is *deterministic*; i.e., the local transition rule yields a unique next state for each cell. Therefore, for fixed initial conditions the future evolution of the automaton is predictable and uniquely determined.

Example: In the model for the "spread of plants" the local rule is defined as follows: Plants do not die during the considered time period; a vacant cell remains vacant if all neighbors are vacant; a vacant cell will be occupied by a plant if at least one of its neighboring cells is occupied by a plant. Fig. 4.12 shows an example of the action of the deterministic local rule in one time step.

In *probabilistic* CA the local transition rule specifies a time- and space-independent probability distribution of next states for each possible neighborhood configuration. It is defined by

$$\mathcal{R}\left(\boldsymbol{s}_{\mathcal{N}(r)}\right) = \begin{cases} z^1 & \text{with probability } W\left(\boldsymbol{s}_{\mathcal{N}(r)} \to z^1\right), \\ \vdots & \\ z^{\|\mathcal{E}\|} & \text{with probability } W\left(\boldsymbol{s}_{\mathcal{N}(r)} \to z^{\|\mathcal{E}\|}\right), \end{cases}$$

where $z^j \in \mathcal{E} := \left\{z^1, \ldots, z^{\|\mathcal{E}\|}\right\}$ and $W\left(\boldsymbol{s}_{\mathcal{N}(r)} \to z^j\right)$ is a time-independent transition probability specifying the probability of reaching an elementary state z^j given by the neighborhood configuration $\boldsymbol{s}_{\mathcal{N}(r)}$. This transition probability has to satisfy the following conditions:

$$W : \mathcal{E}^\nu \times \mathcal{E} \to [0,1] \quad \text{and} \quad \sum_{j=1}^{\|\mathcal{E}\|} W\left(\boldsymbol{s}_{\mathcal{N}(r)} \to z^j\right) = 1, \qquad \boldsymbol{s}_{\mathcal{N}(r)} \in \mathcal{E}^\nu, z^j \in \mathcal{E}.$$

Note that any deterministic local rule can be viewed as a special case of a probabilistic rule with

$$W\left(s_{\mathcal{N}(r)} \to z^j\right) = 1 \quad \text{and} \quad W\left(s_{\mathcal{N}(r)} \to z^l\right) = 0 \qquad \forall\, l \neq j,\ z^j, z^l \in \mathcal{E}.$$

Example: A probabilistic version of the local rule for the "spread of plants" can be defined as follows: Plants do not die during the considered time period; a vacant cell remains vacant if all neighbors are vacant; a vacant cell r is occupied by a plant *with a certain probability* depending on the number $m(r)$ of plants in its interaction neighborhood $\mathcal{N}_b^I(r)$, where each potential "parent" tries to produce a new plant (seed) with probability $\gamma = 1/|\mathcal{N}_b^I|$. This yields a total birth probability at the vacant cell of $\gamma_{\text{total}}(r) = \gamma m(r)$.

Assuming *synchronous* updating of the CA, the local rule is applied simultaneously to each cell of the lattice. Then, the global dynamics is specified by a *global function* $\mathcal{R}_g : \mathcal{S} \to \mathcal{S}$ such that for each global configuration $\boldsymbol{s} \in \mathcal{S}$,

$$\mathcal{R}_g(\boldsymbol{s})(r) = \mathcal{R}\left(\boldsymbol{s}_{\mathcal{N}(r)}\right) \qquad \forall\, r \in \mathcal{L}.$$

Accordingly, global dynamics are completely defined by the local rule $\mathcal{R}$.

We denote the global configuration at subsequent discrete time steps $k \in \mathbb{N}_0$ as $\boldsymbol{s}(k) = (s(r_1, k), \ldots, s(r_{|\mathcal{L}|}, k)) \in \mathcal{S}$, where $s(r_i, k) \in \mathcal{E}$ is the state of cell r_i at time k. The local configuration of cell r at time k is given by $\boldsymbol{s}_{\mathcal{N}(r)}(k) \in \mathcal{S}_{\mathcal{N}(r)}$. For each initial ($k = 0$) global configuration $\boldsymbol{s}(0) \in \mathcal{S}$, the temporal development of the system is determined by

$$\boldsymbol{s}(k+1) = \mathcal{R}_g\left(\boldsymbol{s}(k)\right),$$

where the dynamics of a state $s(r, k)$ is followed by

$$\boxed{s(r, k+1) = \mathcal{R}\left(\boldsymbol{s}_{\mathcal{N}(r)}(k)\right).} \tag{4.2}$$

Example: In mathematical terms the local rule for the model of the "spread of plants" is defined by

$$\mathcal{R}\left(\boldsymbol{s}_{\mathcal{N}(r)}\right) = \begin{cases} 0 & \text{with probability } W\left(\boldsymbol{s}_{\mathcal{N}(r)} \to 0\right), \\ 1 & \text{with probability } W\left(\boldsymbol{s}_{\mathcal{N}(r)} \to 1\right), \end{cases}$$

where the *deterministic* model is specified by

$$W\left(\boldsymbol{s}_{\mathcal{N}(r)} \to 0\right) = (1 - s(r)) \prod_{\tilde{r} \in \mathcal{N}_{2o}^I(r)} (1 - s(\tilde{r})) \in \{0, 1\},$$

$$W\left(\boldsymbol{s}_{\mathcal{N}(r)} \to 1\right) = 1 - W\left(\boldsymbol{s}_{\mathcal{N}(r)} \to 0\right) \in \{0, 1\},$$

and the *probabilistic* model is defined by

$$W\left(s_{\mathcal{N}(r)} \to 0\right) = (1 - s(r))\,(1 - \gamma_{\text{total}}(r)) \in [0, 1],$$
$$W\left(s_{\mathcal{N}(r)} \to 1\right) = 1 - W\left(s_{\mathcal{N}(r)} \to 0\right) \in [0, 1].$$

Here, the total birth probability is

$$\gamma_{\text{total}}(r) := \frac{1}{|\mathcal{N}^I_{2o}(r)|} \sum_{\tilde{r} \in \mathcal{N}^I_{2o}(r)} s(\tilde{r}).$$

The following table summarizes the transition probabilities for an interaction neighborhood template $\mathcal{N}^I_2 = \{-1, 0, 1\}$:

$s_{\mathcal{N}(r)}$	$W\left(s_{\mathcal{N}(r)} \to 0\right)$	$W\left(s_{\mathcal{N}(r)} \to 1\right)$	γ_t
000	1	0	0
100	0.5	0.5	0.5
001	0.5	0.5	0.5
101	0	1	1
111	0	1	1

As the number of possible configurations is finite for a finite lattice, any initial condition must eventually lead to a temporally periodic cycle in a deterministic CA. However, the period can become uninterestingly large. For example, if $|\mathcal{E}| = 2$ and $|\mathcal{L}| = 1000$, a global configuration will certainly be repeated after at most $2^{1000} \approx 10^{300}$ time steps.

In extensions of CA models, *asynchronous* updating is allowed. This can be achieved, for instance, by applying the local rule for each cell only with a certain probability. In (Schönfisch and de Roos 1999), several different asynchronous updating algorithms are presented and analyzed. In many applications involving particle interaction and movement a subdivided local rule is used: the first part (interaction) is applied synchronously, while the second part (movement) changes the values of pairs of cells sequentially. In sec. 5.3.2 we present an algorithm for an asynchronous random particle walk. Note that in a LGCA the local rule is applied synchronously for both particle interaction and movement.

A *cellular automaton* is defined by a regular lattice $\mathcal{L}$, an interaction neighborhood template $\mathcal{N}^I_b$, the finite set of elementary states $\mathcal{E}$, and the local space- and time-independent transition rule $\mathcal{R}$, which is applied synchronously. In a *deterministic* CA the local rule is deterministic while a *probabilistic* CA is characterized by a probabilistic local rule.

Dynamics in Lattice-Gas Cellular Automata. The dynamics of a LGCA arises from repetitive applications of superpositions of local (probabilistic) *interaction* and *propagation* (migration) steps applied simultaneously at all lattice nodes at each discrete time step. The definitions of these steps have to satisfy the exclusion principle; i.e., two or more particles of a given species are not allowed to occupy the same channel.

According to a model-specific *interaction* rule ($\mathcal{R}^{\mathtt{I}}$), particles can change channels (e.g., due to collisions, see fig. 4.13) and/or are created or destroyed. This rule can be *deterministic* or *stochastic*. The dynamics of a state $s(r,k) = \boldsymbol{\eta}(r,k) \in \{0,1\}^{\tilde{b}}$ in a LGCA are determined by the dynamics of the occupation numbers $\eta_i(r,k)$ for each $i \in \{1,\ldots,\tilde{b}\}$ at node r and time k. Therefore, the preinteraction state $\eta_i(r,k)$ is replaced by the postinteraction state $\eta_i^{\mathtt{I}}(r,k)$ determined by

$$\eta_i^{\mathtt{I}}(r,k) = \mathcal{R}_i^{\mathtt{I}}\left(\boldsymbol{\eta}_{\mathcal{N}(r)}(k)\right), \tag{4.3}$$

$$\mathcal{R}^{\mathtt{I}}\left(\boldsymbol{\eta}_{\mathcal{N}(r)}(k)\right) = \left(\mathcal{R}_i^{\mathtt{I}}\left(\boldsymbol{\eta}_{\mathcal{N}(r)}(k)\right)\right)_{i=1}^{\tilde{b}} = z \qquad \text{with probability } W\left(\boldsymbol{\eta}_{\mathcal{N}(r)}(k) \rightarrow z\right),$$

with $z \in \{0,1\}^{\tilde{b}}$ and the time-independent transition probability W.

In the deterministic *propagation* or streaming step ($\mathtt{P}$), all particles are moved simultaneously to nodes in the direction of their velocity; i.e., a particle residing in channel (r, c_i) at time k is moved to another channel $(r + mc_i, c_i)$ during one time step (fig. 4.14). Here, $m \in \mathbb{N}$ determines the *speed* and mc_i the *translocation* of the particle. Because all particles residing at velocity channels move the same number m of lattice units the exclusion principle is maintained. Particles occupying rest channels do not move since they have "zero velocity." In terms of occupation numbers, the state of channel $(r + mc_i, c_i)$ after propagation is given by

$$\eta_i^{\mathtt{P}}(r + mc_i, k+1) = \eta_i(r,k), \quad c_i \in \mathcal{N}_b. \tag{4.4}$$

Hence, if only the propagation step would be applied, then particles would simply move along straight lines in directions corresponding to particle velocities.

Combining interactive dynamics with propagation, (4.3) and (4.4) imply that

$$\eta_i^{\mathtt{IP}}(r + mc_i, k) = \eta_i(r + mc_i, k+1) = \eta_i^{\mathtt{I}}(r,k). \tag{4.5}$$

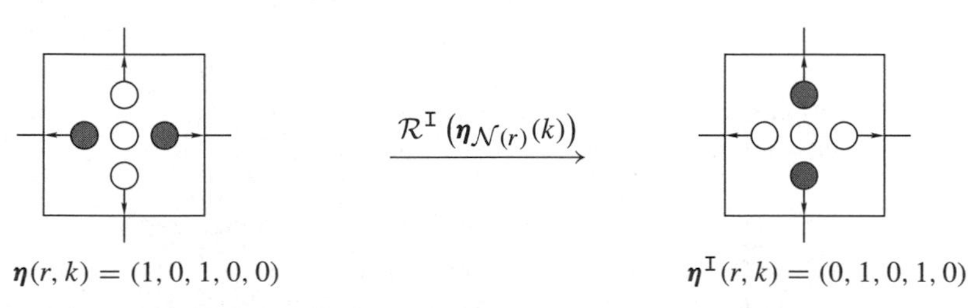

Figure 4.13. Example for collision of particles at two-dimensional square lattice node r; here, the interaction neighborhood is $\mathcal{N}_4^I(r) = \{r\}$ and therefore $\boldsymbol{\eta}_{\mathcal{N}(r)}(k) = \boldsymbol{\eta}(r,k)$. Filled dots denote the presence of a particle in the respective channel. No confusion should arise by the arrows indicating channel directions.

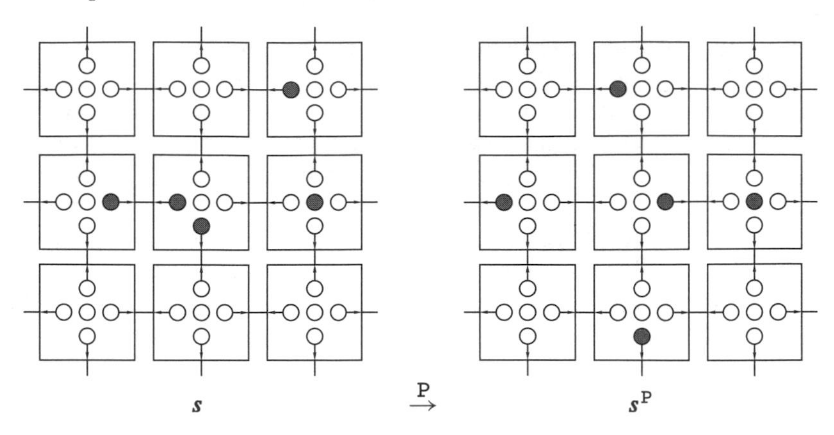

Figure 4.14. Propagation in a two-dimensional square lattice with speed $m = 1$; lattice configurations before ($\mathbf{s}$) and after ($\mathbf{s}^{\mathrm{P}}$) the propagation step; gray dots denote the presence of a particle in the respective channel.

This can be rewritten as the *microdynamical difference equations*

$$\begin{aligned}
&\mathcal{R}_i^{\mathrm{I}}\left(\boldsymbol{\eta}_{\mathcal{N}(r)}(k)\right) - \eta_i(r,k) \\
&\quad = \eta_i^{\mathrm{I}}(r,k) - \eta_i(r,k) = \eta_i(r + mc_i, k+1) - \eta_i(r,k) \\
&\quad =: \mathcal{C}_i\left(\boldsymbol{\eta}_{\mathcal{N}(r)}(k)\right), \qquad i = 1, \dots, \tilde{b},
\end{aligned} \tag{4.6}$$

where the *change in the occupation numbers* due to interaction is given by

$$\mathcal{C}_i\left(\boldsymbol{\eta}_{\mathcal{N}(r)}(k)\right) = \begin{cases} 1 & \text{creation of a particle in channel } (r, c_i), \\ 0 & \text{no change in channel } (r, c_i), \\ -1 & \text{annihilation of a particle in channel } (r, c_i). \end{cases}$$

In a multicomponent system with $\sigma = 1, \dots, \varsigma$ components, (4.6) becomes

$$\eta_{\sigma,i}^{\mathrm{I}}(r,k) - \eta_{\sigma,i}(r,k) = \eta_{\sigma,i}(r + m_\sigma c_i, k+1) - \eta_{\sigma,i}(r,k) = \mathcal{C}_{\sigma,i}\left(\boldsymbol{\eta}_{\mathcal{N}(r)}(k)\right) \tag{4.7}$$

for $i = 1, \dots, \tilde{b}$, with speeds $m_\sigma \in \mathbb{N}$ for each component $\sigma = 1, \dots, \varsigma$. Here, the change in the occupation numbers due to interaction is given by

$$\mathcal{C}_{\sigma,i}\left(\boldsymbol{\eta}_{\mathcal{N}(r)}(k)\right) = \begin{cases} 1 & \text{creation of a particle } X_\sigma \text{ in channel } (r, c_i)_\sigma, \\ 0 & \text{no change in channel } (r, c_i)_\sigma, \\ -1 & \text{annihilation of a particle } X_\sigma \text{ in channel } (r, c_i)_\sigma, \end{cases}$$

where $(r, c_i)_\sigma$ specifies the ith channel associated with node r of the lattice $\mathcal{L}_\sigma$.

4.4 Analysis and Characterization

Two basic questions are underlying the analysis and characterization of cellular automata (see also Wolfram 1985):

1. How can global behavior be predicted from the knowledge of the local rules? (bottom-up approach)
2. How do specific local rules have to be designed in order to yield a preselected global behavior? (top-down approach, inverse problem)

The inverse problem of deducing the local rules from a given global behavior is extremely difficult (Capcarrere 2002; please see Ganguly et al. 2004 for a short review of the inverse problem in CA and further references). Most methods depend on evolutionary computation techniques (Mitchell, Hraber, and Crutchfield 1993; Packard 1988; Richards, Thomas, and Packard 1990). For example, evolutionary algorithms have been suggested to produce rules for random number generators, for solving the density classification,[25] the synchronization task,[26] and the firing squad synchronization problem[27](Crutchfield and Mitchell 1995; Crutchfield, Mitchell, and Das 2003; Das, Mitchell, and Crutchfield 1994; Das, Crutchfield, Mitchell, and Hanson 1995; Umeo, Maeda, and Fujiwara 2003).

Despite the simple construction of CA, they are capable of highly complex behavior. For most CA models the only general method to determine the qualitative dynamic properties of the system is to *run simulations* on a computer for various initial configurations (Jackson 1991; Wolfram 1984). Then, methods from dynamical system analysis and statistical mechanics can be applied. Guided by these concepts, Wolfram 1984 undertook an extensive computer-based search through the properties of a specific group of one-dimensional deterministic CA. Viewing CA as discrete, spatially extended dynamical systems, Wolfram proposed a qualitative classification of CA long-term behavior into four classes, intending to capture all CA behaviors. The first class contains automata evolving into a constant configuration (fixed homogeneous state); the second class comprises automata generating sets of separated simple stable or periodic structures (limit cycles), while the third class consists of chaotic configurations whose behavior is not predictable; the fourth and remaining class includes CA developing complex localized structures, sometimes long lived. Note that these classes are phenomenological; i.e., CA are classified only by visual inspection of space-time diagrams. Based upon Wolfram's classification a detailed

[25] Density classification problem: design a CA in which the initial state of the CA (containing 1s and 0s) will converge to the all-1s state if the number of 1s in the initial configuration is larger and to the all-0s state in the other case.

[26] Synchronization task: design a CA that will reach an oscillating state; i.e., configurations periodically change between all 0s and all 1s in subsequent time steps, after a finite number of time steps.

[27] Firing squad synchronization problem: given a one-dimensional array of n identical CA, including a "general" at the left end which is activated at time $t = 0$, we want to design the automata so that, at some future time, all the cells will simultaneously and, for the first time, enter a special "firing" state.

characterization of different classes has been proposed (see, e.g., Gutowitz 1990; Li, Packard, and Langton 1990). A classification into five disjoint groups based on attractor structure was suggested by Kurka (1997). Walker (1990) has used connected Boolean nets for the characterization of CA. A number of authors have contributed to the characterization of CA with different rule sets (Barbe 1990; Jen 1990; McIntosh 1990; Voorhees 1990; Wootters and Langton 1990).

If a system is capable of universal computation, then with appropriate initial conditions, its evolution can carry out any finite computational process. A computationally universal system can thus mimic the behavior of any other system, and so can in a sense exhibit the most complicated possible behavior. Several specific CA are known to be capable of universal computation. The two-dimensional nearest-neighbor CA with two possible values at each site, known as the "Game of Life" has been proven computationally universal (Gardner 1983). The proof was carried out by showing that the CA could support structures that correspond to all the components of an idealized digital electronic computer, and that these components could be connected so as to implement any algorithm. Some one-dimensional nearest-neighbor cellular automata have been demonstrated to be computationally equivalent to the simplest known universal Turing machines and are thus capable of universal computation (Smith 1971).

Various order/chaos measures have been introduced to characterize global CA dynamics. The topology of the state space is particularly important. For example, characterizations have been proposed that are based on the analysis of Garden of Eden states (nonreachable states) (Amoroso and Patt 1972; Kari 1990; Kari 1994; Myhill 1963). Kaneko 1986 introduced an information-theoretic approach to characterize the complexity of Garden of Eden states in terms of their volumes, stability against noise, information storage capacity, etc.. Lyapunov exponents measure the rate of divergence of trajectories in the space of configurations. Accordingly, the "Lyapunov exponent" is a measure for how much a dynamical system depends on initial conditions (Kaplan and Glass 1995). A comparable concept for deterministic CA is the *Hamming distance*, which is simply the number of cells that are in different states at two successive time steps (e.g., damage spreading; Bagnoli, Rechtman, and Ruffo 1992; Wolfram 1984). *Local irreversibility* is an important feature of many CA, which implies that different initial global configurations may eventually evolve to the same final configuration. Certain global configurations may occur more frequently with respect to initial conditions. This behavior is a property of systems that are capable of self-organization (Hurley 1990). Geometrical aspects of self-similar spatial patterns generated by CA evolution can be investigated, e.g., by the *Hausdorff-Basicovitch* or *fractal dimension* of the pattern. As a measure for the irreversible behavior of a cellular automaton, various kinds of entropy may be defined for CA (Wolfram 1985). Each counts the number of possible sequences of site values corresponding to some space-time region. For example, the spatial entropy gives the dimension of the set of configurations that can be generated at some time step in the evolution of the CA, starting from all possible initial states. There are in general $N(X) \leq k^X$ (k is the number of possible values for each site) possible sequences of values for a block of X sites in this set of configurations. The spatial topological entropy is given by $\lim_{X\to\infty}(1/X)\log_k N(X)$. One may also define a spatial measure entropy formed from the probabilities of possible sequences. Temporal entropies may then be defined to count the number of sequences that occur in the time series of values taken on by each site. Topological entropies reflect the possible configurations of a system; measure entropies reflect those that are probable and are insensitive to phenomena that occur with zero probability.

Beyond entropies and Lyapunov exponents, dynamical systems theory suggests that zeta functions may give a characterization of the global behavior of CA. Zeta functions measure the density of periodic sequences in CA configurations and may possibly be related to Fourier transforms (Wolfram 1985).

The crucial question remains: How can the global behavior be deduced from the local rules? The theory of interacting particle systems yields analytic answers to this problem and can be applied to asynchronous (continuous-time) CA (Liggett 1985, 1999). However, the results are so far limited to a few selected models (e.g., voter's model) and cannot easily be extended to other systems. For synchronous binary CA Langton 1990 suggested the λ parameter. λ is defined as the probability that a particular CA cell will have 1 as its next state; i.e., λ indicates the fraction of 1s in the binary rule table of an automaton cell. For certain rules it has been shown that with the increase of λ the global CA behavior changes from order to chaos (Langton 1990). Since then, various other local parameters have been proposed (Wuensche and Lesser 1992; Wuensche 1996; Zwick and Shu 1995). In this book we focus on a different characterization which can be applied particularly to CA rules mimicking cellular interaction. Starting with the definition of local rules (and local parameters), under certain approximations (especially *mean-field approximation*) a set of (Boltzmann) equations can be derived. Stability analysis of these equations (including only local parameters) allows then to predict the global CA behavior.

4.4.1 Chapman-Kolmogorov Equation

In a previous sec. 4.3.4 we provided a description of CA models in terms of microdynamical discrete variables (eqns. (4.2) and (4.6)), discrete in space and time. Our aim in this section is to derive difference equations that remain discrete in time (and in space) but have continuous state variables. Then, in order to gain more insight into the automaton dynamics, standard analytical techniques can be applied to these equations. Such a description will typically be statistical, not specifying an exact configuration, but merely the probabilities for the appearance of different configurations.

Recall from the definition of a probabilistic local rule (cp. p. 70) that any deterministic CA can be viewed as a probabilistic one. Then, instead of specifying the configuration $\mathbf{s}(k) \in \mathcal{S}$, we analyze the *probability distribution* of each configuration at each time step. At time $k \in \mathbb{N}_0$ let $\boldsymbol{\xi}_k \in \mathcal{S}$ be a realization of the stochastic process $\{\boldsymbol{\xi}_k\}_{k\in\mathbb{N}_0}$ defined in the state space $\mathcal{S}$. In the following, given an arbitrary random distribution of initial states $\boldsymbol{\xi}_0 \in \mathcal{S}$,

$$P(\xi_k(r_1) = s(r_1), \ldots, \xi_k(r_m) = s(r_m))$$

$$=: P_k(s(r_1), \ldots, s(r_m)) \qquad r_i \in \mathcal{M} \subseteq \mathcal{L},\ m = |\mathcal{M}|$$

specifies the probability of observing configuration $\boldsymbol{s}_{\mathcal{M}} \in \mathcal{S}|_{\mathcal{M}}$ at time k.

By construction of the CA, the stochastic process $\{\boldsymbol{\xi}_k\}_{k \in \mathbb{N}_0}$ is Markovian; i.e., there is no memory effect. It is fully characterized by its transition probability matrix with elements $P(\boldsymbol{\xi}_{k+1} = \boldsymbol{s} \mid \boldsymbol{\xi}_k = \tilde{\boldsymbol{s}})$ defining the probability of finding the system in a state $\boldsymbol{s}$ at time $k+1$, given state $\tilde{\boldsymbol{s}}$ at the previous time k. Using the fact that the local CA rule (cp. p. 74) is applied to each site simultaneously and that it specifies the next state of a single site as a function of the interaction neighborhood configuration, we get

$$\begin{aligned} & P\left(\boldsymbol{\xi}_{k+1} = \boldsymbol{s} \mid \boldsymbol{\xi}_k = \tilde{\boldsymbol{s}}\right) \\ & = \prod_{r \in \mathcal{L}} P\left(\xi_{k+1}(r) = s(r) \mid \boldsymbol{\xi}_k|_{\mathcal{N}_b^I(r)} = \tilde{\boldsymbol{s}}|_{\mathcal{N}_b^I(r)}\right), \end{aligned} \tag{4.8}$$

and since the local rule is time-independent

$$= \prod_{r \in \mathcal{L}} W\left(\tilde{\boldsymbol{s}}_{\mathcal{N}(r)} \to s(r)\right) =: W\left(\tilde{\boldsymbol{s}} \to \boldsymbol{s}\right) \tag{4.9}$$

for $\boldsymbol{s}, \tilde{\boldsymbol{s}} \in \mathcal{S}$, $r_i \in \mathcal{N}_b^I(r)$, $i = 1, \ldots, \nu$. $W(\boldsymbol{s}_{\mathcal{A}} \to \boldsymbol{s}_{\mathcal{B}})$ specifies the *time-independent* transition probability of reaching a configuration $\boldsymbol{s}_{\mathcal{B}} = \boldsymbol{s}|_{\mathcal{B}}$ given the configuration $\boldsymbol{s}_{\mathcal{A}} = \boldsymbol{s}|_{\mathcal{A}}$ for any $\mathcal{A}, \mathcal{B} \subseteq \mathcal{L}$. The stochastic process $\{\boldsymbol{\xi}_k\}_{k \in \mathbb{N}_0}$ is a *stationary Markov chain* (Gardiner 1983), and the time evolution of the probability distribution is given by the *Chapman-Kolmogorov equation* or *master equation* (i.e., Gardiner 1983), which becomes, with (4.9),

$$\begin{aligned} P_{k+1}(\boldsymbol{s}) &= \sum_{\tilde{\boldsymbol{s}} \in \mathcal{S}} P_k\left(\tilde{\boldsymbol{s}}\right) W(\tilde{\boldsymbol{s}} \to \boldsymbol{s}) \\ &= \sum_{\tilde{\boldsymbol{s}} \in \mathcal{S}} P_k\left(\tilde{\boldsymbol{s}}\right) \prod_{r \in \mathcal{L}} W\left(\tilde{\boldsymbol{s}}_{\mathcal{N}(r)} \to s(r)\right). \end{aligned} \tag{4.10}$$

Furthermore, using (4.10) the probability for a site to be in a specific elementary state $z^j \in \mathcal{E}$ is given by

$$\begin{aligned} P\left(\xi_{k+1}(r) = z^j\right) &= \sum_{\substack{\boldsymbol{s} \in \mathcal{S} \\ s(r) = z^j}} P_{k+1}(\boldsymbol{s}) \\ &= \sum_{\substack{\boldsymbol{s} \in \mathcal{S} \\ s(r) = z^j}} \sum_{\tilde{\boldsymbol{s}} \in \mathcal{S}} P_k\left(\tilde{\boldsymbol{s}}\right) \prod_{\bar{r} \in \mathcal{L}} W\left(\tilde{\boldsymbol{s}}_{\mathcal{N}(\bar{r})} \to s(\bar{r})\right) \\ &= \sum_{\tilde{\boldsymbol{s}} \in \mathcal{S}} P_k\left(\tilde{\boldsymbol{s}}\right) W\left(\tilde{\boldsymbol{s}}_{\mathcal{N}(r)} \to z^j\right) \sum_{\boldsymbol{s} \in \mathcal{S}|_{\mathcal{L} \setminus \{r\}}} \prod_{\substack{\bar{r} \in \mathcal{L} \\ \bar{r} \neq r}} W\left(\tilde{\boldsymbol{s}}_{\mathcal{N}(\bar{r})} \to s(\bar{r})\right). \end{aligned} \tag{4.11}$$

Taking advantage of

$$\sum_{s \in \mathcal{S}|_{\mathcal{L}\setminus\{r\}}} \prod_{\substack{\bar{r} \in \mathcal{L} \\ \bar{r} \neq r}} W\left(\tilde{s}_{\mathcal{N}(\bar{r})} \to s(\bar{r})\right)$$

$$= \sum_{s \in \mathcal{S}|_{\mathcal{L}\setminus\{r\}}} W\left(\tilde{s}_{\mathcal{N}(\bar{r})} \to s(\bar{r})\right) \cdot \left\{ \sum_{s \in \mathcal{S}|_{\mathcal{L}\setminus\{r,\bar{r}\}}} \prod_{\substack{r' \in \mathcal{L} \\ r' \neq r, \bar{r}}} W\left(\tilde{s}_{\mathcal{N}(r')} \to s(r')\right) \right\}$$

$$= \sum_{s \in \mathcal{S}|_{\mathcal{L}\setminus\{r\}}} W\left(\tilde{s}_{\mathcal{N}(\bar{r})} \to s(\bar{r})\right) \cdot \left\{ \cdots \sum_{s_{\mathcal{N}(r^*)} \in \mathcal{S}_{\mathcal{N}(r^*)}} W\left(\tilde{s}_{\mathcal{N}(r^*)} \to s(r^*)\right) \right\}$$

$$= 1, \qquad \text{where } \bar{r} \neq r, \quad r' \neq r, \bar{r} \quad \text{and} \quad r^* \neq r, \bar{r}, r',$$

and

$$\sum_{\tilde{s} \in \mathcal{S}} P_k(\tilde{s})\, W\left(\tilde{s}_{\mathcal{N}(r)} \to z^j\right)$$

$$= \sum_{\tilde{s}_{\mathcal{N}(r)} \in \mathcal{S}_{\mathcal{N}(r)}} \left(\sum_{\substack{s' \in \mathcal{S} \\ s'_{\mathcal{N}(r)} = \tilde{s}_{\mathcal{N}(r)}}} P_k(s') \right) W\left(\tilde{s}_{\mathcal{N}(r)} \to z^j\right)$$

$$= \sum_{\tilde{s}_{\mathcal{N}(r)} \in \mathcal{S}_{\mathcal{N}(r)}} P_k\left(\tilde{s}_{\mathcal{N}(r)}\right) W\left(\tilde{s}_{\mathcal{N}(r)} \to z^j\right),$$

(4.11) simplifies to

$$\boxed{P\left(\xi_{k+1}(r) = z^j\right) = \sum_{\tilde{s}_{\mathcal{N}(r)} \in \mathcal{S}_{\mathcal{N}(r)}} P_k\left(\tilde{s}_{\mathcal{N}(r)}\right) W\left(\tilde{s}_{\mathcal{N}(r)} \to z^j\right).} \tag{4.12}$$

Although it is straightforward to write down (4.10) and (4.12), which keep track of all information concerning the state of the system, the complete analytical solution is not feasible in most cases. However, it turns out that insight into CA dynamics is possible if some approximation is made.

4.4.2 Cellular Automaton Mean-Field Equations

The simplest approximation is known as the *mean-field theory* (Schulman and Seiden 1978; Wolfram 1983). It is based on the assumption that at any time the states of sites are independent of the states of other sites in the lattice. Although this assumption is generally not true, a simple formula for an estimate of the limit density of each possible state of a cell can be derived from (4.12). Thus, the states of all sites in the lattice are assumed to be independent at all times and therefore the probability of a local configuration $s_{\mathcal{M}}$ is the product of the probabilities of the states of the sites in $\mathcal{M}$, i.e.,

$$P(\mathbf{s}_{\mathcal{M}}) = P\left(s(r_1),\ldots,s(r_{|\mathcal{M}|})\right) = \prod_{r_i\in\mathcal{M}} P(s(r_i)), \tag{4.13}$$

with $\mathcal{M} \subseteq \mathcal{L}$.

For each $k \in \mathbb{N}_0$, $j = 1,\ldots,|\mathcal{E}|$, let $x_k^j \in \{0,1\}$ be a Boolean random variable which equals 1, if the random variable $\xi_k(r)$ is in the elementary state $z^j \in \mathcal{E} := \{z^1,\ldots,z^{|\mathcal{E}|}\}$. For $j = 1,\ldots,|\mathcal{E}|$ it is defined as[28]

$$x_k^j(r) := \delta_{\xi_k(r),z^j} = \begin{cases} 1 & \text{if} \quad \xi_k(r) = z^j, \\ 0 & \text{otherwise.} \end{cases}$$

Then, using (4.12) and (4.13), the temporal evolution of the expected value $E\left(x_k^j(r)\right)$ is given by

$$\begin{aligned} E\left(x_{k+1}^j(r)\right) &= P\left(\xi_{k+1}(r) = z^j\right) \\ &= \sum_{\tilde{\mathbf{s}}_{\mathcal{N}(r)}\in\mathcal{S}_{\mathcal{N}(r)}} W\left(\tilde{\mathbf{s}}_{\mathcal{N}(r)} \to z^j\right) P_k\left(\tilde{\mathbf{s}}_{\mathcal{N}(r)}\right) \\ &= \sum_{\tilde{\mathbf{s}}_{\mathcal{N}(r)}\in\mathcal{S}_{\mathcal{N}(r)}} W\left(\tilde{\mathbf{s}}_{\mathcal{N}(r)} \to z^j\right) \prod_{r_i\in\mathcal{N}_b^I(r)} P_k(\tilde{s}(r_i)), \end{aligned} \tag{4.14}$$

with $r_i \in \mathcal{N}_b^I(r)$, $i = 1,\ldots,\nu$ and $\tilde{\mathbf{s}}_{\mathcal{N}(r)} = (\tilde{s}(r_1),\ldots,\tilde{s}(r_\nu))$. According to the spatial homogeneity of the local transition rule (cp. p. 73), the transition probabilities $W(\tilde{\mathbf{s}}_{\mathcal{N}(r)} \to z^j)$ are translation invariant; i.e., they depend only on the site states which define the interaction neighborhood but not on their location r. Therefore, with $\tilde{s}(r_i) = z_{r_i}$, $P_k(\tilde{s}(r_i))$ can be represented as

$$P_k(\tilde{s}(r_i)) = P\left(\xi_k(r_i) = z_{r_i}\right) = \sum_{l=1}^{|\mathcal{E}|} \delta_{z_{r_i},z^l} E\left(x_k^l(r_i)\right). \tag{4.15}$$

Furthermore, spatially averaged values $x_j(k) \in [0,1]$, which for each $j \in \{1,\ldots,|\mathcal{E}|\}$ denote the expected density of an elementary state z^j on the lattice at time k, are defined by

$$x_j(k) := \frac{1}{|\mathcal{L}|}\sum_{r\in\mathcal{L}} E\left(x_k^j(r)\right) = E\left(x_k^j(r')\right) \in [0,1] \qquad \text{for some } r' \tag{4.16}$$

under the mean-field assumption. Then, combining (4.14)–(4.16), mean-field equations for each $x_j(k)$ are given by

$$\begin{aligned} x_j(k+1) &= \sum_{(z_1,\ldots,z_\nu)\in\mathcal{E}^\nu} W\left((z_1,\ldots,z_\nu) \to z^j\right) \prod_{i=1}^{\nu}\sum_{l=1}^{|\mathcal{E}|} \delta_{z_i,z^l} x_l(k) \\ &=: H_j(\mathbf{x}(k)), \qquad \mathbf{x}^T(k) = \left(x_1(k),\ldots,x_{|\mathcal{E}|}(k)\right) \end{aligned} \tag{4.17}$$

[28] $\delta_{u,v}$ is the Kronecker delta; i.e., $\delta_{u,v} = 1$ if $u = v$ and $\delta_{u,v} = 0$ if $u \neq v$.

for $j = 1, \ldots, |\mathcal{E}|$.

Example: Here, we derive the mean-field equation for the model of plant spread with an interaction neighborhood template $\mathcal{N}_2^I = \{-1, 0, 1\}$. Let $x_0(k)$ denote the expected density of vacant sites and $x_1(k)$ the expected density of plants. Then, according to (4.17), with $\mathcal{E} = \{0, 1\}$ and $x_0(k) + x_1(k) = 1$,

$$x_1(k+1) = \sum_{(z_1,z_2,z_3)\in\{0,1\}^3} W\left((z_1, z_2, z_3) \to 1\right) \prod_{i=1}^{3} \left(\delta_{z_i,0}x_0(k) + \delta_{z_i,1}x_1(k)\right).$$

Using the probabilistic local rule defined on p. 76, we get $W\left((z_1, z_2, z_3) \to 1\right) = 1 - (1 - z_2)(1 - \gamma_t)$, where $\gamma_t = (1/2)\,(z_1 + z_3)$. Therefore,

$$\begin{aligned} x_1(k+1) &= \frac{1}{2}x_1(k)x_0^2(k) + x_0(k)x_1(k)x_0(k) + \frac{1}{2}x_0^2(k)x_1(k) + x_1^2(k)x_0(k) \\ &\quad + x_1(k)x_0(k)x_1(k) + x_0x_1^2(k) + x_1^3(k) \\ &= 2x_1(k)x_0^2(k) + 3x_1^2(k)x_0(k) + x_1^3(k) = 2x_1(k) - x_1^2(k) \qquad \text{(E.1)} \\ x_0(k+1) &= 1 - x_1(k+1) = 1 - 2x_1(k) + x_1^2(k) = x_0^2(k). \end{aligned}$$

Note that (4.17) only encodes the *combinatorial information* contained in the local CA rule, which maps from an interaction neighborhood configuration to the state of a single site, and that it does not reflect the structure of the lattice on which the automaton operates. Therefore, the mean-field theory does not distinguish between CA models that have the same rule with the same number of neighbors, but are defined on different (i.e., one- or two-dimensional) lattices (Gutowitz and Victor 1989). Eqn. (4.17) is called a "mean-field equation" because each site state only depends on the average value of the states of the other sites in the interaction neighborhood. Eqn. (4.17) is *exact* in the case in which

- the lattice is infinitely large, and
- the site states are randomly reallocated after updating.

Although the mean-field approach is a very crude approximation, it often yields a picture of the CA dynamics that is qualitatively correct. This will be discussed in more detail in sec. 6.2.

Lattice-Gas Cellular Automaton Mean-Field (Boltzmann) Equations

Recall that in a LGCA the state of node r is composed of Boolean occupation numbers η_i, i.e.,

$$s(r) = \boldsymbol{\eta}(r) = \left(\eta_1(r), \ldots, \eta_{\tilde{b}}(r)\right) \in \mathcal{E} = \{0, 1\}^{\tilde{b}},$$

where $\tilde{b}$ is the number of channels on a node (cp. p. 72).

Since quantities of interest are not so much the Boolean variables η_i but macroscopic quantities such as densities, we look for an expression similar to (4.14) in terms of average values of the η_i's. These values are given by *single particle distribution functions*

$$f_i(r,k) := E\left(\eta_i(r,k)\right) = P_k\left(\eta_i(r) = 1\right) \in [0,1] \qquad \forall\, r \in \mathcal{L},\ i = 1,\ldots,\tilde{b}, \tag{4.18}$$

where $E\left(\eta_i(r,k)\right)$ is the expected value with respect to the initial particle distribution $\boldsymbol{\xi}_0 \in \mathcal{S}$ of the Markov stochastic process. Note that because of the Boolean nature of the occupation variable $\eta_i(r,k)$, $f_i(r,k)$ equals the probability of finding a particle at channel (r, c_i) at time k. The expected *local number of particles* $\varrho(r,k) \in [0,\tilde{b}]$ at a node is obtained by summing over the expected occupation numbers, i.e.,

$$\varrho(r,k) := E\left(\mathrm{n}(r,k)\right) = E\left(\sum_{i=1}^{\tilde{b}} \eta_i(r,k)\right) = \sum_{i=1}^{\tilde{b}} f_i(r,k),$$

and the expected *total mass* at time k will be denoted by $\varrho(k)$, which is

$$\varrho(k) := \sum_{r\in\mathcal{L}} \varrho(r,k).$$

Furthermore, expected *local* and *total particle densities* are defined as

$$\rho(r,k) := \frac{1}{\tilde{b}}\varrho(r,k) \in [0,1],$$

$$\rho(k) := \frac{1}{|\mathcal{L}|}\sum_{r\in\mathcal{L}} \rho(r,k) = \frac{1}{\tilde{b}}\varrho(k) \in [0,1].$$

Extending the mean-field assumption (4.13) to on-node configurations $s(r) = \boldsymbol{\eta}(r) = \left(\eta_1(r),\ldots,\eta_{\tilde{b}}(r)\right)$, the single node distribution function at time k factorizes to

$$P_k\left(s(r)\right) = P_k\left(\eta_1(r),\ldots,\eta_{\tilde{b}}(r)\right) = \prod_{l=1}^{\tilde{b}} P_k\left(\eta_l(r)\right)$$

$$= \prod_{l=1}^{\tilde{b}} f_l(r,k)^{\eta_l(r)} \left(1 - f_l(r,k)\right)^{(1-\eta_l(r))}. \tag{4.19}$$

Note that with this assumption the Boolean occupation numbers are considered as independent random variables. Therefore, it is possible to replace the average of products by the product of averages. In particular, the following holds true

$$E\left(\eta_i(r,k)\eta_j(r,k)\right)$$

$$= \begin{cases} E\left(\eta_i(r,k)\right) E\left(\eta_j(r,k)\right) = f_i(r,k) f_j(r,k) & \text{if } i \neq j, \quad \text{(a)} \\ E\left(\eta_i(r,k)^2\right) = E\left(\eta_i(r,k)\right) = f_i(r,k) & \text{if } i = j. \quad \text{(b)} \end{cases} \tag{4.20}$$

The physical (statistical mechanics) interpretation of this approach concentrates on *ensembles*, in which each possible configuration occurs with a particular probability. One assumes that appropriate space or time averages of an individual configuration agree with averages obtained from an ensemble of different configurations. In this sense, the single particle distribution function $f_i(r, k)$ is formally treated as the average over an arbitrary distribution of initial occupation numbers $\{\eta_i(r, 0)\}$. Then, the mean-field assumption (4.19) states that different occupation numbers are statistically independent (Chopard and Droz 1998; Rothman and Zaleski 1997). This assumption is based on the famous *molecular chaos* hypothesis for (ideal) gases that was introduced by Rudolf Clausius in 1857: due to the numerous gas molecule collisions, the effects of individual intermolecular forces are relatively weak and fluctuations are negligible; they can be approximated by the average force[29] (Stowe 1984).

Under the mean-field assumption, according to (4.5), the dynamical equation for the single particle distribution function is

$$\begin{aligned} f_i(r + mc_i, k + 1) &= E\left(\eta_i(r + mc_i, k + 1)\right) \\ &= E\left(\eta_i^{\mathrm{I}}(r, k)\right) = P_k\left(\eta_i^{\mathrm{I}}(r) = 1\right) = \sum_{z \in \{0,1\}^{\tilde{b}}} z_i P_k\left(\boldsymbol{\eta}^{\mathrm{I}}(r) = z\right) \\ &= \sum_{z \in \{0,1\}^{\tilde{b}}} \sum_{\tilde{\boldsymbol{\eta}}_{\mathcal{N}(r)} \in \mathcal{S}_{\mathcal{N}(r)}} z_i W\left(\tilde{\boldsymbol{\eta}}_{\mathcal{N}(r)} \to z\right) P_k\left(\tilde{\boldsymbol{\eta}}(r_1), \ldots, \tilde{\boldsymbol{\eta}}(r_\nu)\right), \end{aligned}$$

which, by neglecting off-node correlations[9] (cp. (4.13)), becomes

$$= \sum_{z \in \{0,1\}^{\tilde{b}}} \sum_{\tilde{\boldsymbol{\eta}}_{\mathcal{N}(r)} \in \mathcal{S}_{\mathcal{N}(r)}} z_i W\left(\tilde{\boldsymbol{\eta}}_{\mathcal{N}(r)} \to z\right) \prod_{r_i \in \mathcal{N}_b^I(r)} P_k\left(\tilde{\boldsymbol{\eta}}(r_i)\right),$$

and, by additionally neglecting on-node correlations[10] (cp. (4.19)),

$$\begin{aligned} &= \sum_{z \in \{0,1\}^{\tilde{b}}} \sum_{\tilde{\boldsymbol{\eta}}_{\mathcal{N}(r)} \in \mathcal{S}_{\mathcal{N}(r)}} z_i W\left(\tilde{\boldsymbol{\eta}}_{\mathcal{N}(r)} \to z\right) \prod_{r_i \in \mathcal{N}_b^I(r)} \prod_{l=1}^{\tilde{b}} P_k\left(\tilde{\eta}_l(r_i)\right) \\ &= \sum_{z(r) \in \{0,1\}^{\tilde{b}}} \sum_{\tilde{\boldsymbol{\eta}}_{\mathcal{N}(r)} \in \mathcal{S}_{\mathcal{N}(r)}} z_i(r) W\left(\tilde{\boldsymbol{\eta}}_{\mathcal{N}(r)} \to z\right) \\ &\quad \cdot \prod_{r_i \in \mathcal{N}_b^I(r)} \prod_{l=1}^{\tilde{b}} f_l(r_i, k)^{\tilde{\eta}_l(r_i)} \left(1 - f_l(r_i, k)\right)^{(1 - \tilde{\eta}_l(r_i))}. \end{aligned} \tag{4.21}$$

---

[29] More generally, the hypothesis relies on the rapid randomization of microscopic configurations in many-particle systems.

[30] Off-node correlations are correlations between states of lattice nodes in the lattice.

[31] On-node correlations are correlations between states of occupation numbers at a lattice node.

The standard notation of the mean-field approximation for LGCA is the (nonlinear) *lattice-Boltzmann equation* (Frisch et al. 1987) given by

$$
\boxed{
\begin{aligned}
& f_i(r+mc_i, k+1) - f_i(r,k) \\
& \quad = E\left(\eta_i^{\mathrm{I}}(r,k) - \eta_i(r,k)\right) \\
& \quad = \sum_{z \in \{0,1\}^{\tilde{b}}} \sum_{\tilde{\eta}_{\mathcal{N}(r)} \in \mathcal{S}_{\mathcal{N}(r)}} \Bigg\{ (z_i - \tilde{\eta}_i(r))\, W\left(\tilde{\eta}_{\mathcal{N}(r)} \to z\right) \\
& \qquad\qquad \cdot \prod_{r_i \in \mathcal{N}_b^I(r)} \prod_{l=1}^{\tilde{b}} f_l(r_i,k)^{\tilde{\eta}_l(r_i)} \left(1 - f_l(r_i,k)\right)^{(1-\tilde{\eta}_l(r_i))} \Bigg\} \\
& \quad =: \tilde{C}_i\left(f_{\mathcal{N}(r)}(k)\right),
\end{aligned}
}
\tag{4.22}
$$

where $f^T_{\mathcal{N}(r)}(k) = (f(r_i,k))_{r_i \in \mathcal{N}_b^I(r)} = (f_1(r_i,k), \ldots, f_{\tilde{b}}(r_i,k))_{r_i \in \mathcal{N}_b^I(r)}$. Here, $\tilde{C}_i\left(f_{\mathcal{N}(r)}(k)\right) \in [0,1]$ expresses how the average number of particles with a given translocation mc_i changes, due to interparticle interactions and propagation. Note that the lattice-Boltzmann equation can be derived from the microdynamical difference equation (4.6) by replacing occupation numbers η_i by average occupation numbers f_i, i.e.,

$$
\begin{aligned}
& E\left(\eta_i(r+mc_i,k+1) - \eta_i(r,k)\right) = E\left(C_i\left(\eta_{\mathcal{N}(r)}(k)\right)\right) \\
& \quad = \boxed{f_i(r+mc_i,k+1) - f_i(r,k) = \tilde{C}_i\left(f_{\mathcal{N}(r)}(k)\right),} \qquad i = 1,\ldots,\tilde{b},
\end{aligned}
\tag{4.23}
$$

in which the average is taken under consideration of (4.20)(b). Thus, for a multicomponent system it follows from (4.7) that

$$
f_{\sigma,i}(r+m_\sigma c_i,k+1) - f_{\sigma,i}(r,k) = \tilde{C}_{\sigma,i}\left(f_{\mathcal{N}(r)}(k)\right), \tag{4.24}
$$

for $i = 1,\ldots,\tilde{b}$, $\sigma = 1,\ldots,\varsigma$, where

$$
f^T_{\mathcal{N}(r)}(k) = \left((f_\sigma(r_i,k))_{r_i \in \mathcal{N}^I_{b,\sigma}(r)} \right)_{\sigma=1}^{\varsigma},
$$

with

$$
f_\sigma(r_i,k) = \left(f_{\sigma,1}(r_i,k), \ldots, f_{\sigma,\tilde{b}}(r_i,k) \right).
$$

The obtained mean-field equations (4.17) and (4.23) are deterministic equations that describe the dynamics of the automaton average concentrations and have been derived neglecting all the correlations in the automaton. They encode information contained in the *local* CA rule. In addition, the lattice-Boltzmann equation (4.23) of a LGCA model also *keeps information* about the *structure of the lattice* on which the automaton operates (for $m \neq 0$). The analysis of the automaton (stochastic) dynamics in terms of these time- and space-dependent difference equations can provide spatio-temporal information. But how "good" is the mean-field approximation? Can we properly predict important aspects of the (stochastic) automaton dynamics, such as the wavelength of an observed pattern? We will try to explore answers to this question in this book.

Higher Order Correlations. Some research to improve the mean-field approximation by considering higher order correlations has been carried out on individual-based models with discrete space and continuous time (e.g., Durrett and Levin 1994a). In order to improve the mean-field approximation given by

$$P\left(\mathbf{s}_{\mathcal{M}}\right) = P\left(s(r_1), \ldots, s(r_{|\mathcal{M}|})\right) = \prod_{r_i \in \mathcal{M}} P\left(s(r_i)\right),$$

probabilities of large local configurations of length $|\mathcal{M}| > 2$ can be estimated by taking into account probabilities of smaller blocks with maximal length $2 \leq l < |\mathcal{M}|$ which they contain (this approach is sometimes called local structure approximation; Gutowitz, Victor, and Knight 1987); i.e.,

$$P\left(s(r_1), s(r_2), \ldots, s(r_l), s(r_{l+1})\right)$$
$$= \begin{cases} \dfrac{P\left(s(r_2), \ldots, s(r_{l+1})\right) P\left(s(r_1), \ldots, s(r_l)\right)}{P\left(s(r_2), \ldots, s(r_l)\right)} & \text{if } P\left(s(r_2), \ldots, s(r_l)\right) > 0, \\ 0 & \text{otherwise.} \end{cases} \tag{4.25}$$

Note that the resulting equations involve rational fractions, whereas the mean-field equations are polynomial equations.

Example: An improved mean-field approximation for the model of the plant spread with an interaction neighborhood template $\mathcal{N}_2^I = \{-1, 0, 1\}$ is derived including probabilities of local configurations of length 1 and 2 ($l = 2$). From (4.12) we get for the expected density of plants

$$x_1(k+1) = \sum_{(z_1,z_2,z_3) \in \{0,1\}^3} W\left((z_1, z_2, z_3) \to 1\right) P_k(z_1, z_2, z_3).$$

Using (4.25) this becomes

$$= \sum_{(z_1,z_2,z_3) \in \{0,1\}^3} W\left((z_1, z_2, z_3) \to 1\right) \frac{1}{P_k(z_2)} P_k(z_2, z_3) P_k(z_1, z_2),$$

and with the probabilistic local rule defined on p. 76,

$$
\begin{aligned}
&= \sum_{(z_1,1,z_3)\in\{0,1\}^3} \frac{1}{P_k(1)} P_k(1,z_3)P_k(z_1,1) \\
&\quad + \sum_{(z_1,0,z_3)\in\{0,1\}^3} \frac{1}{2}(z_1+z_3)\frac{1}{P_k(0)} P_k(0,z_3)P_k(z_1,0) \\
&= \frac{P_k(1,0)P_k(0,1)+P_k(1,0)P_k(1,1)+P_k(1,1)P_k(0,1)+P_k(1,1)^2}{P_k(1)} \\
&\quad + \frac{\frac{1}{2}P_k(0,0)P_k(1,0)+\frac{1}{2}P_k(0,1)P_k(0,0)+P_k(0,1)P_k(1,0)}{P_k(0)} \\
&= 2P_k(1)-P_k(1,1) = 2x_1(k)-P_k(1,1), \qquad \text{(E.2)}
\end{aligned}
$$

where we have used the assumption that $P_k(1,0) = P_k(0,1)$ and the fact that $P_k(0) = 1 - P_k(1)$, $P_k(1,0) = P_k(0) - P_k(0,0) = P_k(1) - P_k(1,1)$. In order to derive a closed expression for the dynamics of the expected density of plants, the time evolution equation for the two-block probability $P_k(1,1) \in \{0,1\}^2$ has to be analyzed in a subsequent step. Note that (E.2) reduces to the mean-field equation (E.1) when we assume that the state of a site is independent of all its neighboring sites, i.e., $P_k(1,1) = P_k(1)^2$.

This or very similar approximation techniques are known under various names, e.g., probability path method (Kikuchi 1966), local structure theory (Gutowitz, Victor, and Knight 1987; Gutowitz and Victor 1989), l-step Markov approximation (Gutowitz 1990), cluster approximation (Ben-Avraham and Köhler 1992; Rieger, Schadschneider, and Schreckenberg 1994; Schreckenberg, Schadschneider, Nagel, and Ito 1995), or the so-called BBGKY[32] hierarchy (Bussemaker 1995). An approximation with blocks of size $l \geq 2$ explicitly takes into account spatial short-range correlations between the sites, and therefore the quality of the approximation improves with increasing block size l. Unfortunately, even in one dimension, approximations for $l \geq 2$ are very hard to obtain (Rieger, Schadschneider, and Schreckenberg 1994).

4.4.3 Linear Stability Analysis

In order to analyze the automaton dynamics in terms of the mean-field equations (4.17) or (4.23), standard mathematical tools, such as linear stability analysis, can be applied (e.g., Murray 1989; Kelley and Peterson 1991). We begin with a short outline of the linear stability analysis of the system of time-discrete mean-field equations (4.17), which is given in vectorial notation by

$$\boldsymbol{x}(k+1) = \boldsymbol{H}(\boldsymbol{x}(k)),$$

where $\boldsymbol{x}^T = (x_i)_{i=1}^{|\mathcal{E}|}$ and $\boldsymbol{H}^T = (H_i)_{i=1}^{|\mathcal{E}|}$. The H_i's are nonlinear functions of polynomial type. Of special interest are *steady-state solutions* in which x_i keep the same value for all times, i.e.,

[32] Bogoliubov, Born, Green, Kirkwood, and Yvon.

$$x_i(k) = \bar{x}_i \qquad \forall\, k \in \mathbb{N}_0,\ i = 1, \ldots, |\mathcal{E}|.$$

These can be obtained by solving the fixed-point equation for the system

$$\boldsymbol{x}(k+1) = \boldsymbol{H}\,(\boldsymbol{x}(k)) = \boldsymbol{x}(k).$$

In order to characterize the behavior of the system close to a fixed point $\bar{\boldsymbol{x}}^T = (\bar{x}_i)_{i=1}^{|\mathcal{E}|}$, the time evolution of a small perturbation $\delta\boldsymbol{x}^T = (\delta x_i)_{i=1}^{|\mathcal{E}|}$ around $\bar{\boldsymbol{x}}$ is analyzed.

$$\begin{aligned} \delta\boldsymbol{x}(k) &:= \boldsymbol{x}(k) - \bar{\boldsymbol{x}}, \\ \delta\boldsymbol{x}(k+1) &= \boldsymbol{x}(k+1) - \bar{\boldsymbol{x}} = \boldsymbol{H}\,(\boldsymbol{x}(k)) - \bar{\boldsymbol{x}} = \boldsymbol{H}\,(\bar{\boldsymbol{x}} + \delta\boldsymbol{x}(k)) - \bar{\boldsymbol{x}} \\ &\simeq \boldsymbol{J}\delta\boldsymbol{x}(k), \end{aligned} \tag{4.26}$$

where a first-order Taylor expansion of $\boldsymbol{H}\,(\bar{\boldsymbol{x}} + \delta\boldsymbol{x}(k))$ is used to obtain a linear approximation of $\delta\boldsymbol{x}(k+1)$. The Jacobian matrix $\boldsymbol{J}$ is defined as

$$J_{ij} = \left.\frac{\partial H_i\,(\boldsymbol{x}(k))}{\partial x_j(k)}\right|_{\bar{\boldsymbol{x}}}, \qquad i, j = 1, \ldots, |\mathcal{E}|.$$

Eqn. (4.26) is a linear first-order homogeneous difference equation with constant coefficients, whose general solution is

$$\delta\boldsymbol{x}(k) = \boldsymbol{J}^k \delta\boldsymbol{x}(0) \quad \text{or} \quad \delta x_i(k) = \sum_{l=1}^{n} p_{li}(k)\lambda_l^k \qquad i = 1, \ldots, |\mathcal{E}|. \tag{4.27}$$

Here, $\Lambda_{\boldsymbol{J}} = \{\lambda_1, \ldots, \lambda_n\}$ is the set of *distinct* eigenvalues (spectrum) of $\boldsymbol{J}$ and $p_{li}(k)$ is a polynomial of degree less than α_l which is given by the minimal polynomial $\prod_{l=1}^{n} (x - \lambda_l)^{\alpha_l}$ of $\boldsymbol{J}$ (cp., e.g., Kelley and Peterson 1991). If $\boldsymbol{J}$ has $|\mathcal{E}|$ linearly independent eigenvectors v_l, which means that $\boldsymbol{J}$ is *diagonalizable*, then (4.27) reduces to

$$\delta x_i(k) = \sum_{l=1}^{|\mathcal{E}|} d_l v_{li} \lambda_l^k, \qquad i = 1, \ldots, |\mathcal{E}|, \tag{4.28}$$

where v_{li} is the ith component of v_l and coefficients $d_l \in \mathbb{C}$ are constants that are uniquely specified by the initial condition, i.e., $\delta\boldsymbol{x}(0) = d_1 v_1 + \cdots + d_{|\mathcal{E}|} v_{|\mathcal{E}|}$.

Hence, the dynamics of the perturbation $\delta\boldsymbol{x}(k)$ are determined by the eigenvalues of $\boldsymbol{J}$, especially by the *spectral radius*

$$\mu := \max\left\{|\lambda| : \lambda \in \Lambda_{\boldsymbol{J}}\right\}.$$

Typical behaviors[33] are summarized as follows:

- If $\mu < 1$, then all perturbations $\delta\boldsymbol{x}(k)$ decrease, i.e., $\lim_{k\to\infty} \delta\boldsymbol{x}(k) = 0$, which means the stationary state $\bar{\boldsymbol{x}}$ is locally *stable*.

[33] For details and proofs see, e.g., Kelley and Peterson (1991).

- If $\mu = 1$ and real or complex conjugate eigenvalues λ with $|\lambda| = 1$ are simple,[34] then there is a constant C such that $|\delta \boldsymbol{x}(k)| \leq C|\delta \boldsymbol{x}(0)|$ for all times $k \in \mathbb{N}$. Hence, all perturbations are *bounded*, which is a *weaker form of stability* of the stationary state $\bar{\boldsymbol{x}}$. $\mu = 1$ implies that $\bar{\boldsymbol{x}}$ is not hyperbolic; hence stability can only be decided by considering the nonlinear terms of the system.
- If $\mu \geq 1$, then perturbations $\delta \boldsymbol{x}(k)$ exist that do not decay. If $\mu > 1$, then perturbations can become very large in time such that the approximate equation (4.26) will no longer be valid. In this case the stationary state $\bar{\boldsymbol{x}}$ is called locally *unstable*.
- A real eigenvalue less than zero, $\lambda < 0$, implies a converging ($\lambda > -1$) or diverging ($\lambda < -1$) sawtooth oscillation in one of the expressions on the right-hand side of (4.27), since $\lambda^k = (-1)^k|\lambda|^k$. As all eigenvalues can have different signs, a variety of possibilities exists for the linear combination of solutions. In the linear system (4.27), the eigenvalue corresponding to the spectral radius determines the qualitative behavior for $k \to \infty$.

Next, we outline the linear stability analysis of the lattice-Boltzmann equation (4.23) which is not only time- but also space-dependent.

Stability Analysis of the Lattice-Boltzmann Equation. In a first attempt to understand the solutions of the time- and space-discrete lattice-Boltzmann equation (4.23), we analyze the stability of *spatially uniform steady states*

$$f_i(r,k) = \bar{f}_i, \qquad \forall\, r \in \mathcal{L},\ \forall\, k \in \mathbb{N}_0,$$

with respect to small, spatially heterogeneous local fluctuations

$$\delta f_i(r,k) := f_i(r,k) - \bar{f}_i, \qquad i = 1,\ldots,\tilde{b}.$$

The homogeneous stationary states $\bar{f}_i$ are obtained under the assumption that the automaton is in a local equilibrium state; i.e., they satisfy the equations

$$\tilde{\mathcal{C}}_i\left(\bar{\boldsymbol{f}}_{\mathcal{N}}\right) = 0, \qquad i = 1,\ldots,\tilde{b},$$

where

$$\bar{\boldsymbol{f}} = \left(\bar{f}_1,\ldots,\bar{f}_{\tilde{b}}\right) \quad \text{and} \quad \bar{\boldsymbol{f}}_{\mathcal{N}}^T = \left(\underbrace{\bar{\boldsymbol{f}},\ldots,\bar{\boldsymbol{f}}}_{\nu}\right), \qquad \nu = |\mathcal{N}_b^I|.$$

If (4.23) contains *nonlinearities*, linearization of "$\tilde{\mathcal{C}}_i\left(\boldsymbol{f}_{\mathcal{N}(r)}(k)\right)$" around $\bar{\boldsymbol{f}}$ yields the *linear lattice-Boltzmann equation*

$$\boxed{\begin{aligned}&\delta f_i(r+mc_i,k+1)\\ &\quad = \delta f_i(r,k) + \sum_{j=1}^{\tilde{b}} \Omega_{ij}^0 \delta f_j(r,k) + \sum_{r_n \in \mathcal{N}_{bo}^I(r)} \sum_{j=1}^{\tilde{b}} \Omega_{ij}^n \delta f_j(r_n,k),\end{aligned}} \tag{4.29}$$

[34] This means that the multiplicity of λ as a root of the characteristic equation is 1.

where the elements of the $\tilde{b} \times \tilde{b}$-matrices $\mathbf{\Omega}^n$ are defined as

$$\Omega_{ij}^0 := \left. \frac{\partial \tilde{C}_i\,(\delta f\, v N r(k))}{\partial \delta f_j(r,k)} \right|_{\bar{f}}, \qquad i,j = 1,\ldots,\tilde{b}$$

and

$$\Omega_{ij}^n := \left. \frac{\partial \tilde{C}_i\,(\delta f\, v N r(k))}{\partial \delta f_j(r_n,k)} \right|_{\bar{f}}, \qquad i,j = 1,\ldots,\tilde{b}.$$

Note that all $\mathbf{\Omega}^n$, $n = 0, \ldots$, and $\nu_o = |\mathcal{N}_{bo}^I(r)|$, are independent of node and time since the term is evaluated in a spatially homogeneous and temporarily stationary situation. Note further, that $\mathcal{N}_{bo}^I(r)$ denotes the outer interaction neighborhood; i.e., the node r is not included here.

To determine the stability of the steady state $\bar{\boldsymbol{f}}$, we have to find solutions $\delta f_i(r,k)$ of the set of spatially coupled linear difference equations (4.29). If fluctuations decrease to zero for $k \to \infty$, then $\bar{\boldsymbol{f}}$ is stable with respect to spatially heterogeneous local fluctuations $\delta f_i(r,0)$. Eqn. (4.29) can be solved explicitly under the assumption that the fluctuations have a spatial dependence of sinoidal shape (Press et al. 1988).

In particular, all solutions for a *one-dimensional* periodic lattice with L nodes are of the form

$$\delta f_i(r,k) \propto \lambda^k \cos(\hat{q} \cdot r). \tag{4.30}$$

As $\lambda^k \cos(\hat{q} \cdot r) = \delta f_i(r,k) = \delta f_i(r+L,k) = \lambda^k \cos(\hat{q} \cdot (r+L))$, we find that $\hat{q} = (2\pi/L)q$, where q is an integer, $q = 0, \ldots, L-1$; L/q gives the *wavelength* of the fluctuation, and from now on we shall refer to q as the *wave number*. In other words, q is the number of sine or cosine waves that "fit" into a domain of a given size.

Similarly, for a *two-dimensional* lattice with $|\mathcal{L}| = L_1 \cdot L_2$ nodes $r = (r_1, r_2)$, we consider fluctuations of the form

$$\delta f_i(r,k) \propto \lambda^k \cos(\hat{q}_1 \cdot r_1) \cos(\hat{q}_2 \cdot r_2). \tag{4.31}$$

Again, implying periodic boundary conditions we find that

$$\hat{q} = (\hat{q}_1, \hat{q}_2) = \left(\frac{2\pi}{L_1} q_1, \frac{2\pi}{L_2} q_2 \right)$$

with the wave number $q = (q_1, q_2)$, where $q_1 = 0, \ldots, L_1 - 1$ and $q_2 = 0, \ldots, L_2 - 1$. The wavelength is given by $L/|q| = L/\sqrt{q_1^2 + q_2^2}$. Here and in the remainder, we solely consider two-dimensional lattices with $L_1 = L_2 = L$.

Note that fluctuations associated with the wave number $q = 0$ are *spatially homogeneous*.

Fourier series theory allows the representation of arbitrary fluctuations as a generalized sum of sinusoidal terms. In order to find all solutions of equation (4.29), we apply the discrete finite Fourier transform $\mathcal{F}$ and its inverse $\mathcal{F}^{-1}$, which are defined as

$$\boxed{\begin{aligned} \mathcal{F}\{\delta f_i(r,k)\} &= \sum_r e^{\mathbf{i}(2\pi/L)q\cdot r}\,\delta f_i(r,k) =: F_i(q,k), \\ \mathcal{F}^{-1}\{F_i(q,k)\} &= \frac{1}{|\mathcal{L}|}\sum_q e^{-\mathbf{i}(2\pi/L)q\cdot r}\,F_i(q,k) = \delta f_i(r,k), \end{aligned}} \tag{4.32}$$

where $|\mathcal{L}| = L$, $q = 0, \ldots, L-1$, in one dimension. In two dimensions $|\mathcal{L}| = L^2$, $q \cdot r = q_1 r_1 + q_2 r_2$, and $q_1, q_2 = 0, \ldots, L-1$.

Transforming the linear Boltzmann equations (4.29) into Fourier space we obtain

$$\begin{aligned} &\mathcal{F}\{\delta f_{i,}(r + mc_i, k+1)\} \\ &\quad = e^{\mathbf{i}(2\pi/L)q\cdot mc_i}\,F_i(q,k+1) \\ &\quad = F(q,k) + \sum_{j=1}^{\tilde{b}} \Omega^0_{ij} F_j(q,k) + \mathcal{F}\left\{\sum_{n=1}^{\nu_o}\sum_{j=1}^{\tilde{b}} \Omega^n_{ij}\delta f_j(r + c_n, k)\right\}. \end{aligned} \tag{4.33}$$

According to the translation invariance of the outer interaction neighborhood $\mathcal{N}^I_{bo}(r)$ (cp. definition on p. 70), (4.33) can be simplified to

$$\begin{aligned} &F_i(q,k+1) \\ &\quad = e^{-\mathbf{i}(2\pi/L)q\cdot mc_i} \\ &\qquad \cdot \left\{F(q,k) + \sum_{j=1}^{\tilde{b}} \Omega^0_{ij} F_j(q,k) + \sum_{n=1}^{\nu_o}\sum_{j=1}^{\tilde{b}} \Omega^n_{ij} e^{\mathbf{i}(2\pi/L)q\cdot\tilde{c}_n} F_j(q,k)\right\}, \end{aligned} \tag{4.34}$$

with $\tilde{c}_n \in \mathcal{N}^I_{bo}$, $n = 1, \ldots, \nu_o$. Let $\boldsymbol{F}^T(q,k) = \big(F_1(q,k), \ldots, F_{\tilde{b}}(q,k)\big)$; then each mode $\boldsymbol{F}(q,k)$ evolves according to the vector equation

$$\boldsymbol{F}(q,k+1) = \Gamma(q)\,\boldsymbol{F}(q,k), \tag{4.35}$$

where the elements of the $\tilde{b} \times \tilde{b}$ matrix $\Gamma(q)$ are given by

$$\Gamma_{ij}(q) = e^{-\mathbf{i}(2\pi/L)q\cdot mc_i}\left\{\delta_{ij} + \Omega^0_{ij} + \sum_{n=1}^{\nu_o} \Omega^n_{ij} e^{\mathbf{i}(2\pi/L)q\cdot\tilde{c}_n}\right\},$$

with the Kronecker symbol δ_{ij}. In matrix notation this becomes

$$\boxed{\Gamma(q) = T\left\{\boldsymbol{I} + \boldsymbol{\Omega}^0 + \sum_{n=1}^{\nu_o} \boldsymbol{\Omega}^n e^{\mathbf{i}(2\pi/L)q\cdot\tilde{c}_n}\right\},} \tag{4.36}$$

with $\tilde{c}_n \in \mathcal{N}^I_{bo}$, $n = 1, \ldots, \nu_o$, the identity matrix $\boldsymbol{I}$, and the "transport matrix" $T = \mathrm{diag}\left(e^{-\mathbf{i}(2\pi/L)q\cdot mc_1}, \ldots, e^{-\mathbf{i}(2\pi/L)q\cdot mc_{\tilde{b}}}\right)$.

$\Gamma(q)$ is known as the *Boltzmann propagator* (Frisch et al. 1987). The solution of the linear discrete-time equation (4.35) for each q becomes

$$\boldsymbol{F}(q,k) = \Gamma(q)^k \boldsymbol{F}(q,0) \qquad \forall\, k \in \mathbb{N}. \tag{4.37}$$

Therefore, each Fourier component $\boldsymbol{F}(q,k)$ corresponding to wave number q evolves independently of the components corresponding to other wave numbers. The general solution (4.37) can be rewritten in the form

$$F_i(q,k) = \sum_{l=1}^{n} p_{li}(q,k)\lambda_l(q)^k, \tag{4.38}$$

where the spectrum of $\Gamma(q)$ is given by $\Lambda_{\Gamma(q)} = \{\lambda_1(q), \ldots, \lambda_n(q)\}$ and $p_{li}(q,k)$ is a polynomial in k with coefficients depending on q defined in the same way as the polynomial in (4.27). Taking the inverse Fourier transform $\mathcal{F}^{-1}$ of (4.38), we obtain that the complete solution for the linear lattice-Boltzmann equations (4.29) is given by a superposition of single q-mode solutions, which are the modes corresponding to the wave number q, i.e.,

$$\delta f_i(r,k) = \frac{1}{|\mathcal{L}|} \sum_q \sum_{l=1}^{n} p_{li}(q,k) e^{-\mathbf{i}(2\pi/L)q\cdot r} \lambda_l(q)^k.$$

Again, if $\Gamma(q)$ is *diagonalizable*, which means that it has $\tilde{b}$ linearly independent eigenvectors $v_l(q) = (v_{li}(q))_{i=1}^{\tilde{b}}$, then

$$\boxed{F_i(q,k) = \sum_{l=1}^{\tilde{b}} d_l(q) v_{li}(q) \lambda_l(q)^k,} \tag{4.39}$$

where the constants $d_l(q) \in \mathbb{C}$ are specified by the *initial condition*

$$\boxed{F_i(q,0) = \sum_{l=1}^{\tilde{b}} d_l(q) v_{li}(q) = \sum_r e^{\mathbf{i}(2\pi/L)q\cdot r} \delta f_i(r,0),} \qquad i = 1, \ldots, \tilde{b}, \tag{4.40}$$

and hence

$$\boxed{\begin{aligned} \delta f_i(r,k) &= \frac{1}{|\mathcal{L}|} \sum_q e^{-\mathbf{i}(2\pi/L)q\cdot r} F_i(q,k) \\ &= \frac{1}{|\mathcal{L}|} \sum_q e^{-\mathbf{i}(2\pi/L)q\cdot r} \sum_{l=1}^{\tilde{b}} d_l(q) v_{li}(q) \lambda_l(q)^k. \end{aligned}} \tag{4.41}$$

Thus, from (4.37) it follows that equations (4.29) are linearly stable with respect to perturbations of wave number q if the vector $\boldsymbol{F}(q,k)$ remains bounded. As in space-independent discrete systems (cp. p. 91), this is the case if $\Gamma(q)$ has no eigenvalue with an absolute value larger than 1. Hence, by analyzing the spectrum $\Lambda_{\Gamma(q)}$ of the Boltzmann propagator $\Gamma(q)$, we can predict which modes can grow with time, leading to the possible emergence of spatial patterns. The corresponding set of *critical wave numbers* is denoted by

$$Q^c := \left\{ q = (q_i)_{i=1}^d , \; q_i \in \{0, \ldots, L_i - 1\} : \mu(q) \geq 1 \right\},$$

where d is the dimension of the underlying lattice. The *spectral radius* $\mu(q)$ for each wave number q is given by

$$\mu(q) := \max\{|\lambda(q)| : \lambda(q) \in \Lambda_{\Gamma(q)}\}.$$

Consequently, the dynamics of the system (4.29) are determined by

$$\delta f_i(r,k) \sim \sum_{q \in Q^c} e^{-\mathrm{i}(2\pi/L) q \cdot r} F_i(q,k). \tag{4.42}$$

Let $q_* = (q_{i*})_{i=1}^d$ be the *dominant critical wave number* for which the spectral radius is maximal, i.e.,

$$\mu(q_*) = \max_{q \in Q^c} \mu(q).$$

For this wave number q_* the corresponding mode $\boldsymbol{F}(q_*, k)$ grows fastest. Furthermore, we introduce a decomposition of the set of critical wave numbers in $Q^+ \cup Q^- \subset Q^c$, where for $q = (q_i)_{i=1}^d, q_i \in \{0, \ldots, L_i - 1\}$,

$$Q^+ := \left\{ q : \exists\, \lambda(q) \in \Lambda_{\Gamma(q)}\; \lambda(q) \geq 1 \wedge |\lambda(q)| \equiv \mu(q) \right\}$$

and

$$Q^- := \left\{ q : \exists\, \lambda(q) \in \Lambda_{\Gamma(q)}\; \lambda(q) \leq -1 \wedge |\lambda(q)| \equiv \mu(q) \right\}.$$

Certain cases may be distinguished:

- *No spatial pattern.* The spatially homogeneous stationary solution $\bar{\boldsymbol{f}}$ is stable to any perturbation if $\mu(q) \leq 1$ for each wave number q. The LGCA model for simultaneous random movement studied in sec. 5.4 belongs to this category.
- *Stationary spatial pattern.* If a critical wave number $q \in Q^c$ exists such that $|q| \neq 0$ and if the corresponding eigenvalue is real, then these q-modes grow and there is an indication of a prevailing spatial wavelength $L/|q|$. The q-directions (in two dimensions) determine the structure of the pattern. Pattern formation of this kind is exemplarily studied in a one- and two-dimensional LGCA model based on activator–inhibitor interactions introduced in ch. 11. An indication of a spatial mode is also given in a model of adhesive interaction (see ch. 7).

- *Nonstationary spatial pattern.* Again, a critical wave number $q \in Q^c$ exists such that $|q| \neq 0$, but the corresponding eigenvalue has a nonzero imaginary part. In sec. 11.2 we introduce a two-dimensional LGCA model of an excitable medium which can exhibit spatial oscillating spirals.
- *Orientation-induced pattern formation.* In ch. 8 we present a two-dimensional LGCA as a model for the formation of propagating swarms, i.e., aligned cell patches. Here, the dominant unstable eigenvalue corresponds to a critical wave number $|q| = 0$ and the structure of the corresponding eigenspace indicates a global drift of cells: horizontal and vertical momentum both become unstable.

It is important to stress here that it is a property of *linear* theory that each of the various modes grows or decays independently of the other modes.

4.5 From Cellular Automata to Partial Differential Equations

Here, we are interested in the relationship between microscopic CA models and macroscopic partial differential equation systems. In particular, in this section we show that for appropriate space and time scaling the lattice-Boltzmann equation corresponding to a LGCA (4.24) can be transformed to LGCA macroscopic *partial differential equations* (cp. also Chopard and Droz 1998). In order to derive these equations we assume that the LGCA dynamics takes place on a lattice $\mathcal{L} = \mathcal{L}_\epsilon$ with periodic boundary conditions, and that $\mathcal{L}_\epsilon$ has $|\mathcal{L}_\epsilon| = L = l/\epsilon$ nodes in order to cover the interval $[0, l]$, where $\epsilon \in \mathbb{R}^+$ is a small parameter. On each lattice $\mathcal{L}_\epsilon$ we impose an initial concentration profile of particles as follows. We assume that the initial distribution $\xi_0 = \xi_{0,\epsilon}$ is a product distribution, such that

$$E\left(\eta_{\sigma,i}(r,0)\right) = g_{\sigma,i}(\epsilon r), \qquad i = 1,\ldots,\tilde{b},\ \sigma = 1,\ldots,\varsigma,$$

where $g_{\sigma,i}$ are nonnegative periodic smooth functions on the interval $[0, l]$ independent of ϵ and bounded by 1. Since the microscopic space variable r has been scaled by ϵ we must also scale time to obtain finite velocities in the LGCA model. Hence, the microscopic lattice-node $r \in \mathcal{L}_\epsilon$ and the time $k \in \mathbb{N}$ correspond to the macroscopic variables x and t, such that

$$\begin{aligned} x &= r\epsilon \in [0, l], \\ t &= k\delta \in \mathbb{R}^+, \qquad \epsilon, \delta \in \mathbb{R}^+. \end{aligned}$$

Additionally, we assume that for each lattice $\mathcal{L}_\epsilon$ the interaction process scales by $h(\epsilon, \delta)$. This means that the change of the average number of particles[35] in a short time interval of duration δ in a small space interval of length of order ϵ is approximately $h(\epsilon, \delta)\tilde{\mathcal{C}}_{\sigma,i}(x, t)$. Note that $h(\epsilon, \delta)$ has to be decreasing with the arguments to obtain finite rates of change of the particle numbers in the macroscopic regime.

[35] Recall that $\tilde{\mathcal{C}}_{\sigma,i}(r, k)$ (cp. p. 88) is the change of the average number of σ-particles at channel (r, c_i) at time k.

The relationship between δ and ϵ can be chosen according to a concrete model in a suitable way later on. For the sake of clarity, we restrict the technical derivation of the macroscopic LGCA equations to the case of a one-dimensional lattice ($b = 2$) with one rest channel ($\beta = 1$). Extensions to higher dimensions are straightforward. Under the discussed scaling the lattice-Boltzmann equations (4.24) become the continuous space and time finite-difference equations

$$f_{\sigma,1}(x + m_\sigma\epsilon, t + \delta) - f_{\sigma,1}(x,t) = h(\epsilon,\delta)\tilde{\mathcal{C}}_{\sigma,1}\left(\boldsymbol{f}_{\mathcal{N}(x)}(t)\right), \tag{4.43}$$

$$f_{\sigma,2}(x - m_\sigma\epsilon, t + \delta) - f_{\sigma,2}(x,t) = h(\epsilon,\delta)\tilde{\mathcal{C}}_{\sigma,2}\left(\boldsymbol{f}_{\mathcal{N}(x)}(t)\right), \tag{4.44}$$

$$f_{\sigma,3}(x, t + \delta) - f_{\sigma,3}(x,t) = h(\epsilon,\delta)\tilde{\mathcal{C}}_{\sigma,3}\left(\boldsymbol{f}_{\mathcal{N}(x)}(t)\right). \tag{4.45}$$

Taylor expansion of (4.43), (4.44) and (4.45) in powers of ϵ and δ up to second order yields

$$\begin{aligned} h(\epsilon,\delta)\tilde{\mathcal{C}}_{\sigma,1}\left(\boldsymbol{f}_{\mathcal{N}}\right) &= \delta\partial_t f_{\sigma,1} + \frac{\delta^2}{2}\partial_{tt} f_{\sigma,1} + m_\sigma\epsilon\partial_x f_{\sigma,1} + \frac{m_\sigma^2\epsilon^2}{2}\partial_{xx} f_{\sigma,1} \\ &\quad + m_\sigma\epsilon\delta\partial_{tx} f_{\sigma,1} + \mathcal{O}(\delta^2\epsilon) + \mathcal{O}(\delta\epsilon^2), \end{aligned} \tag{4.46}$$

$$\begin{aligned} h(\epsilon,\delta)\tilde{\mathcal{C}}_{\sigma,2}\left(\boldsymbol{f}_{\mathcal{N}}\right) &= \delta\partial_t f_{\sigma,2} + \frac{\delta^2}{2}\partial_{tt} f_{\sigma,2} - m_\sigma\epsilon\partial_x f_{\sigma,2} + \frac{m_\sigma^2\epsilon^2}{2}\partial_{xx} f_{\sigma,2} \\ &\quad - m_\sigma\epsilon\delta\partial_{tx} f_{\sigma,2} + \mathcal{O}(\delta^2\epsilon) + \mathcal{O}(\delta\epsilon^2), \end{aligned} \tag{4.47}$$

$$h(\epsilon,\delta)\tilde{\mathcal{C}}_{\sigma,3}\left(\boldsymbol{f}_{\mathcal{N}}\right) = \delta\partial_t f_{\sigma,3} + \frac{\delta^2}{2}\partial_{tt} f_{\sigma,3} + \mathcal{O}(\delta^3), \tag{4.48}$$

where $f_{\sigma,i} = f_{\sigma,i}(x,t)$ and $\boldsymbol{f}_{\mathcal{N}} = \boldsymbol{f}_{\mathcal{N}(x)}(t)$. Let $J_\sigma(x,t) := m_\sigma(f_{\sigma,1}(x,t) - f_{\sigma,2}(x,t))$ be the *flux* and $\varrho_\sigma(x,t) := f_{\sigma,1}(x,t) + f_{\sigma,2}(x,t) + f_{\sigma,3}(x,t)$ be the local number of particles of each component $\sigma = 1, \ldots, \varsigma$. Then from (4.46)–(4.48) we get

eqn. (4.46) + eqn. (4.47) + eqn. (4.48):

$$\begin{aligned} &\delta\partial_t\varrho_\sigma + \frac{\delta^2}{2}\partial_{tt}\varrho_\sigma + \epsilon\partial_x J_\sigma + \frac{m_\sigma^2\epsilon^2}{2}\partial_{xx}\varrho_\sigma + \epsilon\delta\partial_{tx} J_\sigma \\ &\quad + \mathcal{O}(\delta^2\epsilon) + \mathcal{O}(\delta\epsilon^2) + \mathcal{O}(\delta^3) \\ &= h(\epsilon,\delta)\sum_{i=1}^{3}\tilde{\mathcal{C}}_{\sigma,i}\left(\boldsymbol{f}_{\mathcal{N}}\right), \end{aligned} \tag{4.49}$$

∂_x\{eqn. (4.46) – eqn. (4.47)\}:

$$\begin{aligned} &\frac{\delta}{m_\sigma}\partial_{tx} J_\sigma + \frac{\delta^2}{2m_\sigma}\partial_{ttx} J_\sigma + m_\sigma\epsilon\partial_{xx}\varrho_\sigma + \frac{m_\sigma\epsilon^2}{2}\partial_{xxx} J_\sigma + m_\sigma\epsilon\delta\partial_{txx}\varrho_\sigma \\ &\quad + \mathcal{O}(\delta^2\epsilon) + \mathcal{O}(\delta\epsilon^2) \end{aligned}$$

$$= h(\epsilon, \delta) \left(\partial_x \tilde{C}_{\sigma,1} (\boldsymbol{f}_{\mathcal{N}}) - \partial_x \tilde{C}_{\sigma,2} (\boldsymbol{f}_{\mathcal{N}}) \right), \tag{4.50}$$

where we used that $\partial_x \varrho_\sigma = \partial_x (f_{\sigma,1} + f_{\sigma,2}) + \partial_x f_{\sigma,3} = \partial_x (f_{\sigma,1} + f_{\sigma,2})$. Note that $\partial_x f_{\sigma,3} = 0$, since the index 3 refers to the nonmoving rest particle. Solving (4.50) for $\partial_{tx} J_\sigma$ and substituting into (4.49) yields, after rearranging terms,

$$\begin{aligned}
\partial_t \varrho_\sigma = & \frac{h(\epsilon, \delta)}{\delta} \sum_{i=1}^{3} \tilde{C}_{\sigma,i} (\boldsymbol{f}_{\mathcal{N}}) + \frac{m_\sigma^2 \epsilon^2}{2\delta} \partial_{xx} \varrho_\sigma \\
& - \frac{h(\epsilon, \delta)}{\delta} m_\sigma \epsilon \left(\partial_x \tilde{C}_{\sigma,1} (\boldsymbol{f}_{\mathcal{N}}) - \partial_x \tilde{C}_{\sigma,2} (\boldsymbol{f}_{\mathcal{N}}) \right) - \frac{\delta}{2} \partial_{tt} \varrho_\sigma \\
& - \frac{\epsilon}{\delta} \partial_x J_\sigma + \frac{m_\sigma^2 \epsilon^3}{2\delta} \partial_{xxx} J_\sigma + m_\sigma^2 \epsilon^2 \partial_{txx} \varrho_\sigma \\
& - \mathcal{O}(\delta^2 \epsilon) - \mathcal{O}(\delta \epsilon^2) - \mathcal{O}(\delta^3) + \mathcal{O}(\delta^2 \epsilon^2) + \mathcal{O}(\delta \epsilon^3).
\end{aligned} \tag{4.51}$$

Hence, in the "diffusion limit," i.e., if

$$\delta \to 0, \ \epsilon \to 0 \quad \text{and} \quad \lim_{\substack{\delta \to 0 \\ \epsilon \to 0}} \frac{\epsilon^2}{\delta} = \text{const} := a,$$

and with appropriate scalings of the rates of change of the number of particles, i.e.,

$$h(\epsilon, \delta) = \hat{a}\delta, \quad \hat{a} \in \mathbb{R}^+,$$

(4.51) reduces to

$$\boxed{\partial_t \varrho_\sigma (x, t) = \hat{a} \sum_{i=1}^{3} \tilde{C}_{\sigma,i} \left(\boldsymbol{f}_{\mathcal{N}(x)}(t) \right) + D_\sigma \partial_{xx} \varrho_\sigma (x, t),} \tag{4.52}$$

where

$$\boxed{D_\sigma := \lim_{\substack{\delta \to 0 \\ \epsilon \to 0}} \frac{m_\sigma^2 \epsilon^2}{2\delta} = \frac{1}{2} a m_\sigma^2}$$

is the *diffusion coefficient* of species $\sigma = 1, \ldots, \varsigma$. Note that the only approximative assumption in the derivation of (4.52) is the restriction to second order terms in the truncated series leading to (4.46)–(4.48).

The derived partial differential equation system (4.52) is a scaled mean-field approximation of the underlying LGCA. It allows for analysis of spatio-temporal limiting behavior (see examples in ch. 11). These equations describe the limiting behavior that emerges on macroscopic scales from the microscopic automaton dynamics when both the distance between neighboring nodes and the time between transitions tend to zero. The limiting behavior depends accordingly on the precise relationships between considered time, space, and rate scalings. It is reflected through the types of emerging macroscopic partial differential equations, which can be different for different scalings (e.g., Deutsch and Lawniczak 1999; Lawniczak 1997).

Summary. In this chapter we have introduced and formally defined CA, in particular of deterministic, probablistic, and lattice-gas type. Furthermore, possible strategies for the analytic treatment of CA have been presented. The definitions in this chapter provide the basis for specific model developments. In the following chapters we will show how the CA concept can be used to design models for interacting cell systems.

4.6 Further Research Projects[36]

1. **Complexity/evolution/self-reproduction/Game of Life**
   a) To what extent do the different entropy concepts (in particular spatial and temporal entropy) serve as good measures of the complexity of deterministic and stochastic CA? Consider specific CA examples, e.g., the Game of Life.
   b) The Game of Life automaton is capable of self-reproduction. Could it be possible for a machine to produce another machine more complicated than itself? How could this be done? What is the connection between such a complexity-increasing construction and the idea of natural evolution for living organisms (Casti 1989)?
   c) It is usually assumed that Darwinian evolution involves passages from a state of lower to increasingly higher complexity. Are there relationships between this concept and the evolution and complexity of patterns in the Game of Life CA?
   d) How do small changes in the Game of Life rules (e.g., a different threshold for birth) influence the automaton evolution? Can different patterns be expected?
   e) A self-reproduction process must contain a blueprint for constructing offspring, a factory to carry out the construction, a controller to make sure the factory follows the plan, and a duplicating machine to transmit a copy of the blueprint to the offspring. J. von Neumann designed a CA capable of self-reproduction as observed in a biological cell. Are there other systems with self-reproduction capabilities, e.g., ecological networks, dynamical networks (the Internet)?
2. **Irreversibility:** A CA is irreversible if a particular configuration may arise from several distinct predecessors. A CA is reversible if the CA rules are bijective (invertible).
   (a) Give examples of rules for irreversible and reversible CA.
   (b) If the initial configuration for a one-dimensional CA is chosen at random and a reversible rule is applied, what kind of long-term behavior can be expected?

[36] For additional problems see also Casti 1989; Wolfram 1985.

(c) If one chooses a reversible rule that allows information to propagate, what kind of temporal behavior will emerge?
(d) Consider the possibility of using CA with reversible rules to model the processes of Newtonian physics (space, time, locality, microscopic reversibility, conservation laws, etc.).
(e) Will rules for CA modeling biological cell interaction be reversible or irreversible?

3. **Pattern recognition:** The evolution of a CA can be viewed as a pattern recognition process, in which all initial configurations in the basin of attraction of a particular attractor are thought of as instances of some pattern, with the attractor being the archetype of this pattern. Thus, the evolution of the different state trajectories toward this attractor would constitute recognition of the pattern. How could one formalize this idea into a practical pattern recognition device for image processing (Casti 1989)?
4. **Inverse problem:** Design CA rules for the solution of the density clasification problem, which is an example of "distributed computing": the initial state of the CA (containing 1s and 0s) shall converge to the all 1s state if the number of 1s in the initial configuration is larger and to the all 0s state in the other case.
5. **Partial differential/finite-difference equations/scaling**
   (a) We have shown how to derive a partial differential equation from CA rules under appropriate scaling. Are there alternative CA rules to mimic a given partial differential equation, e.g., a reaction-diffusion system.
   (b) Is it possible to develop a partial differential equation whose space-time behavior exactly matches a given CA at the discrete lattice nodes? Could such an equation be used to predict the long-term CA behavior without having to engage in direct simulation?
   (c) We have shown that the approximation by a deterministic lattice-Boltzmann equation can mimic the behavior of stochastic CA. Give examples of CA for which this approximation will not work.
   (d) We have demonstrated in this chapter how to derive a partial differential equation system as a "diffusively scaled" mean-field approximation of a LGCA. Discuss the influence of "diffusive scaling." Are there alternative realistic scaling possibilities? Of what type are corresponding partial differential equations?

Part III

Applications

5

Random Movement

"That which is static and repetitive is boring. That which is dynamic and random is confusing."

(John A. Locke)

In this chapter we present an overview of random walk models for random particle movement. The notion of a *particle* refers to any identifiable unit such as molecules (genes), biological cells, or individuals. The motion of particles in space[37] can be *directed* or *random*, *independent* or *dependent* (if the motion is biased by particle interaction with themselves or with their environment), and *active* or *passive*. Although it is not always easy to identify essential properties of particle motion, some examples are given in Table 5.1. Mathematically similar models can be applied even if the nature of the components is different. For example, the motion of pollen in a liquid and the spread of mutations in a genetic population can in certain limits both be described by a diffusion equation.

In particular, we introduce approaches to model a random walk in the framework of cellular automata (CA). Rules usually defined for random movement in probabilistic CA models are either time- and space-dependent (cp. sec. 5.3.1) or asynchronous (cp. sec. 5.3.2), while rules defined for lattice-gas cellular automata (LGCA) models (cp. sec. 5.4) are time- and space-independent and applied synchronously. We show that the stochastic random walk model (cp. equation (5.3)) and the LGCA model (cp. equation (5.32)) are described by identical equations for the evolution of the average total mass of particles, if the jump probability and the number of rest particles are adjusted appropriately. The diffusive dynamics of the LGCA is well captured by the corresponding linear lattice-Boltzmann equations, which can be solved explicitly. Moreover, the existence of an artificial (geometrical) invariant, known as the checkerboard invariant, can be identified with the help of the lattice-Boltzmann description. This artefact, not necessarily obvious in simulations can also be observed in stochastic random walk models without exclusion principle.

---

[37] The term *space* refers not only to Euclidean space, but to any abstract space such as ecological niche space or fitness landscapes.

Phenomenon	"Particle"	Movement		
		Directed/ random	Independent/ dependent	Active/ passive
Calcium ions in cytoplasm[a]	Molecule (ion)	Random	Independent	Passive
Pollen grains in liquid[b]	Cell	Rrandom	Independent	Passive
Spread of plants[c]	Cell	Random	Independent	Passive
Flock of birds[d]	Organism	Directed	Dependent	Active
Evolution (mutation)[e]	Genotype	Random	Independent	Passive

Table 5.1. Examples of "particle motion;" references: [a]Hasslacher et al. 1993, [b]Brown 1828, [c]cp. sec. 4.3, [d]Bussemaker et al. 1997, [e]Ebeling 1990.

5.1 Brownian Motion

Biological motion typically involves "interaction," and, in fact, in the chapters to follow we focus on the analysis of various forms of biologically motivated interactions. But there are important examples of *independent motion* of single cells or organisms, e.g., in the preaggregation feeding phase of myxobacteria (cp. the life cycle in fig. 1.2). Furthermore, diffusive motion can underlie growth processes; a well-known example is diffusion-limited aggregation (DLA) (Witten and Sander 1981; cp. also sec. 5.5).

Random walk as endless *irregular motion* was first observed with the aid of a microscope in the motion of pollen grains in liquid by the English botanist **Robert Brown** (1773–1858) (Brown 1828). He interpreted this movement as a general phenomenon characterizing small physical particles. Later, it was recognized that the random motion observed by Brown was the result of uncountable collisions of water molecules, which are thermally driven (Einstein 1905; Smoluchowski 1916). This explanation provided a link to the (macroscopic) *physical* process of molecular *diffusion* based on Fick's phenomenological law (formulated by the physiologist **Adolf Fick** (1829–1901) in 1855). Fick's law states that particle flux due to random motion is approximately proportional to the local gradient in particle concentration. A more general definition of a diffusion concept is given by Okubo 1980, who stated that the collective behavior of many particles is a diffusive phenomenon, "*when the microscopic irregular motion of each particle gives rise to a regularity of motion of the total particle group (macroscopic regularity).*" In his book Okubo provided an excellent review of diffusion models with *biological* applications,[38] e.g., the spread of genes (Fisher 1937) or the random dispersal of biological populations (Skellam

[38] See also Britton 1986.

1951, 1973). Other applications of diffusion models are innovation and expenditure diffusion, worker migration, and price waves.[39]

If particle motion is not random it is *biased*, either by an external force, e.g., flow of the medium, or internal force, e.g., motivation to move to some "attractive position" in space. In this case the collective movement of particles has a resulting directional component.

Research on random walk strongly triggered the development of such important fields as random processes, random noise, spectral analysis, and stochastic equations, just to name a few. "Certainly, Brown could not have dreamed that the motion of pollen in a single drop of water provided the stimulus for such achievements" (quoted from Okubo 1980).

In the next section we give a brief formal description of a simple model that describes Brownian motion of a *single particle* as a stochastic process (Feller 1968; Haken 1978b). For simplicity, we restrict our attention to the one-dimensional random motion. The common name for stochastic displacement is *random walk*. In the random walk model introduced here, time and space are assumed to be discrete. Furthermore, the relationship to the macroscopic diffusion equation will be described.

5.2 Discrete Random Walk and Diffusion

Here, we assume that a single particle walks randomly on a one-dimensional periodic lattice $\mathcal{L}_\epsilon$. Space and time are uniformly discretized such that the possible positions $x \in \mathcal{L}_\epsilon$ of the particle at time t are given by

$$\boxed{x = x(t) = r\epsilon} \quad \text{and} \quad \boxed{t = k\delta,} \qquad r \in \{0, \ldots, L-1\},\ k \in \mathbb{N}$$

with $|\mathcal{L}_\epsilon| = L$ spatial units of length $\epsilon \in \mathbb{R}^+$, and temporal units of duration $\delta \in \mathbb{R}^+$. Furthermore, we assume that besides motion, no other processes (e.g., birth/death) occur. The random walk of the particle is described by choosing a new random direction at each time t and moving a step in that direction during the time interval between t and $t + \delta$. The spatial steps are always of equal length ϵ in any direction and we suppose that successive moves are independent of each other. Since the particle should not be directionally affected, we choose equal probabilities $\alpha/2$ for jumps to the left and right. In this case the random walk is called *isotropic*.[40] Furthermore, there is a probability that the particle does not move. Hence, as illustrated in fig. 5.1, the change of the particle position is given by

$$x(t+\delta) = \begin{cases} x(t) - \epsilon & \qquad \alpha/2, \\ x(t) & \text{with probability} \quad 1-\alpha\ , \\ x(t) + \epsilon & \qquad \alpha/2, \end{cases}$$

where $\alpha \in [0, 1]$ (Weimar 1995). Let the particle position x at time t be a discrete

[39] For references and further examples of diffusion models see Banks 1994.

[40] From Greek *isos*, the same, and *tropos*, direction.

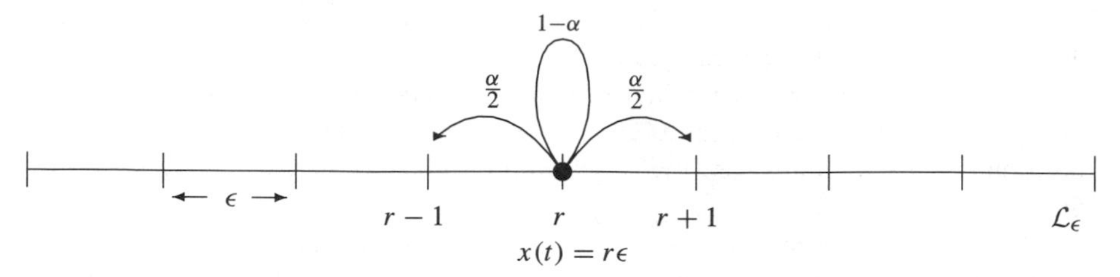

Figure 5.1. Random walk model for a single particle on a one-dimensional lattice $\mathcal{L}_\epsilon$.

random variable and let $P_s(x,t)$ be the probability that the *single particle* has reached point x at time t. Then, the probabilistic description of the random walk model is given by

$$P_s(x, t+\delta) = \frac{\alpha}{2} P_s(x+\epsilon, t) + (1-\alpha) P_s(x,t) + \frac{\alpha}{2} P_s(x-\epsilon, t) \,. \tag{5.1}$$

In order to solve this equation it is convenient to consider the corresponding equation:

$$\begin{aligned} P_s\left(\frac{x}{\epsilon}, \frac{t+\delta}{\delta}\right) &= P_s(r, k+1) \\ &= \frac{\alpha}{2} P_s(r+1, k) + (1-\alpha) P_s(r,k) + \frac{\alpha}{2} P_s(r-1, k). \end{aligned} \tag{5.2}$$

This is a linear difference equation, which can be solved by applying the discrete Fourier transformation and its inverse, as defined by (4.32). The spatial Fourier transform of (5.2) is

$$\begin{aligned} \hat{P}_s(q, k+1) &:= \mathcal{F}\{P_s(r, k+1)\} \\ &= \frac{\alpha}{2} e^{-\mathrm{i}(2\pi/L)q}\, \hat{P}_s(q,k) + (1-\alpha)\hat{P}_s(q,k) + \frac{\alpha}{2} e^{\mathrm{i}(2\pi/L)q}\, \hat{P}_s(q,k) \\ &= \left(1 - \alpha + \alpha\cos\left(\frac{2\pi}{L} q\right)\right) \hat{P}_s(q,k) \end{aligned}$$

with solution

$$\hat{P}_s(q,k) = \left(1 - \alpha + \alpha\cos\left(\frac{2\pi}{L} q\right)\right)^k \hat{P}_s(q,0),$$

where

$$\hat{P}_s(q,0) = \sum_{r=0}^{L-1} e^{\mathrm{i}(2\pi/L)qr}\, P_s(r,0)$$

for wave numbers $q \in \{0, \ldots, L-1\}$. Hence, the probability of finding the particle after k time steps at position r is given by

$$\boxed{\begin{aligned} P_s(r,k) &= \mathcal{F}^{-1}\left\{\hat{P}_s(q,k)\right\} \\ &= \frac{1}{L}\sum_{q=0}^{L-1}\hat{P}_s(q,0)e^{-\mathbf{i}(2\pi/L)qr}\left(1-\alpha+\alpha\cos\left(\frac{2\pi}{L}q\right)\right)^k. \end{aligned}} \tag{5.3}$$

If the particle starts at time step $k=0$ at position r^*, i.e.,

$$P_s(r,0) = \begin{cases} 1 & r = r^*, \\ 0 & \text{otherwise}, \end{cases},$$

the solution (5.3) becomes

$$P_s(r,k) = \frac{1}{L}\sum_{q=0}^{L-1} e^{-\mathbf{i}(2\pi/L)q(r-r^*)}\left(1-\alpha+\alpha\cos\left(\frac{2\pi}{L}q\right)\right)^k,$$

and hence

$$P_s(x,t) = \frac{1}{L}\sum_{q=0}^{L-1} e^{-\mathbf{i}(2\pi/L)q(\frac{x}{\epsilon}-\frac{x^*}{\epsilon})}\left(1-\alpha+\alpha\cos\left(\frac{2\pi}{L}q\right)\right)^{\frac{t}{\delta}} \tag{5.4}$$

for $x \in \{0, \epsilon, 2\epsilon, \ldots, (L-1)\epsilon\}$, $x^* = r^*\epsilon$ and $t = 0, \delta, 2\delta, \ldots$. Fig. 5.2 shows this probability distribution $P_s(x,t)$ at different times.

If a *large number of particles* follows the same random walk rule independently of each other, the probabilities $P_s(x,t)$ for each walker can simply be added. In this sense, $P(x,t) \equiv P_s(x,t)$ also represents the probability density of particles at position $x = r\epsilon$ at time $t = k\delta$.

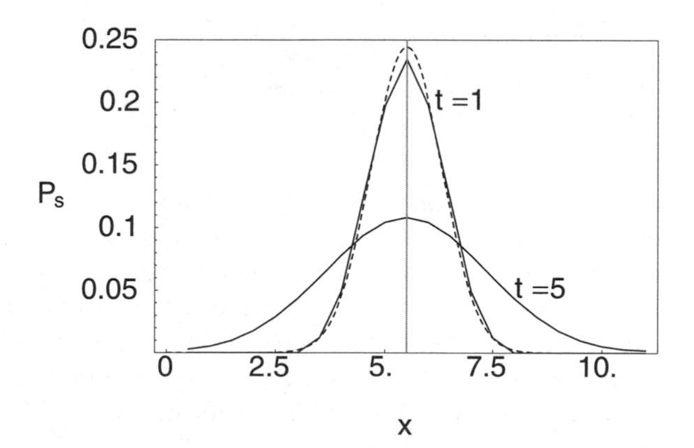

Figure 5.2. Solid lines: probability distribution $P_s(x,t)$ with initial particle position $x^* = 11\epsilon$ and lattice size $L = 22$ scaled with $\epsilon = 0.5$, $\delta = 0.25$: $k\delta = t \in \{1, 5\}$, $x = r\epsilon$, and $\alpha = 2/3$, calculated from equation (5.4). Dashed line: solution of the continuous diffusion equation at time $t = 1$ with diffusion coefficient $D \approx (\alpha/2)(\epsilon^2/\delta) = 1/3$, calculated from (5.11).

In order to derive a description of the random walk in *continuous* time and space we assume that the density of particles within a small spatial interval of length ϵ is large enough to neglect spatial and temporal random fluctuations in density. With this assumption, $P(x,t)$ can be considered as a smooth function in x and t such that it can be approximated using its Taylor expansion. We can rewrite (5.1) as

$$P(x,t) = \frac{\alpha}{2} P(x+\epsilon, t-\delta) + (1-\alpha)P(x, t-\delta) + \frac{\alpha}{2} P(x-\epsilon, t-\delta) \tag{5.5}$$

and approximate $P(x, t-\delta)$ and $P(x \pm \epsilon, t-\delta)$ by

$$P(x, t-\delta) = P(x,t) - \delta \frac{\partial P(x,t)}{\partial t} + O(\delta^2), \tag{5.6}$$

$$\begin{aligned} P(x \pm \epsilon, t-\delta) = P(x,t) &- \delta \frac{\partial P(x,t)}{\partial t} \pm \epsilon \frac{\partial P(x,t)}{\partial x} \\ &+ \frac{\delta^2}{2} \frac{\partial^2 P(x,t)}{\partial t^2} \mp \delta\epsilon \frac{\partial^2 P(x,t)}{\partial t \partial x} + \frac{\epsilon^2}{2} \frac{\partial^2 P(x,t)}{\partial x^2} \\ &+ O(\epsilon^2 \delta, \epsilon \delta^2). \end{aligned} \tag{5.7}$$

Substituting (5.6) and (5.7) in (5.5) and rearranging terms yields

$$\frac{\partial P(x,t)}{\partial t} = \alpha \frac{\epsilon^2}{2\delta} \frac{\partial^2 P(x,t)}{\partial x^2} + O(\delta^2) + O(\epsilon^2 \delta, \epsilon \delta^2). \tag{5.8}$$

Hence, in the limit for $\epsilon \to 0$ and $\delta \to 0$ but with the requirement

$$\lim_{\substack{\epsilon \to 0 \\ \delta \to 0}} \frac{\epsilon^2}{\delta} = \text{constant} \neq 0, \tag{5.9}$$

(5.8) becomes

$$\boxed{\frac{\partial P(x,t)}{\partial t} = D \frac{\partial^2 P(x,t)}{\partial x^2} \qquad \text{with} \qquad D := \frac{\alpha}{2} \lim_{\substack{\epsilon \to 0 \\ \delta \to 0}} \frac{\epsilon^2}{\delta}} \tag{5.10}$$

for $x \in \mathbb{R}$, $t \in \mathbb{R}^+$. This is the (partial differential) equation of *diffusion* for the random walk that results from the limiting process. If $P(x,t)$, the probability density, is multiplied by the total number of particles in the system, then the resulting equation for the particle density is known as Fick's diffusion equation or heat equation. The analytic solvability of (5.10) strongly depends on the imposed initial and boundary conditions. An extensive mathematical treatment of this equation can be found in Crank 1975. For example, if all particles start their walk at time $t = 0$ from the same position x^*, i.e.,

$$P(x,0) = \begin{cases} 1 & x = x^*, \\ 0 & \text{otherwise}, \end{cases}$$

the solution of (5.10) for an *infinite domain* is given by

$$P(x,t) = \frac{1}{\sqrt{4\pi Dt}} e^{-\frac{(x-x^*)^2}{4Dt}}, \qquad t > 0. \tag{5.11}$$

A comparison of this solution with the exact solution (5.4) of the random walk model (cp. p. 109) for $t = 1$ is shown in fig. 5.2. Note that D, which is the *diffusion coefficient*, has dimensions (length)^2/(time) and is *not* the speed of particle motion. The speed $c = \epsilon/\delta$ approaches positive infinity as $\epsilon \to 0$ and $\delta \to 0$. In other words, the continuous diffusion equation predicts that in an arbitrarily short time some particle will be found at an arbitrarily large distance from its starting point, which for most problems is a rather unrealistic assumption. Particles in the discrete random walk model have a finite speed. Note further that the derivation of the diffusion equation relies on ϵ and δ approaching zero in a rather specific way (cp. (5.9)). As Okubo (1980) observed, the use of the continuous diffusion approximation is justified when "the time of observation t is much greater than the duration time δ of each random step, and when the scale of observation x is much greater than the length of each random step ϵ."

5.3 Random Movement in Probabilistic Cellular Automaton Models

The implementation of a random walk of many particles in a CA may cause problems. Since all (independent) particles move *synchronously* in a classical CA, several of them may choose to move simultaneously to the same cell of the lattice. This may lead to an arbitrarily large number of particles in some lattice cells, which conflicts with the requirement that the set of elementary states should be small and finite (cp. the definition on p. 71). In the following we present some solutions to this problem based on space-time dependent local rules (sec. 5.3.1) and asynchronous update (sec. 5.3.2).

5.3.1 Random Walk Rule According to Toffoli and Margolus

The random walk rule of Toffoli and Margolus (1987) is based on the assumptions that (i) Brownian motion does not occur in a vacuum, i.e., pollen grain molecules interact with water molecules; and (ii) particles are impenetrable. Hence, when a particle moves, it swaps positions with other particles. In order to deal with a *synchronous* swapping of all particles, the lattice is partitioned into nonoverlapping "blocks" (pairs) of two adjacent cells. As shown in fig. 5.3, two partitions of lattice nodes are defined, one for even and one for odd time steps. Each cell can be empty or can host one particle. At each time step k, the contents of each block of cells is randomly swapped according to the actual partitioning of the lattice. This swapping rule is data blind since, if particles are present in both cells of a block, both will exchange places. Note that this random walk algorithm leads to a *time-* and

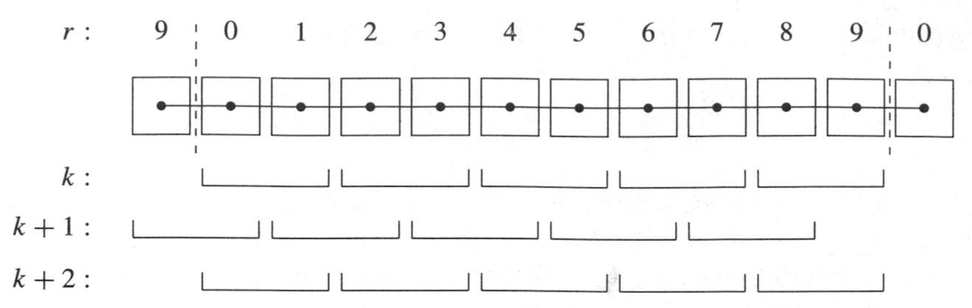

Figure 5.3. Example of time-dependent (k) block-partitioning of cells (r) for a one-dimensional periodic lattice $\mathcal{L}$ with $L = 10$ nodes.

space-dependent local transition rule. Thus the Toffoli and Margolus model is not a probabilistic cellular automaton according to the definition given previously (sec. 4.3).

The model is formally defined as follows: the state of each cell is made up of two components $s(r, k) = (s_1(r, k), s_2(r, k)) \in \mathcal{E} = \{0, 1\}^2$, where $s_1(r, k)$ is 1 or 0 with equal probability and $s_2(r, k)$ denotes the presence ($s_2(r, k) = 1$) or absence ($s_2(r, k) = 0$) of a particle in cell r at time k. The random bit s_1 is used to model the randomness of swapping in such a way that whenever two cells r_1 and r_2 belonging to the same block have equal random bits, i.e., $s_1(r_1, k) = s_1(r_2, k)$, then they exchange their contents, i.e., $s_2(r_1, k + 1) = s_2(r_2, k)$ and $s_2(r_2, k + 1) = s_2(r_1, k)$. With an interaction neighborhood template given by $\mathcal{N}_2^I = \{-1, 0, 1\}$, the local interaction rule $\mathcal{R}$ (cp. (4.2)) can be defined as

$$s(r, k + 1) = \mathcal{R}\left(\boldsymbol{s}_{\mathcal{N}(r)}(k)\right) = \left(\mathcal{R}_1(r, k), \mathcal{R}_2(\boldsymbol{s}_{\mathcal{N}(r)}(k))\right)$$

with

$$s_1(r, k + 1) = \mathcal{R}_1\,(r, k) = \xi(r, k + 1)$$

and

$$\begin{aligned} s_2&(r, k + 1) \\ &= \mathcal{R}_2\left((s(r - 1, k), s(r, k), s(r + 1, k))\,, r, k\right) \\ &= ((r + k) \bmod 2) \left\{ \left(1 - (s_1(r, k) - s_1(r - 1, k))^2\right) s_2(r - 1, k) \right. \\ &\qquad\qquad \left. + (s_1(r, k) - s_1(r - 1, k))^2 s_2(r, k) \right\} \\ &\quad + (1 - (r + k) \bmod 2)) \left\{ \left(1 - (s_1(r, k) - s_1(r + 1, k))^2\right) s_2(r + 1, k) \right. \\ &\qquad\qquad \left. + (s_1(r, k) - s_1(r + 1, k))^2 s_2(r, k) \right\}, \end{aligned}$$

where $\xi(r, k) \in \{0, 1\}$ are time- and cell-independent random variables taken from a uniform distribution, i.e., $P\,(\xi(r, k) = 1) = 1/2$. Note that the resulting random walk of a *single particle* is *not isotropic*, as can be seen from the example given in fig. 5.4.

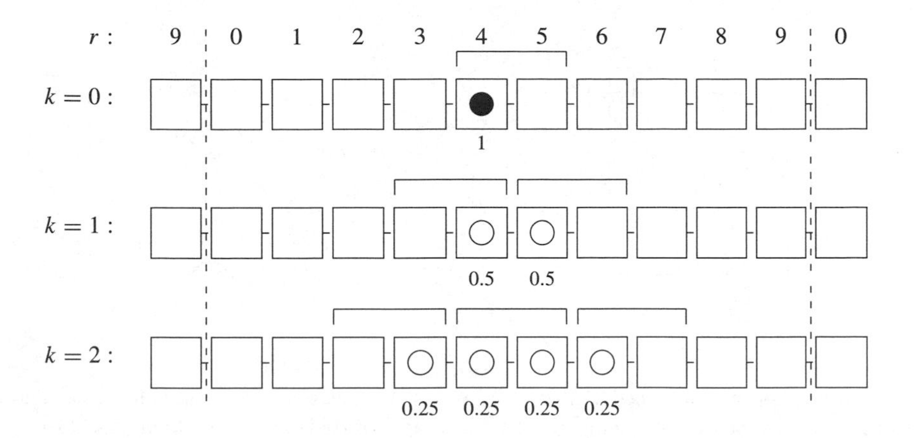

Figure 5.4. Example of a random walk of one particle; a number below a lattice cell r denotes the probability of the particle starting from position $r^* = 4$ to reach position r at times $k = 1$ and $k = 2$; the brackets above the lattice indicate the block-partitioning at time k. It can be seen that the random walk is not isotropic since the particle has a drift to the right.

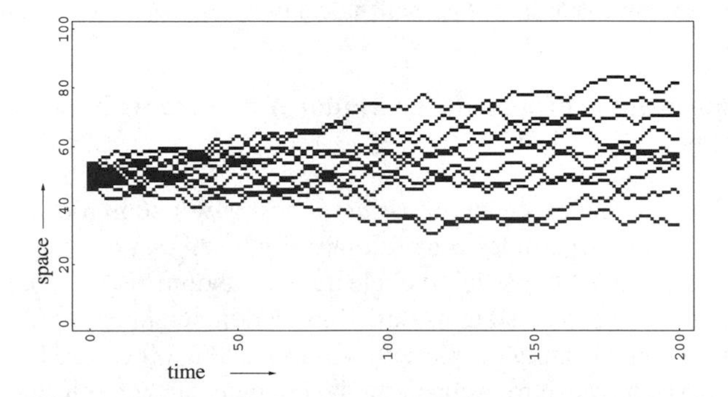

Figure 5.5. Space-time dynamics of the one-dimensional random walk model on a periodic lattice according to Toffoli and Margolus (1987). Parameters: $L = 100$, $k = 0, \ldots, 200$, and initial particle distribution $s_2(r, 0) = 1$ for $r = 45, \ldots, 55$ and $s_2(r, 0) = 0$ otherwise.

The repeated application of this cellular automaton rule leads to a random walk of many particles, as shown in fig. 5.5.

An extension of this model to *two dimensions* is straightforward (see Toffoli and Margolus 1987). The resulting partitioning of the two-dimensional lattice is known as the *Margolus neighborhood* (see fig. 5.6(a)).

The contents of such a 2×2 block of adjacent cells can be shuffled in 4! different ways. Chopard and Droz (1990) showed that a shuffling algorithm exclusively based on clockwise and counterclockwise block-rotations with equal probability is sufficient to model diffusive behavior on appropriate continuum space and time scales (cp. fig. 5.6(b)).

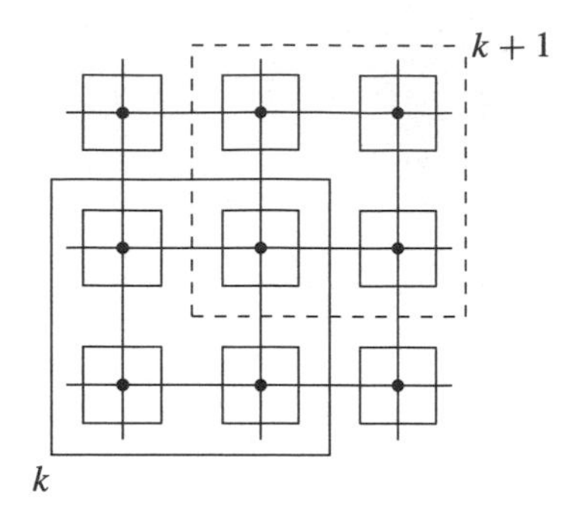

(a) "Margolus neighborhood": Part of time-dependent block-partitioning of cells for a two-dimensional lattice. The solid frame belongs to the partition at time k and the dashed frame to the partition at time $k+1$.

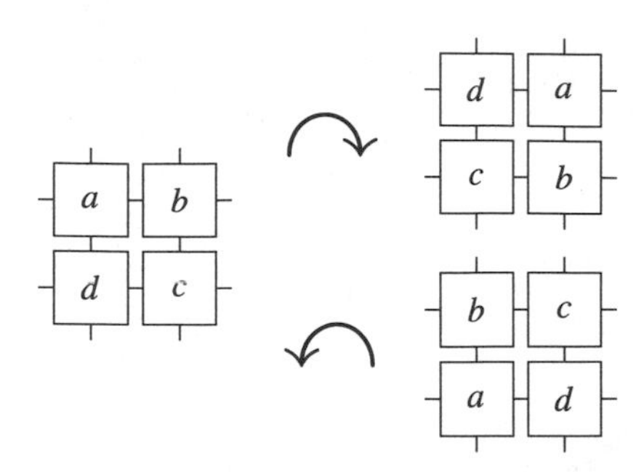

(b) Clockwise and counter-clockwise rotation of the contents of a Margolus block.

Figure 5.6. Two-dimensional random walk model using the "Margolus neighborhood."

5.3.2 Random Walk in Probabilistic Cellular Automata with Asynchronous Updating

If we allow the updating procedure to be *asynchronous*, a common way[41] to model a random walk of many particles is as follows: Each cell can be empty or occupied by at most one particle. A particle is selected at random (equal chance for each particle) and may move to a cell in its outer interaction neighborhood,[42] also chosen at random. If this cell is empty the particle will move to it, otherwise the particle will not move. This is equivalent to exchanging the contents of each selected pair of cells since the particles are indistinguishable.[43] The interaction neighborhood defines the *range* of the move. Note that this sequential procedure allows some particles to move more than others.[44]

Let N be the total number of particles in the system. In *one time step* $\tau\mathrm{N}$, $\tau > 0$, particles are sequentially selected at random to perform a move. The probability that a specific particle is not selected at all in this time step is $\left(1-\bar{\mathrm{N}}/\mathrm{N}\right)^{\tau\mathrm{N}}$. For a large

[41] Boccara and Cheong 1992, 1993; Boccara, Nasser, and Roger 1994, Brieger and Bonomi 1991; and McCauley, Wilson, and de Roos 1996.

[42] We use the term "interaction" in a rather general way; cp. the definition on p. 70.

[43] Automata, in which the local rule assigns new states to *two* cells—depending on their former states—are known as "dimer automata" (Schönfisch and Hadeler 1996). Such systems are also called "artificial ecologies" (Rand and Wilson 1995).

[44] This is not the case if all particles are chosen one by one (but in a random order) in every time step. This procedure is, e.g., applied in Berec, Boukal, and Berec 2001.

number of particles we get

$$\lim_{N\to\infty}\left(1-\frac{1}{N}\right)^{\tau N}=e^{-\tau}.$$

Hence, the probability that a single particle is updated at least once in a time step is given by $1-e^{-\tau}$ in this limit. Hence, τ represents the average number of *tentative* moves per particle during a unit of time. These considerations are the link to *continuous time* interacting particle systems (cp. p. 52). Herein, the probability that an event occurs with a rate α at least once between times t and $t+dt$, where $t, dt \in \mathbb{R}^+$ and dt is small, follows $1-e^{-\alpha dt}$ (Durrett 1995). Then, the rate α corresponds to $\alpha \hat{=} \tau\rho$, where ρ is the density of particles. One can identify $\tau N = (\alpha/\rho)N = \alpha L$ steps with one unit of time, where L is the number of cells in the lattice.[45]

From now on we set the range of the move to 1 (outer von Neumann interaction neighborhood) and measure the degree of mixing with the parameter τ. Fig. 5.7 shows a space-time plot of this asynchronous random walk for $\tau = 1$ and $\tau = 5$. Note that models of this type are also named "probabilistic automaton networks" (Boccara and Cheong 1992).

In the following sections, random movement in deterministic and probabilistic cellular automaton models will always be defined according to this asynchronous random walk implementation (sec. 5.3.2). Random movement in LGCA models follows the rules defined in the next section (5.4).

5.4 Random Movement in Lattice-Gas Cellular Automaton Models

The independent random walk of many particles within the framework of LGCA[46] is modeled by a shuffling (mixing) operator M which acts as follows (Lawniczak 1997): Before particles move from one node to a neighboring node according to the propagation step (defined on p. 77), each particle randomly selects a new velocity among the values permitted by the lattice. In order to avoid two or more particles occupying the same channel, a random permutation of the velocity vectors is performed at each lattice node at each time step, independently of the node configuration (data-blind). This interaction step, called the *shuffling step*, does not take into account the configuration of neighboring nodes, i.e., $\mathcal{N}_b^I(r) = \{r\}$. Since $\tilde{b}$ channels are assigned to each node, they can be shuffled in $\tilde{b}!$ ways, which is the number of permutations of $\tilde{b}$ objects. Fig. 5.8 shows all possible node configurations for a one-dimensional lattice with two velocity channels and one rest channel, i.e., $\tilde{b} = 3$. Note that the shuffling operation conserves the number of particles but not the momentum[47].

[45] Note that algorithms for (discrete) simulations of continuous interacting particle systems (Markov chains) are defined similar to the asynchronous updating rule.

[46] Similar models have been analyzed by various authors (see, for example, Boon, Dab, Kapral, and Lawniczak 1996; Chopard and Droz 1998).

[47] That is, the product of particle mass and velocity.

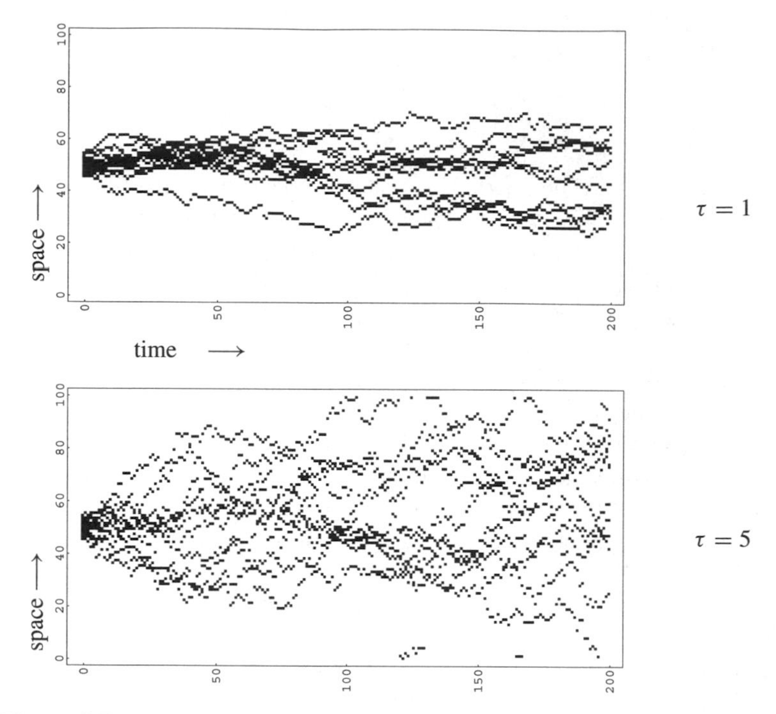

Figure 5.7. Space-time dynamics of a one-dimensional asynchronous random walk model on a periodic lattice. Parameters: $L = 100, k = 0, \ldots, 200$ and initial particle distribution $s_2(r, 0) = 1$ for $r = 45, \ldots, 55$ and $s_2(r, 0) = 0$ otherwise.

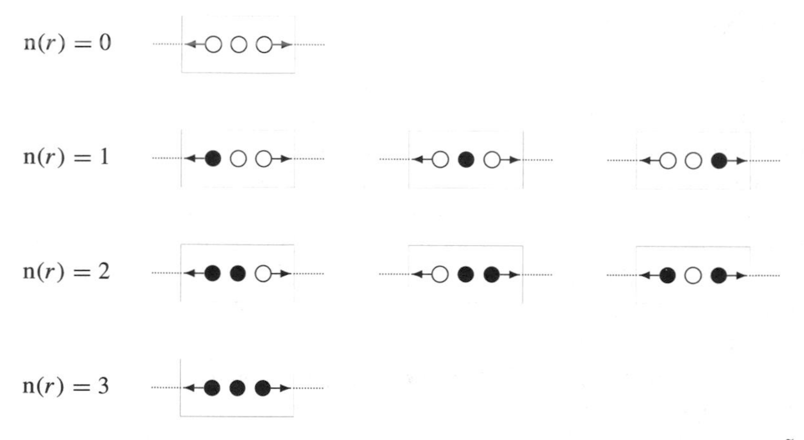

Figure 5.8. Possible node configurations for nodes of a one-dimensional lattice with $\tilde{b} = 3$; $\mathrm{n}(r)$ denotes the total number of particles present at node r.

The local shuffling process can be formally described as follows: If $\boldsymbol{u}^j$, $j = 1, \ldots, \tilde{b}$, are unit vectors with $\tilde{b}$ components, $\mathcal{A}_{\tilde{b}}$ denotes the set of all orthonormal permutation matrices A, given by

$$\mathcal{A}_{\tilde{b}} := \left\{ A \in \mathbb{R}^{\tilde{b}\times\tilde{b}} : \exists\, \pi \in \Pi_{\tilde{b}}\ \boldsymbol{a}^i = \boldsymbol{u}^{\pi(i)}\ \forall\, i = 1,\ldots,\tilde{b} \right\} = \{A_1, \ldots, A_{\tilde{b}!}\},$$

where $\boldsymbol{a}^i$ is the ith column vector of A and $\Pi_{\tilde{b}}$ is the set of all permutations of $\tilde{b}$ elements. In order to select a permutation matrix $A_j \in \mathcal{A}_{\tilde{b}}$, an independent sequence $\{\xi_j(r,k)\}_{r\in\mathcal{L},k\in\mathbb{N}}$ of independent, identically distributed Bernoulli-type random variables $\xi_j \in \{0,1\}$ is defined for each $j = 1,\ldots,\tilde{b}!$, such that for every k the random variables $\{\xi_j(r,k)\}_{r\in\mathcal{L}}$ are independent of the past evolution of the automaton,

$$p_j := P\left(\xi_j(r,k) = 1\right), \qquad j = 1,\ldots,\tilde{b}! \tag{5.12}$$

and

$$\sum_{j=1}^{\tilde{b}!} \xi_j(r,k) = \sum_{j=1}^{\tilde{b}!} p_j = 1 \qquad \forall\, r \in \mathcal{L},\ k \in \mathbb{N}. \tag{5.13}$$

Then, the microdynamics can be described by

$$\eta_i^{\mathrm{M}}(r,k) = \mathcal{R}_i^{\mathrm{M}}\left(\boldsymbol{\eta}(r,k)\right) = \sum_{j=1}^{\tilde{b}!} \xi_j(r,k) \sum_{l=1}^{\tilde{b}} \eta_l(r,k) a_{li}^j, \tag{5.14}$$

where a_{li}^j is a matrix element of $A_j \in \mathcal{A}_{\tilde{b}}$. Note that at a given node r and time step k only one of the ξ_j's is equal to 1.

As an *example* consider again a LGCA with $\tilde{b} = 3$. Then, $\mathcal{A}_3$ is given by

$$\mathcal{A}_3 = \left\{ \begin{pmatrix} 1&0&0\\0&1&0\\0&0&1 \end{pmatrix}, \begin{pmatrix} 1&0&0\\0&0&1\\0&1&0 \end{pmatrix}, \begin{pmatrix} 0&1&0\\1&0&0\\0&0&1 \end{pmatrix}, \begin{pmatrix} 0&0&1\\1&0&0\\0&1&0 \end{pmatrix}, \begin{pmatrix} 0&0&1\\0&1&0\\1&0&0 \end{pmatrix}, \begin{pmatrix} 0&1&0\\0&0&1\\1&0&0 \end{pmatrix}, \right\}$$

and therefore

$$\begin{aligned} \eta_1^{\mathrm{M}} &= \xi_1\eta_1 + \xi_2\eta_1 + \xi_3\eta_2 + \xi_4\eta_2 + \xi_5\eta_3 + \xi_6\eta_3, \\ \eta_2^{\mathrm{M}} &= \xi_1\eta_2 + \xi_2\eta_3 + \xi_3\eta_1 + \xi_4\eta_3 + \xi_5\eta_2 + \xi_6\eta_1, \\ \eta_3^{\mathrm{M}} &= \xi_1\eta_3 + \xi_2\eta_2 + \xi_3\eta_3 + \xi_4\eta_1 + \xi_5\eta_1 + \xi_6\eta_2. \end{aligned} \tag{5.15}$$

where we have neglected the (r,k) dependence for simplicity .

Hence, the complete dynamics is governed by the composition of the shuffling step (eqn. (5.14)) with the propagation step (eqn. (4.4)) and can be described by the microdynamical difference equation

$$\begin{aligned} \eta_i(r + mc_i, k+1) - \eta_i(r,k) &= \eta_i^{\mathrm{M}}(r,k) - \eta_i(r,k) \\ &= \left(\sum_{j=1}^{\tilde{b}!} \xi_j(r,k) \sum_{l=1}^{\tilde{b}} \eta_l(r,k) a_{li}^j \right) - \eta_i(r,k) \\ &= \mathcal{C}_i\left(\boldsymbol{\eta}(r,k)\right). \end{aligned} \tag{5.16}$$

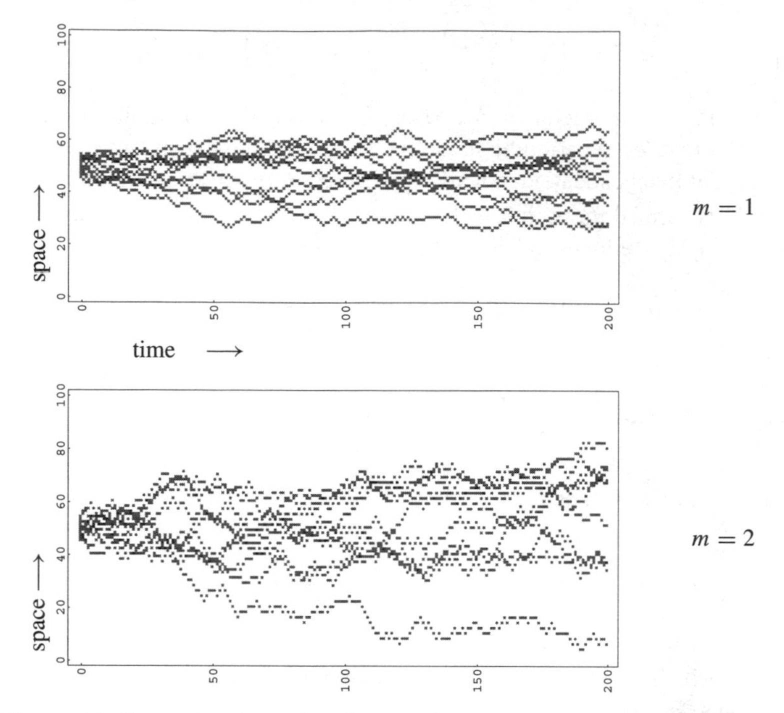

Figure 5.9. Space-time dynamics of a one-dimensional LGCA random walk model with one rest particle on a periodic lattice for different speeds $m \in \{1, 2\}$; parameters: $L = 100, k = 0, \ldots, 200$ and initial particle distribution $P(\eta_i(r, 0) = 1) = 1/3$ for $r = 45, \ldots, 55$ and $P(\eta_i(r, 0) = 1) = 0$ otherwise.

The repeated application of propagation and shuffling operators leads to a random walk for each particle, as shown in fig. 5.9, where each particle moves a distance m in a random direction within each time step.

This random walk performed by different particles is *not completely independent* because the exclusion principle does not allow independent changes of velocities of different particles at the same node. Nevertheless, since the interaction between random walkers is strictly local, the macroscopic behavior of the particle density can be expected to be *diffusive*. For example, Lawniczak (1997) showed that the presented automaton rules describe diffusive behavior in a continuous limit that can be reached using appropriate time- and space-scaling assumptions.[48].

In order to analyze the diffusive dynamics, we study the corresponding lattice-Boltzmann equation, which is obtained by taking the expectation of the microdynamical equation (5.16) as described in sec. 4.4.3. Note that the right-hand side of (5.16) is *linear*. Therefore, the lattice-Boltzmann equation can be solved explicitly.

To be more specific, we shall now consider the one-dimensional ($b = 2$) case with one rest channel ($\beta = 1$), i.e., $\tilde{b} = 3$.

[48] See also sec. 11.1.5, in which a macroscopic description for a LGCA model is derived.

Several other LGCA models for the simultaneous random walk of many particles are conceivable. For example, in the case of isotropic random walk, the shuffling operator `M` can be restricted to cyclic permutations of the velocity vectors (Boon, Dab, Kapral, and Lawniczak 1996; Chopard and Droz 1998; Lawniczak 1997). More isotropy can be obtained by introducing velocity channels that are linked to sites on the lattice beyond nearest neighbors (Wolfram 1986a). For example, Qian, d'Humières, and Lallemand 1992 studied a one-dimensional LGCA model for unsteady flows, in which each node is associated with four velocity channels $c_1 = 1$, $c_2 = 2$, $c_3 = -1$, and $c_4 = -2$. Also here, the shuffling step defines deterministic particle collisions which conserve mass and momentum.

5.4.1 Stability Analysis for the One-Dimensional Random Walk Model with One Rest Channel

By definition, the random variables ξ_i's are mutually independent of the η_l's. Therefore, according to (4.23), the lattice-Boltzmann equation can be derived from (5.16) as

$$\begin{aligned} & f_i(r + mc_i, k+1) - f_i(r,k) \\ & \quad = E\left(C_i\left(\boldsymbol{f}(r,k)\right)\right) = \tilde{C}_i\left(\boldsymbol{f}(r,k)\right) \qquad (5.17) \\ & \quad = \left(\sum_{j=1}^{\tilde{b}!} p_j \sum_{l=1}^{\tilde{b}} f_l(r,k) a_{li}^j \right) - f_i(r,k) \qquad \text{with } \left(a_{li}^j\right) = A_j \in \mathcal{A}_{\tilde{b}} \end{aligned}$$

for $i = 1, \ldots, \tilde{b}$, where we used (5.12) and (5.13). For the one-dimensional random walk with one rest channel, i.e., $\tilde{b} = 3$, this is equivalent to (cp. $\mathcal{A}_3$ given on p. 117)

$$\begin{aligned} & f_1(r+m, k+1) - f_1(r,k) \\ & \quad = (p_1 + p_2 - 1) f_1(r,k) + (p_3 + p_4) f_2(r,k) + (p_5 + p_6) f_3(r,k), \\ & f_2(r-m, k+1) - f_2(r,k) \\ & \quad = (p_3 + p_6) f_1(r,k) + (p_1 + p_5 - 1) f_2(r,k) + (p_2 + p_4) f_3(r,k), \\ & f_3(r, k+1) - f_3(r,k) \\ & \quad = (p_4 + p_5) f_1(r,k) + (p_2 + p_6) f_2(r,k) + (p_1 + p_3 - 1) f_3(r,k). \end{aligned}$$

Note that in the special case in which all configurations have equal probabilities, i.e., $p_i = p_j, i, j = 1, \ldots, \tilde{b}!$ and therefore $p_i = 1/\tilde{b}!$ due to constraint (5.13), the following expression can be obtained from (5.17):

$$\boxed{\begin{aligned} f_i(r + mc_i, k+1) - f_i(r,k) &= \left(\frac{(\tilde{b}-1)!}{\tilde{b}!} \sum_{l=1}^{\tilde{b}} f_l(r,k) \right) - f_i(r,k) \\ &= \left(\frac{1}{\tilde{b}} \sum_{l=1}^{\tilde{b}} f_l(r,k) \right) - f_i(r,k). \end{aligned}} \qquad (5.18)$$

Since the right-hand side of (5.17) is linear and the interaction neighborhood of a node r was defined to be the node itself (i.e., the outer interaction neighborhood is empty, $\nu_o = 0$), the Boltzmann propagator (4.36) simply becomes

$$\Gamma(q) = T\{\boldsymbol{I} + \boldsymbol{\Omega}^0\}$$
$$= \begin{pmatrix} e^{-\mathbf{i}(2\pi/L)q\cdot m} & 0 & 0 \\ 0 & e^{\mathbf{i}(2\pi/L)q\cdot m} & 0 \\ 0 & 0 & 1 \end{pmatrix} \begin{pmatrix} p_1 + p_2 & p_3 + p_4 & p_5 + p_6 \\ p_3 + p_6 & p_1 + p_5 & p_2 + p_4 \\ p_4 + p_5 & p_2 + p_6 & p_1 + p_3 \end{pmatrix}, \tag{5.19}$$

with $|\mathcal{L}| = L$. In the following, we will consider only *isotropic diffusion* processes, i.e., $p_i = 1/6, i = 1, \ldots, 6$. Under this assumption, $\Gamma(q)$ has a very simple structure such that we can determine the spectrum $\Lambda_{\Gamma(q)} = \{\lambda_1(q), \lambda_2(q), \lambda_3(q)\}$ as

$$\lambda_1(q) = \frac{1}{3} + \frac{2}{3}\cos\left(\frac{2\pi}{L}qm\right)$$
$$\approx 1 - \frac{1}{3}\left(\frac{2\pi}{L}qm\right)^2 + \frac{1}{36}\left(\frac{2\pi}{L}qm\right)^4 + O\left(\left(\frac{q}{L}\right)^6\right),$$
$$\lambda_2(q) = \lambda_3(q) = 0,$$

and corresponding eigenvectors as

$$v_1^T(q) = \left(e^{-\mathbf{i}(2\pi/L)q\cdot m}, e^{\mathbf{i}(2\pi/L)q\cdot m}, 1\right),$$
$$v_2^T(q) = (1, -1, 0)$$

and

$$v_3^T(q) = (1, 0, -1).$$

$\Gamma(q)$ is diagonalizable because the dimension of the eigenspace of the multiple eigenvalue $\lambda_2(q) = \lambda_3(q) = 0$ is $3 - \text{rank}(\Gamma(q)) = 2$, and therefore, according to eqn. (4.41) (cp. p. 95), the general solution of the lattice-Boltzmann equation (5.17) is given by

$$\delta f_i(r, k) = \frac{1}{L}\sum_{q=0}^{L-1} e^{-\mathbf{i}(2\pi/L)q\cdot r} F_i(q, k), \tag{5.20}$$

with

$$F_i(q, k) = d_1(q)v_{1i}(q)\lambda_1(q)^k \tag{5.21}$$

and

$$F_i(q, 0) = d_1(q)v_{1i}(q) + d_2(q)v_{2i}(q) + d_3(q)v_{3i}(q) \tag{5.22}$$

$$= \sum_{r=0}^{L-1} e^{\mathbf{i}(2\pi/L)q\cdot r} \delta f_i(r, 0) .$$

Hence, solutions corresponding to the eigenvalues $\lambda_2(q) = \lambda_3(q) = 0$ decrease to zero in one time step independently of the value of q. Because $\mu(q) = |\lambda_1(q)| \leq 1$ for all q, no modes can grow with time; i.e., pattern formation is not possible since all instabilities (with $q \neq 0$) are damped out. A spectral radius with $\mu(q) = 1$ indicates the existence of *invariants* in the system. Here, the spectral radius $\mu(q) = |\lambda_1(q)|$ converges to 1 if the wavelength L/q ($q \neq 0$) becomes very large, which is the case for large lattices $L \to \infty$ and small wave numbers q. In this case, there exists a dominant projection of the eigenvector $v_1(q)$ on the "mass vector" $v_1^T(0) = (1, 1, 1)$ since

$$\begin{pmatrix} 1 & 1 & 1 \\ 1 & -1 & 0 \\ 1 & 0 & -1 \end{pmatrix}^{-1} v_1(q) \approx \left(1 + O\left(\left(\frac{q}{L}\right)^2\right), -\mathbf{i}O\left(\frac{q}{L}\right), O\left(\left(\frac{q}{L}\right)^2\right)\right).$$

According to (5.21), we find that for $q = 0$,

$$F_1(0, k) = F_2(0, k) = F_3(0, k) = d_1(0) \qquad \forall\, k \in \mathbb{N},$$

since $\lambda_1(0) = 1$ and $v_1^T(0) = (1, 1, 1)$. $d_1(0)$ is determined by (5.22), i.e.,

$$\begin{aligned} & & d_1(0) + d_2(0) + d_3(0) &= \sum_{r=0}^{L-1} \delta f_1(r, 0), \\ &\wedge & d_1(0) - d_2(0) &= \sum_{r=0}^{L-1} \delta f_2(r, 0), \\ &\wedge & d_1(0) - d_3(0) &= \sum_{r=0}^{L-1} \delta f_3(r, 0), \\ &\Rightarrow & 3d_1(0) &= \sum_{r=0}^{L-1} \sum_{i=0}^{3} \delta f_i(r, 0), \end{aligned}$$

and hence

$$\sum_{i=1}^{3} F_i(0, k) = 3d_1(0) = \sum_{r=0}^{L-1} \sum_{i=0}^{3} \delta f_i(r, 0). \tag{5.23}$$

Therefore, persisting modes, which are associated with long wavelengths, describe diffusion of the particles initially present in the system (diffusive modes).

In particular, a closed equation for the *average local number of particles* at node r at time k, denoted by $\varrho(r, k) = \delta f_1(r, k) + \delta f_2(r, k) + \delta f_3(r, k) \in [0, 3]$, can be derived as (cp. (5.20))

$$\varrho(r,k) = \frac{1}{L}\sum_{q=0}^{L-1}\left(\sum_{i=1}^{3} F_i(q,0)\right) e^{-\mathbf{i}(2\pi/L)q\cdot r}\left(\frac{1}{3}+\frac{2}{3}\cos\left(\frac{2\pi}{L}qm\right)\right)^k, \quad (5.24)$$

with the Fourier-transformed initial conditions $\sum_{i=1}^{3} F_i(q,0)$. Note that the solution of the one-dimensional simultaneous random walk LGCA model with one rest particle given by (5.24) is equivalent to the solution (5.3), which we derived for the random walk model in sec. 5.2 with $\alpha = 2/3$ and $m = 1$.

Next, we explain the importance of rest particles in LGCA models by means of the one-dimensional simultaneous random walk.

5.4.2 Checkerboard Artefact

The linear stability analysis for the one-dimensional isotropic random walk model with an arbitrary number ($\beta > 0$) of zero-velocity channels is straightforward. A special situation is the LGCA random walk model *without rest particles*, i.e., $\beta = 0$. The corresponding Boltzmann propagator (cp. (5.19)) is given by

$$\Gamma(q) = \frac{1}{2}\begin{pmatrix} e^{-\mathbf{i}(2\pi/L)q\cdot m} & e^{-\mathbf{i}(2\pi/L)q\cdot m} \\ e^{\mathbf{i}(2\pi/L)q\cdot m} & e^{\mathbf{i}(2\pi/L)q\cdot m} \end{pmatrix},$$

with eigenvalues

$$\lambda_1(q) = \cos\left(\frac{2\pi}{L}qm\right), \quad \lambda_2(q) = 0, \quad (5.25)$$

and linear independent eigenvectors

$$v_1^T(q) = \left(e^{-\mathbf{i}\frac{4\pi}{L}q\cdot m}, 1\right), \quad v_2^T(q) = (-1, 1)\,. \quad (5.26)$$

Due to possible simplifications in the mathematical derivations, we examine the special case $m = 1$ in the following.

Here, another invariant is present in the system, because

$$\mu(\tilde{q}) = |\lambda_1(\tilde{q})| = |-1| = 1 \qquad \text{for } \tilde{q} = \frac{L}{2}.$$

Note that $\tilde{q}$ is an integer only for *even lattice sizes* L. The corresponding eigenvector is $v_1^T(\tilde{q}) = (1, 1)$, and hence modes

$$F_i(\tilde{q}, k) = (-1)^k d_1(\tilde{q})$$

perform undamped oscillations over time with period 2. Using the initial condition we get

$$d_1(\tilde{q}) - d_2(\tilde{q}) = \sum_{r\in\mathcal{L}} e^{\mathbf{i}\pi r}\delta f_1(r,0),$$

$$\wedge \qquad d_1(\tilde{q}) + d_2(\tilde{q}) = \sum_{r \in \mathcal{L}} e^{\mathbf{i}\pi r} \delta f_2(r, 0),$$

$$\Rightarrow \qquad 2d_1(\tilde{q}) = \sum_{r \in \mathcal{L}} e^{\mathbf{i}\pi r} \left(\delta f_1(r, 0) + \delta f_2(r, 0)\right).$$

Hence,

$$\begin{aligned}
\sum_{i=1}^{2} F_i(\tilde{q}, k) &= (-1)^k 2d_1(\tilde{q}) \\
&= (-1)^k \sum_{r \in \mathcal{L}} e^{\mathbf{i}\pi r} \varrho(r, 0) \\
&= (-1)^k \left(\sum_{\substack{r \in \mathcal{L} \\ r \text{ even}}} e^{\mathbf{i}\pi r} \varrho(r, 0) + \sum_{\substack{r \in \mathcal{L} \\ r \text{ odd}}} e^{\mathbf{i}\pi r} \varrho(r, 0) \right) \\
&= (-1)^k \left(\sum_{\substack{r \in \mathcal{L} \\ r \text{ even}}} \varrho(r, 0) - \sum_{\substack{r \in \mathcal{L} \\ r \text{ odd}}} \varrho(r, 0) \right),
\end{aligned}$$

where $\tilde{q} = L/2$. This invariant, which obviously does not have any interpretable meaning in the real system (i.e., it is artificial (spurious)[49]) is known as the *checkerboard invariant* in the context of LGCA models (d'Humières et al. 1989) and as *mesh-drift instability* in the context of numerical treatment of partial differential equations (Press et al. 1988). Note that it is straightforward to find checkerboard invariants in higher dimensional systems with square symmetries (Boon, Dab, Kapral, and Lawniczak 1996; Deutsch 1999a; d'Humières, Qian, and Lallemand 1989). This invariant with wavelength $2 = L/\tilde{q}$ is related to the fact that odd and even lattice nodes are completely decoupled; i.e., there is no interaction at any time between particles that are not located on the same even or odd sublattice. This can be seen more clearly in Table 5.2, which we derive next. According to (4.41) we get

$$\begin{aligned}
\varrho(r, k) &= \frac{1}{L} \sum_{i=1}^{2} \sum_{q=0}^{L-1} e^{-\mathbf{i}(2\pi/L)q \cdot r} F_i(q, k) \qquad (5.27) \\
&= \frac{1}{L} \sum_{i=1}^{2} \left(F_i(0, k) + e^{-\mathbf{i}\pi r} F_i(\tilde{q}, k) + \sum_{\substack{q=1 \\ q \neq \tilde{q}}}^{L-1} e^{-\mathbf{i}(2\pi/L)q \cdot r} F_i(q, k) \right) \\
&= \frac{1}{L} \left(\sum_{\tilde{r} \in \mathcal{L}} \varrho(\tilde{r}, 0) + (-1)^k e^{-\mathbf{i}\pi r} \left(\sum_{\substack{\tilde{r} \in \mathcal{L} \\ \tilde{r} \text{ even}}} \varrho(\tilde{r}, 0) - \sum_{\substack{\tilde{r} \in \mathcal{L} \\ \tilde{r} \text{ odd}}} \varrho(\tilde{r}, 0) \right) \right.
\end{aligned}$$

[49] In contrast to the mass conservation invariant.

	k even	k odd
r even	$\varrho(r,k) \approx \frac{2}{L} \sum_{\substack{\tilde{r}\in\mathcal{L} \\ \tilde{r} \text{ even}}} \varrho(\tilde{r},0)$	$\varrho(r,k) \approx \frac{2}{L} \sum_{\substack{\tilde{r}\in\mathcal{L} \\ \tilde{r} \text{ odd}}} \varrho(\tilde{r},0)$
r odd	$\varrho(r,k) \approx \frac{2}{L} \sum_{\substack{\tilde{r}\in\mathcal{L} \\ \tilde{r} \text{ odd}}} \varrho(\tilde{r},0)$	$\varrho(r,k) \approx \frac{2}{L} \sum_{\substack{\tilde{r}\in\mathcal{L} \\ \tilde{r} \text{ even}}} \varrho(\tilde{r},0)$

Table 5.2. Time- and (even/odd) sublattice-dependence of the expected local mass $\varrho(r,k)$; this table is based on eqn. (5.28).

$$
\left. + \sum_{\substack{q=1 \\ q\neq\tilde{q}}}^{L-1} e^{-\mathbf{i}(2\pi/L)q\cdot r} \sum_{i=1}^{2} F_i(q,k) \right),
$$

where $\tilde{q} = L/2$ and $F_1(0,k) + F_2(0,k) = \sum_{r\in\mathcal{L}} \varrho(r,0)$, which follows with arguments similar to those used in (5.23) (cp. p. 121). Furthermore, since the spectral radius $\mu(q) < 1$ for every $q \notin \{0, L/2\}$, all corresponding modes $F_i(q,k)$ will decay with time and accordingly their sum. Therefore, using the fact that

$$
\sum_{\tilde{r}\in\mathcal{L}} \varrho(\tilde{r},0) = \sum_{\substack{\tilde{r}\in\mathcal{L} \\ \tilde{r} \text{ even}}} \varrho(\tilde{r},0) + \sum_{\substack{\tilde{r}\in\mathcal{L} \\ \tilde{r} \text{ odd}}} \varrho(\tilde{r},0),
$$

we get

$$
\varrho(r,k) \approx \frac{1}{L}\left(\left(1 + (-1)^k e^{-\mathbf{i}\pi r}\right) \sum_{\substack{\tilde{r}\in\mathcal{L} \\ \tilde{r} \text{ even}}} \varrho(\tilde{r},0) + \left(1 - (-1)^k e^{-\mathbf{i}\pi r}\right) \sum_{\substack{\tilde{r}\in\mathcal{L} \\ \tilde{r} \text{ odd}}} \varrho(\tilde{r},0) \right), \tag{5.28}
$$

and hence, as summarized in Table 5.2, at even lattice nodes and even time steps, the expected mass depends solely on the mass of nodes in the even sublattice while at even lattice nodes and odd time steps it depends solely on the nodes on the odd sublattice.

If the *lattice size* L is *odd*, the checkerboard invariant is absent, because in this case no integer solution of the equation $\lambda_1(q) = -1$ (cp. (5.25)) exists. But modes corresponding to wave numbers close to $L/2$ decay very slowly, due to their sign-oscillating nature. Therefore, also in this case, a local decoupling of odd and even lattice nodes can persist for a while (Boon et al. 1996).

One possibility for *avoiding a checkerboard invariant* is the introduction of rest particles.[50] This induces a coupling of the two (even and odd) sublattices. It is easy

[50] For further possibilities for avoiding checkerboard invariants see, e.g., Boon et al. 1996 and Chopard and Droz 1998.

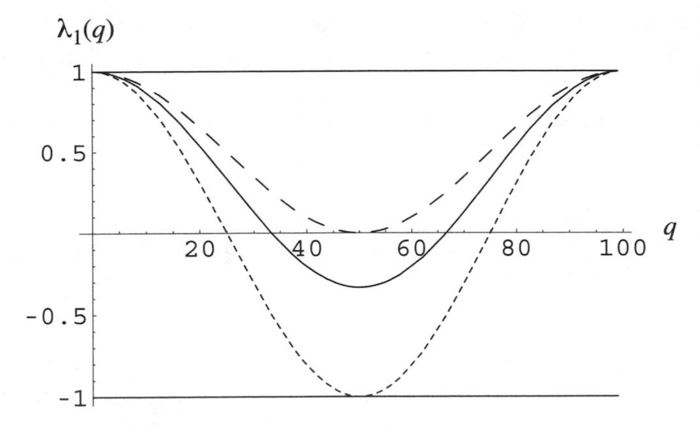

Figure 5.10. Dominant eigenvalues for one-dimensional LGCA random walk models given by (5.30). Small dashed line: no rest particle ($\beta = 0$); solid line: one rest particle ($\beta = 1$); large dashed line: two rest particles ($\beta = 2$); parameters: $L = 100$, $m = 1$.

to show that for any number $\beta > 0$ of rest particles the $(2+\beta) \times (2+\beta)$-Boltzmann propagator is given by

$$\Gamma(q) = \frac{1}{2+\beta} \begin{pmatrix} e^{-\mathbf{i}(2\pi/L)q\cdot m} & \dots & e^{-\mathbf{i}(2\pi/L)q\cdot m} \\ e^{\mathbf{i}(2\pi/L)q\cdot m} & \dots & e^{\mathbf{i}(2\pi/L)q\cdot m} \\ 1 & \dots & 1 \\ \vdots & \ddots & \vdots \\ 1 & \dots & 1 \end{pmatrix}. \tag{5.29}$$

The eigenvalues of (5.29) are

$$\boxed{\lambda_1(q) = \frac{1}{2+\beta}\left(\beta + 2\cos\left(\frac{2\pi}{L}qm\right)\right),} \tag{5.30}$$

$$\lambda_2(q) = \cdots = \lambda_{2+\beta}(q) = 0,$$

and corresponding eigenvectors are

$$v_1^T(q) = \left(e^{-\mathbf{i}(2\pi/L)q\cdot m}, e^{\mathbf{i}2\pi/L)q\cdot m}, 1, \dots, 1\right),$$

$$v_j^T(q) = \left(-1, 0, \dots, 0, \underset{j\text{th position}}{1}, 0\dots, 0\right) \quad \text{for} \quad j = 2, \dots, 2+\beta. \tag{5.31}$$

Hence, $\Gamma(q)$ is always diagonalizable. As illustrated in fig. 5.10, the magnitude of the minimum of the dominant eigenvalue $\lambda_1(q)$ depends on the number β of rest particles. Note that all sign-oscillating modes can be eliminated by introducing at least two zero velocity channels.

Later (e.g., in ch. 7), we present a two-dimensional LGCA model in which unstable modes that change sign at each time step are present and also become visible (cp. fig. 7.7).

If we compare the solution for the expected local mass of particles for $m = 1$, given by

$$\varrho(r,k) = \frac{1}{L}\sum_{q=0}^{L-1}\left(\sum_{i=1}^{2+\beta} F_i(q,0)\right) e^{-\mathrm{i}(2\pi/L)q\cdot r}\left(\frac{1}{2+\beta}\left(\beta + 2\cos\left(\frac{2\pi}{L}q\right)\right)\right)^k, \tag{5.32}$$

with the solution (5.3) for the stochastic random walk model derived in sec. 5.2, we get the following relationship between the number of rest particles β and the jump probability α (cp. p. 107), shown in Table 5.3. In the interpretation of the stochastic

No. rest particles	Jump probability (α)
0	1
1	2/3
2	1/2
$\vdots$	$\vdots$
β	$2/(2+\beta)$

Table 5.3. Relationship between the number of test particles β and the jump probability α.

random walk model introduced in sec. 5.2, a jump probability of $\alpha = 1$ implies that a particle jumps with probability $1/2$ either to the left or to the right, which implies that particles starting at even lattice nodes can only reach even lattice nodes after an even number of time steps (checkerboard artefact, Weimar 1995).

5.5 Growth by Diffusion-Limited Aggregation

Our analysis of the diffusive LGCA (cp. sec. 5.4.1) has shown in particular that there are no unstable modes in "diffusive" CA; i.e., we cannot expect any pattern formation: all initial disturbances will either be conserved (modes with corresponding eigenvalue $|\lambda| = 1$) or eventually die out (modes with $|\lambda| \leq 1$). Nevertheless the diffusive LGCA may be used as models of pattern formation if coupled with an antagonistic growth process. Turing showed that an interplay of diffusion and appropriately chosen reaction may destabilize an initially homogeneous situation and yield spatial patterns (Turing 1952). The reaction he proposed is a complicated chemical

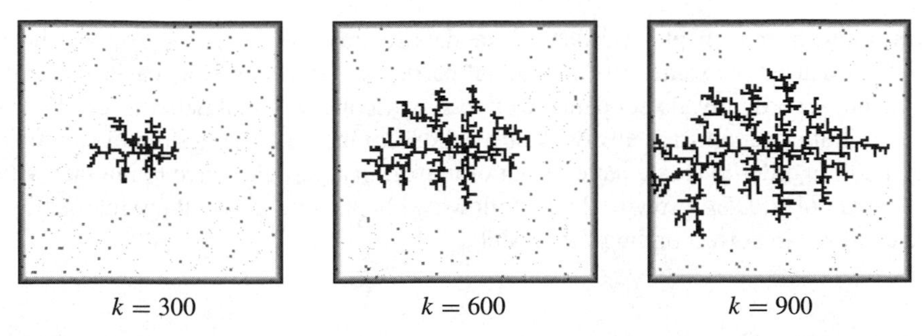

Figure 5.11. Growth pattern formation in two-dimensional diffusive LGCA (DLA model as described in the text). Temporal development initiated by a "resting particle nucleus" in a homogeneous distribution of "diffusing particles." Parameters: aggregation threshold: one particle; average density per velocity channel: 0.2; lattice size: $L = 50$.

reaction. It turns out that a much simpler "reaction," namely a *sticking condition*, can induce spatial anisotropies. Corresponding patterns are growth or aggregation patterns (cp. fig. 4.3). The prototype of such models is *diffusion-limited aggregation* (DLA) (Witten and Sander 1981). Note that the lattice model presented here is a stochastic model. There are also deterministic lattice models for diffusion-limited crystal growth (Liu and Goldenfeld 1991).

Diffusion-Limited Aggregation Interaction. The DLA-LGCA is defined on a two-dimensional lattice (coordination number $b = 4$) with one rest channel ($\beta = 1$). In this model, the existence of a rest channel is essential for modeling "stationary structures" (aggregates). The "reaction" may be implemented in various ways: it must be specified how moving particles are transformed into rest particles. Typically, if a migrating particle is the nearest neighbor of a rest particle, it sticks to the rest particle; i.e., it transforms into a rest particle itself. The *aggregation threshold* can be influenced by varying the minimum number of resting particles in the von Neumann interaction neighborhood of a moving particle which is a prerequisite for sticking. Growth speed can be regulated by varying the *sticking probability*. Because up to b particles may be simultaneous candidates for sticking, one can also require a threshold density to induce aggregation. The "diffusion" of the migrating particles is modeled by the rule for random movement, i.e., by shuffling of particles residing on velocity channels (see sec. 5.4).

Initially, in a spatially homogeneous distribution of migrating particles, one rest particle is introduced as an aggregation seed in the center. We show a temporal sequence of simulations (fig. 5.11). In order to characterize the simulated patterns one can determine fractal dimensions and growth rates (Chopard and Droz 1998). Introducing rest particles corresponds to a disturbance of the spatially homogeneous distribution. The consequences can be analyzed by means of the corresponding (linearized) Boltzmann equation, which yields indications for the behavior of unstable solutions (see further research projects below).

Summary. In this chapter various CA models for random particle movement have been introduced. Movement of individual particles was assumed as basically independent. Accordingly, no cooperative effects, particularly spatial patterns, can be expected. Initial patterns are always destroyed due to the underlying diffusive particle behavior. In order to create patterns out of uniformity, particles have to interact with each other (e.g., DLA growth). In the following chapters various cellular interactions are analyzed in corresponding CA models.

5.6 Further Research Projects

1. **Lattice and dimension**
   (a) Formally describe the independent random walk of many particles within the framework of "diffusive" LGCA for two- ($d = 2, b = 4$ and $b = 6$) and three- ($d = 3, b = 6$) dimensional lattices *without* resting particles ($\beta = 0$).
   (b) Develop and analyze a three-dimensional DLA-LGCA. Analyze the influence of modifications of the DLA rule.
2. **Analysis:** Perform a linear stability analysis for the two- and three-dimensional "diffusive" LGCA models and for the DLA-LGCA (defined in sec. 5.5).
3. **Artefacts:** Compare the dominant eigenvalues in the different "diffusive" LGCA models. Does the choice of lattice and dimension have an influence on the model capability of generating checkerboard artefacts?
4. **Deterministic versus stochastic system:** We have presented CA models for random walk that satisfy deterministic or stochastic rules. In addition, initial states typically have a random form. Discuss the influences of random initial conditions and stochastic dynamics and how stochastic influences can be captured in deterministic systems.
5. **Inverse problem:** Are there systems in which checkerboard patterns naturally occur? Try to design corresponding CA rules that would lead from any initial configuration to checkerboard patterns.
6. **Invariants/conservation laws:** The checkerboard artefact is an example of an artificial (geometrical) invariant: the CA evolves as if operating on several disjoint spatial lattices. In the random walk models introduced in this chapter, there is also a conservation of particles. Further indication of invariants can be found in CA, which support a set of persistent structures or "particles" that interact in simple ways. Find examples of systems with natural invariants and design corresponding CA rules.
7. **Movement of biological cells:** In this chapter we have not distinguished the random walk of physical particles and biological cells. In which situations are these models appropriate? How could the models be refined in order to yield more realistic models of biological cell movement?

6

Growth Processes

"The trees that are slow to grow bear the best fruit."
(Moliere, 1622–1673)

The term "growth" is used to indicate both *increase in size*, i.e., volume increase of an individual organism or cell, and *increase in numbers*, i.e., number of organismic or cellular individuals. Volume growth can, e.g., arise from the translocation of cells as, e.g., in tumor cell invasion. In this chapter we focus on models of "particle populations" growing in number. Starting from a short overview of historic growth concepts, we provide examples of probabilistic and LGCA growth models.

6.1 Classical Growth Models

A very detailed study of various growth models together with applications can be found in Banks (1994) and Sloep (1992). Following these authors, we briefly present some basic aspects of growth processes. In the history of growth models the English clergyman Thomas R. Malthus, together with the two mathematicians Benjamin Gompertz and Pierre F. Verhulst, are three especially noteworthy researchers whose basic frameworks are still adequate for handling many kinds of practical problems. They all studied growth processes of (human) populations. In modern language, the underlying assumptions are as follows: let $N(t)$ be the number of individuals in a population at continuous time t. Then Malthus assumed that the *relative growth rate*, or per capita growth rate, is constant,[51] i.e.,

$$\boxed{\frac{N'(t)}{N(t)} = \gamma,} \tag{6.1}$$

[51] A synonym for the relative growth rate is *specific* growth rate. But this term might be misleading, since the term "specific" carries the connotation of being constant, which is not true in general, as the upcoming examples show.

where $N'(t) = dN(t)/dt$. In other words, (6.1) expresses that the rate at which the quantity N changes in time is directly proportional to the amount of N present. The solution of (6.1) is

$$N(t) = N(0)e^{\gamma t}.$$

Accordingly, this simple law of growth is also known as the *exponential growth* model because the quantity of interest grows exponentially. For example, under suitable conditions, the growth of bacterial cells follows Malthusian growth, at least for short times, since it is characterized by independent cell divisions (proliferation). But exponential growth cannot go on forever. A more realistic approach takes limitations of the growth process into account, for example, a shortage of food and other resources or "crowding effects." A suggestion to account for such effects was originally proposed by Verhulst. He stated that the relative growth rate is density-dependent in such a way that it decreases linearly with the size of the population, i.e.,

$$\boxed{\frac{N'(t)}{N(t)} = \gamma \left(1 - \frac{N(t)}{K}\right),} \tag{6.2}$$

where K (carrying capacity) is the population size that will be reached after a long time. The solution of (6.2) is

$$N(t) = K\left(1 + \left(\frac{K}{N(0)} - 1\right)e^{-\gamma t}\right)^{-1}.$$

Verhulst's growth model is also known as the *logistic growth model*. The solution curve is S-shaped and it is almost exponential for "short" times and low values of $N(0)$. The third growth law that frequently appears in models of single component growth is the Gompertz law. Gompertz proposed that the relative growth rate decreases exponentially with *time*, i.e.,

$$\boxed{\frac{N'(t)}{N(t)} = \gamma e^{-\alpha t}.} \tag{6.3}$$

Using the solution given by

$$N(t) = N(0)e^{\frac{\gamma}{\alpha}(1-e^{-\alpha t})} \underset{t\to\infty}{\longrightarrow} K = N(0)e^{\frac{\gamma}{\alpha}},$$

a formally equivalent version of the relative growth rate can be derived as

$$\frac{N'(t)}{N(t)} = \alpha\left(\ln(K) - \ln(N)\right). \tag{6.4}$$

This growth model is widely used to describe the growth of solid tumors in mice, rats, and rabbits (e.g., Adam and Bellomo 1996; Casey 1934).

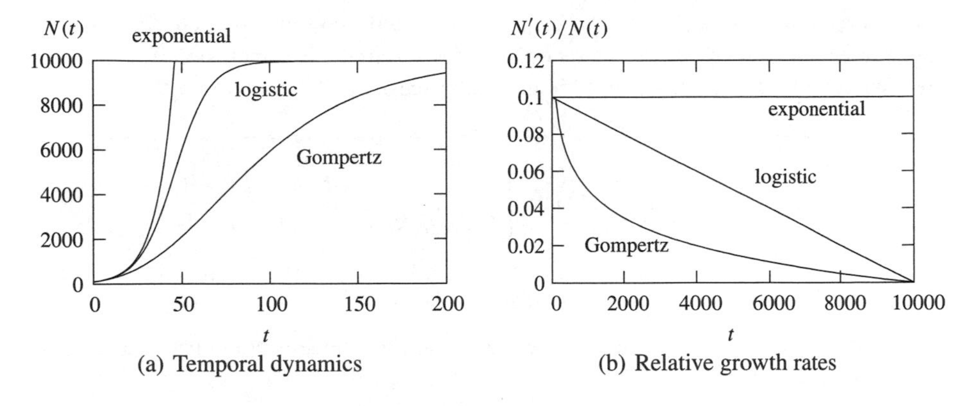

Figure 6.1. Comparison of exponential (eqn. (6.1)), logistic (eqn. (6.2)), and Gompertz (eqn. (6.4)) growth curves. Parameters: $N(0) = 100$, $K = 10000$, $\gamma = 0.1$, and $\alpha = 0.0217$.

In fig. 6.1 a comparison of exponential, logistic, and Gompertz growth curves is shown. Note that for all these models it is assumed that the relative growth rate is maximal if no other individuals are around.

In the models presented so far, space is not considered explicitly. But, *space* plays an important role in the growth of *aggregates* viewed as a collection of particles in space. Two frequent mechanims of aggregate growth can be distinguished (Williams, Desjardins, and Billings 1998):

1. growing aggregates due to a particle reproduction limited to points adjacent to the aggregate (e.g., Eden 1961; Richardson 1973) and
2. growing aggregates by sticking together diffusing particles if they meet. A prototype model for this process is the so-called "diffusion-limited aggregation," introduced by Witten and Sander 1983 (cp. sec. 5.5).

Originally, these growth models have been formulated as continuous-time stochastic Markov processes. But recently, (discrete-time) cellular automaton (CA) models have been constructed to study these phenomena (e.g., Chopard and Droz 1998; Eloranta 1997; Packard and Wolfram 1985; Williams, Desjardins, and Billings 1998).

In the following, we consider particle growth processes based on the first mechanism (particle reproduction) modeled in the framework of CA. We restrict our analysis to CA defined on a two-dimensional square lattice with periodic boundary conditions.

6.2 Growth Processes in Cellular Automata

A particle population is represented by particles occupying cells of the lattice. Each cell is either vacant ($s(r, k) = 0$) or occupied ($s(r, k) = 1$) by one particle, i.e., $\mathcal{E} = \{0, 1\}$. An *empty cell* becomes occupied with a certain time-independent transition probability $W(\boldsymbol{s}_{\mathcal{N}(r)} \to 1)$ which depends on the local configuration $\boldsymbol{s}_{\mathcal{N}(r)}$ of the interaction neighborhood $\mathcal{N}_4^I(r)$. In simple growth processes particles do not disappear; i.e., *occupied cells* remain occupied for all times. In conclusion, the general formal definition of a local rule for a growth process in a CA is given by

$$
\begin{aligned}
s(r, k+1) = \mathcal{R}\left(\mathbf{s}_{\mathcal{N}(r)}(k)\right) &= \begin{cases} 1 & \text{if} \quad s(r,k) = 1, \\ \zeta\left(\mathbf{s}_{\mathcal{N}(r)}(k)\right) & \text{if} \quad s(r,k) = 0, \end{cases} \\
&= s(r,k) + (1 - s(r,k))\,\zeta\left(\mathbf{s}_{\mathcal{N}(r)}(k)\right),
\end{aligned}
\tag{6.5}
$$

where $\zeta\left(\mathbf{s}_{\mathcal{N}(r)}(k)\right) \in \{0, 1\}$ is a space- and time-independent random variable with probability distribution

$$
\begin{aligned}
P\left(\zeta\left(\mathbf{s}_{\mathcal{N}(r)}(k)\right) = 1\right) &= W\left(\mathbf{s}_{\mathcal{N}(r)}(k) \to 1\right), \\
P\left(\zeta\left(\mathbf{s}_{\mathcal{N}(r)}(k)\right) = 0\right) &= 1 - W\left(\mathbf{s}_{\mathcal{N}(r)}(k) \to 1\right).
\end{aligned}
\tag{6.6}
$$

We will consider only growth processes that are not directionally biased. Hence, the transition probability W should depend solely on the *total number* of particles in the outer interaction neighborhood, which is denoted by

$$
\mathrm{n}_{\mathcal{N}(r)}(k) := \sum_{\tilde{r} \in \mathcal{N}_{4o}^I(r)} s(\tilde{r}, k) \in [0, \nu_o], \qquad \nu_o = |\mathcal{N}_{4o}^I|.
$$

Next we give some examples for the transition probability. Boccara, Roblin, and Roger (1994b) considered a generalized Game of Life model (see also sec. 4.1), in which the interaction neighborhood is R-axial (cp. p. 70). The growth part of their automaton rule is defined as follows: A vacant cell becomes occupied with probability $\gamma = 1$, if at least $B_{\min}$ and maximal $B_{\max}$ particles are in the outer interaction neighborhood; i.e., if $[B_{\min}, B_{\max}]$ represents an interval of "fertility." Hence,

$$
W\left(\mathbf{s}_{\mathcal{N}(r)}(k) \to 1\right) = \begin{cases} \gamma & \text{if} \quad \mathrm{n}_{\mathcal{N}(r)}(k) \in [B_{\min}, B_{\max}], \\ 0 & \text{else}. \end{cases}
\tag{6.7}
$$

Note that the growth part of the classical Game of Life model[52] suggested by John Conway corresponds to $R = 1$ and $B_{\min} = B_{\max} = 3$, $\gamma = 1$. A comprehensive treatment of Game of Life–like models can be found, for example, in Chaté and Manneville (1992) or Torre and Mártin (1997). Fig. 6.2(a) shows an example for the spatial distribution of particles resulting from a growth process after $k = 200$ steps with initially two seeds in the center. Since the growth process described by this rule

[52] The complete rule for the Game of Life model includes a death part, according to which a particle "dies" if $\mathrm{n}_{\mathcal{N}(r)}(k) \neq 3$.

is very restrictive and strongly dependent on the initial conditions, we do not go into further details.

For many infection or excitation processes[53] a lower bound B is defined, such that a susceptible/resting cell ("empty" cell) becomes infected/excited (cell occupied by a particle) with probability γ, if at least B cells in state 1 are in the outer interaction neighborhood, i.e.,

$$W\left(s_{\mathcal{N}(r)}(k) \to 1\right) = \begin{cases} \gamma & \text{if } \ \mathrm{n}_{\mathcal{N}(r)}(k) \geq B, \\ 0 & \text{else.} \end{cases} \tag{6.8}$$

For example, Greenberg et al. (1978) studied an excitation process. They used a von Neumann interaction neighborhood (cp. p. 70), i.e., $\nu = |\mathcal{N}_4^I| = 5$ and $\nu_o = 4$, and took the threshold to be $B = 1$ with $\gamma = 1$. Further examples are forest fire models (e.g., Chen, Bak, and Tang 1990). A "green tree" ("empty" cell) becomes a burning tree (cell occupied by a particle) if at least one of its nearest neighbors is burning, i.e., $B = 1$ and $\gamma = 1$.

An automaton for the spread of innovations has been introduced and studied by Boccara and Fukś (1998). They studied a one-dimensional automaton with R-axial interaction neighborhood. Here, we state the rule for the two-dimensional lattice, i.e., $\nu = |\mathcal{N}_4^I| = (2R+1)^2$ and $\nu_o = \nu - 1$. A neutral individual ("empty" cell) becomes an adopter (cell occupied by a particle) with a probability depending on the local density of adopters, i.e.,

$$W\left(s_{\mathcal{N}(r)}(k) \to 1\right) = \frac{\gamma \mathrm{n}_{\mathcal{N}(r)}(k)}{\nu_o}. \tag{6.9}$$

Another approach assumes that each particle in the outer interaction neighborhood produces an "offspring" with probability γ. Then, the probability for an empty cell to become occupied is given by

$$W\left(s_{\mathcal{N}(r)}(k) \to 1\right) = 1 - (1-\gamma)^{\mathrm{n}_{\mathcal{N}(r)}(k)}. \tag{6.10}$$

In the context of epidemics, γ is interpreted as the probability of becoming infected by contact with one individual (Boccara and Cheong 1993; Duryea et al. 1999). This rule is also used for predator–prey systems (Boccara, Roblin, and Roger 1994a; Hiebeler 1997; Rozenfeld and Albano 1999).

Fig. 6.2 shows snapshots of the particle distribution in space for each growth model.

In order to study the *temporal evolution* of the number of particles $N(k)$ in the described growth processes, we focus on the model defined by (6.8) for $B = 1$. First,

[53] Reviews of epidemic models and excitable media (compare sec. 11.2) in the framework of CA are given in Schönfisch (1993) and Kapral, Lawniczak, and Masiar (1991).

(a) Growth according to rule (6.7): Moore interaction neighborhood, two seeds in the center at $k = 0$ and $B_{\text{min}} = 2$, $B_{\text{max}} = 3$.

(b) Growth according to rule (6.8): von Neumann interaction neighborhood, one seed in the center at $k = 0$ and $B = 1$.

(c) Growth according to rule (6.9): von Neumann interaction neighborhood, one seed in the center at $k = 0$.

(d) Growth according to rule (6.10): von Neumann interaction neighborhood, one seed in the center at $k = 0$.

Figure 6.2. Two-dimensional growth processes in CA models shown after $k = 200$ steps. Parameters: $L = 100 \times 100$, $\gamma = 0.1$.

let us consider the growth process triggered by a *single seed* in the center of the lattice, i.e., $N(0) = 1$. A snapshot of the particle distribution in space after $k = 200$ steps with the von Neumann interaction neighborhood and $\gamma = 0.1$ is shown in fig. 6.2(b). A crude approximation of this growth process is given by the following difference equation:

$$\begin{aligned} & N(k+1) = N(k) + \nu_o \gamma k \\ \Rightarrow \quad & N(k) = 1 + \frac{\nu_o}{2} \gamma k \,(k-1)\,, \end{aligned} \tag{6.11}$$

where $\nu_o = 4$ for the outer von Neumann neighborhood and $\nu_o = 8$ for the outer Moore neighborhood. Eqn. (6.11) is exact for $\gamma = 1$ and an infinite lattice. If we place $N_0 > 1$ seeds on the lattice and assume independent growth for all seeds, then

$$N(k) = N_0 + N_0 \frac{\nu_o}{2} \gamma k \,(k-1)\,,$$

or in terms of spatial averages $\rho := \frac{1}{L} N \in [0, 1]$,

$$\rho(k) = \rho(0) + \rho(0) \frac{\nu_o}{2} \gamma k \,(k-1) \tag{6.12}$$

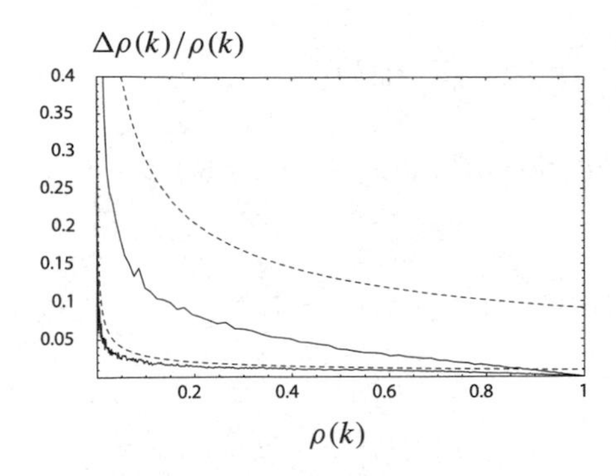

(a) Relative growth rates. The two curves at the bottom correspond to $\rho(0) = 1/L = 0.0001$ and the others to $\rho(0) = 0.01$.

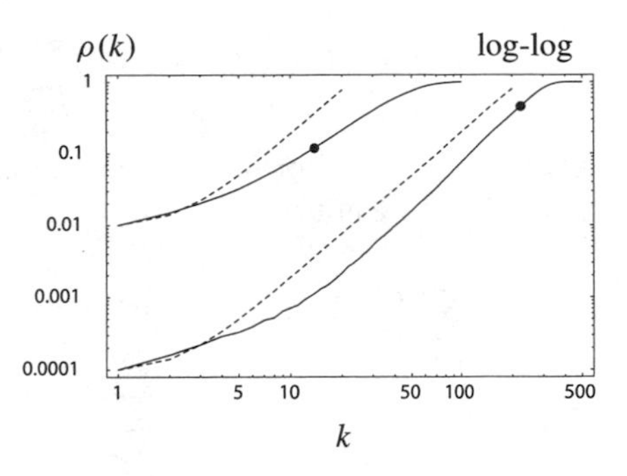

(b) Double logarithmic plot of the particle density $\rho(k)$. The dots indicate the inflection points of the curves.

Figure 6.3. Two-dimensional growth according to rule (6.8). Two solid lines: simulated growth process with two different initial conditions; dashed lines: growth process according to (6.12); parameters: $L = 100 \times 100$, $\gamma = 0.1$ and $B = 1$ (von Neumann interaction neighborhood).

is an upper bound for the growth process. As shown in fig. 6.3(a) the relative growth rate predicted by (6.12) for $\rho(0) = 1/L$ (single seed) is in good agreement with simulation data. Furthermore, the simulated growth process is qualitatively similar to the predicted equation (6.12), because the double logarithmic plot in fig. 6.3(b) shows a linear relationship for intermediate times, which is typical for polynomial processes. For higher initial densities, i.e., $\rho(0) > 1/L$, the prediction given by (6.12) is not useful. The predicted relative growth rate overestimates the growth process (cp. fig. 6.3(a)), since we did not take into account crowding effects of the agglomerations growing from each seed. In addition, although the shape of the predicted relative growth rate is similar to the relative growth rate obtained from simulation data, the simulated growth process is qualitatively different, as can be seen in fig. 6.3(b).

Mean-Field Approximation. We continue with an approximation of the given "growth" rules by mean-field theory (cp. sec. 4.4.2, p. 83). In particular, the mean-field equation for rule (6.8) is derived, while we simply state the mean-field equations for the other rules. According to (4.17) the evolution equation for the spatially averaged density of particles $x_1(k)$ is given by

$$x_1(k+1) = \sum_{(z_1,\ldots,z_\nu)\in\{0,1\}^\nu} W\left((z_1,\ldots,z_\nu) \to 1\right) \prod_{i=1}^{\nu} \sum_{l=0}^{1} \delta_{z_i,z^l} x_l(k), \tag{6.13}$$

where $x_0(k)$ denotes the spatially averaged density of empty cells and $z^0 = 0$ and $z^1 = 1$. For rule (6.8) this equation becomes

$$x_1(k+1) = x_1(k) + x_0(k) \sum_{i=B}^{\nu_o} \binom{\nu_o}{i} \gamma \, (x_1(k))^i \, (x_0(k))^{\nu_o - i} , \qquad (6.14)$$

where $\nu_0 = \nu - 1$. For $B = 1$, with the substitution $\rho(k) \equiv x_1(k)$, $1 - \rho(k) \equiv x_0(k)$, and with the identity

$$\sum_{i=0}^{\nu_o} \binom{\nu_o}{i} \rho^i(k) \, (1 - \rho(k))^{\nu_o - i} = (\rho(k) + 1 - \rho(k))^{\nu_o} = 1,$$

(6.14) simplifies to

rule (6.8):

$$\boxed{\rho(k+1) = \rho(k) + \gamma \, (1 - \rho(k)) \, (1 - (1 - \rho(k))^{\nu_o}) .} \qquad (6.15)$$

Mean-field equations for rules (6.9) and (6.10) are obtained in a similar way, i.e.,

rule (6.9):

$$\boxed{\rho(k+1) = \rho(k) + \gamma \rho(k) \, (1 - \rho(k))} \qquad (6.16)$$

and

rule (6.10):

$$\boxed{\rho(k+1) = \rho(k) + (1 - \rho(k)) \, (1 - (1 - \gamma \rho(k))^{\nu_o}) .} \qquad (6.17)$$

For a comparison of these growth laws see fig. 6.4. The basic assumption underlying the derivation of mean-field equations is to disregard any correlations. Since random independent particle movement destroys correlations, we expect that the mean-field equations are good approximations for processes which combine interaction with motion. Therefore, we impose an asynchronous random walk step (cp. sec. 5.3.2) after each growth step. As it is summarized in fig. 6.5, the qualitative evolution of these automata is correctly predicted by mean-field theory. As the degree of mixing (τ) increases, the prediction becomes also quantitatively correct (Boccara, Nasser, and Roger 1994). Note that particle motion leads also to an increase of the relative growth rate, because due to the spread of particles, limitations imposed by local "crowding effects" are reduced. Furthermore, these examples indicate that the evolution of CA models lacking any correlation-destroying mechanism can be neither qualitatively nor quantitatively captured by mean-field approximation.

6.3 Growth Processes in Lattice-Gas Cellular Automata

Lattice-gas cellular automata (LGCA) are especially designed to model processes of interaction (I) *and* movement (propagation operator P) (cp. p. 77). In the case of

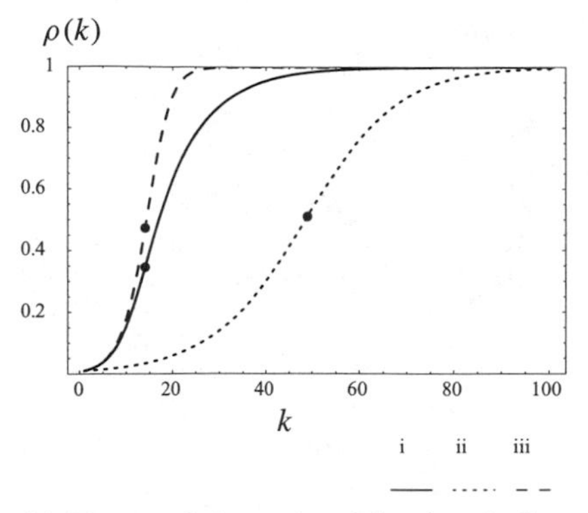

(a) Temporal dynamics; The dots indicate the inflection points of the curves.

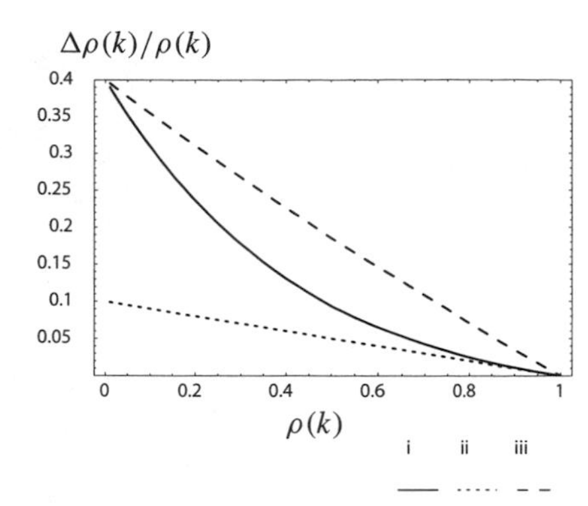

(b) Relative growth rates

Figure 6.4. Comparison of mean-field equations: (i) (6.15), (ii) (6.16), (iii) (6.17). Parameters: von Neumann interaction neighborhood, i.e., $\nu = 5$, $\nu_o = 4$, $\rho(0) = 0.01$, and $\gamma = 0.1$.

growth the interaction step is split into two parts: the "growth" mechanism (operator $\mathtt{G}$) and the shuffling operator ($\mathtt{M}$), which was introduced to model a random walk of many particles (cp. sec. 5.4). Therefore, the dynamics is given by repeated application of operators $\mathtt{P} \circ \mathtt{M} \circ \mathtt{G}$. We consider a two-dimensional lattice with coordination number $b = 4$ and introduce one additional rest particle $\beta = 1$, i.e., $\tilde{b} = 5$. Hence, the state at node r at time k is given by $\boldsymbol{\eta}(r, k) \in \{0, 1\}^5 = \mathcal{E}$. For simplicity, we restrict the interaction neighborhood for the growth and shuffling part to the node itself, i.e., $\mathcal{N}_4^I(r) = \{r\}$. Through the propagation step, particles are distributed and can, therefore, also interact with particles at neighboring nodes. Furthermore, we consider totalistic growth rules, which means that they are independent of the particle distribution in channels and solely depend on the total number of particles $\mathrm{n}(r, k)$ at the node. Hence, the preinteraction state $\eta_i(r, k)$ is replaced by the post-growth-interaction state $\eta_i^{\mathtt{G}}(r, k)$ determined by

$$\eta_i^{\mathtt{G}}(r, k) = \mathcal{R}_i^{\mathtt{G}}(\boldsymbol{\eta}(r, k)) = \eta_i(r, k) + (1 - \eta_i(r, k))\, \zeta_i(\boldsymbol{\eta}(r, k)). \tag{6.18}$$

$\zeta_i(\boldsymbol{\eta}(r, k)) \in \{0, 1\}$, $i = 1, \ldots, 5$, are space- and time-independent random variables with probability distribution

$$\begin{aligned} P(\zeta_i(\boldsymbol{\eta}(r, k)) = 1) &= W_i\left(\boldsymbol{\eta}(r, k) \to (z_1, \ldots, z_5)|_{z_i=1}\right), \\ P(\zeta_i(\boldsymbol{\eta}(r, k)) = 0) &= 1 - W_i\left(\boldsymbol{\eta}(r, k) \to (z_1, \ldots, z_5)|_{z_i=1}\right), \end{aligned} \tag{6.19}$$

where $z_j \in \{0, 1\}$, $j = 1, \ldots, 5$ and $\mathbf{y}|_{y_i=a}$ denotes any vector $\mathbf{y}$ for which the ith component assumes value a. Hence, using (6.18) together with (5.14) and (4.5), the complete dynamics can be described by the microdynamical equation

$$\eta_i(r + mc_i, k + 1) = \mathcal{R}_i^{\mathtt{M}}\left(\boldsymbol{\eta}^{\mathtt{G}}(r, k)\right). \tag{6.20}$$

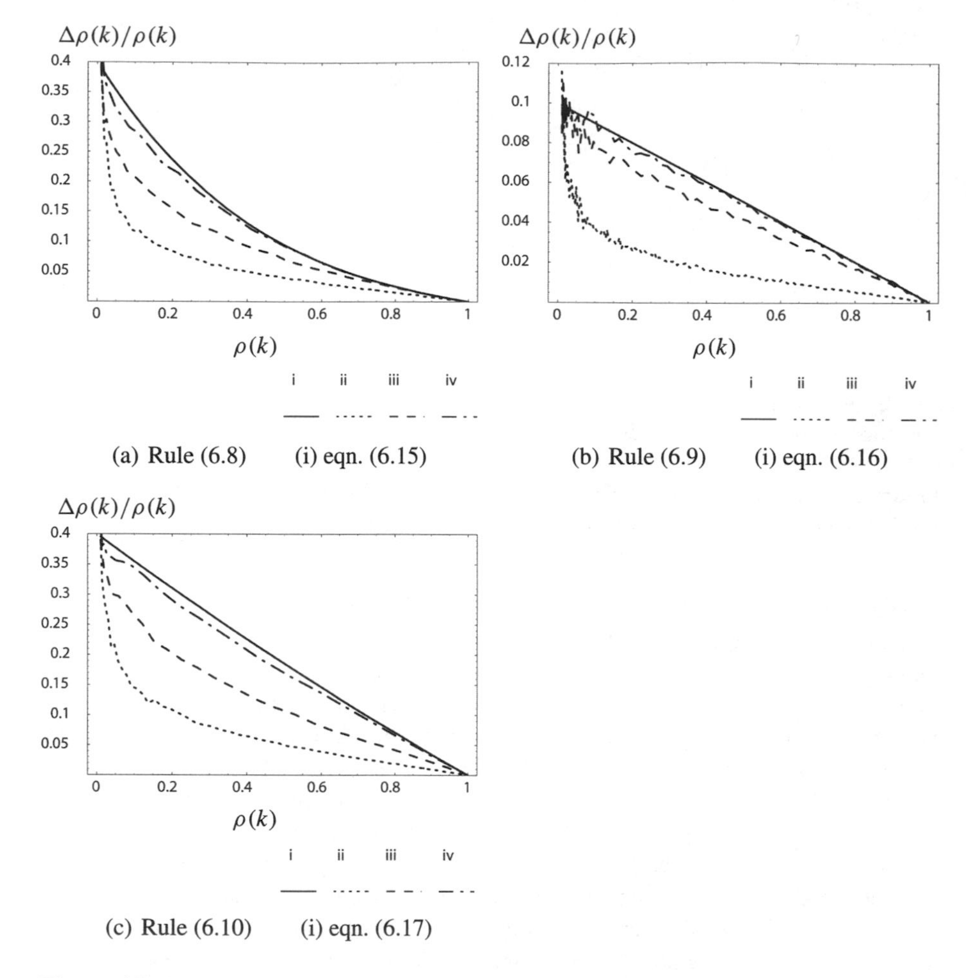

Figure 6.5. Comparison of relative growth rates: (i) mean-field equation, (ii) no movement, (iii) asynchronous movement ($\tau = 1$), (iv) asynchronous movement ($\tau = 10$). Parameters: $L = 100 \times 100$, von Neumann interaction neighborhood, i.e., $\nu = 5$, $\nu_o = 4$, $\rho(0) = 0.01$, and $\gamma = 0.1$.

A growth rule similar to rule (6.8), defined for probabilistic CA, could be formulated as follows: an empty channel (r, c_i) gains a particle with probability γ if at least B particles are present at the node, i.e.,

$$W_i\left(\boldsymbol{\eta}(r,k) \to (z_1,\ldots,z_5)|_{z_i=1}\right) = \begin{cases} \gamma & \text{if} \quad \mathrm{n}(r,k) \geq B, \\ 0 & \text{else.} \end{cases} \tag{6.21}$$

Mean-Field Approximation. We now derive the mean-field equation for the growth process defined by (6.21). Since the growth-interaction operator G depends only on the number of particles at a node ($\mathrm{n}(r, k)$), it is convenient to define *indicator functions* $\Psi_a : \{0, 1\}^{\tilde{b}} \to \{0, 1\}$, for $a = 0, \ldots, \tilde{b}$, such that

$$\boxed{\Psi_a\left(\boldsymbol{\eta}(r,k)\right) = \begin{cases} 1 & \text{if} \quad \mathrm{n}(r,k) = a, \\ 0 & \text{else.} \end{cases}} \tag{6.22}$$

These can be formulated as

$$\Psi_a\left(\boldsymbol{\eta}(r,k)\right) := \sum_{l=1}^{\binom{\tilde{b}}{a}} \prod_{i \in \mathcal{M}_l^a} \eta_i(r,k) \prod_{j \in \mathcal{M}/\mathcal{M}_l^a} \left(1 - \eta_j(r,k)\right), \tag{6.23}$$

with the index set $\mathcal{M} := \{1, \ldots, \tilde{b}\}$ and with $\mathcal{M}_l^a$ denoting the ath subset of $\mathcal{M}$ with l elements. Now, the transition probability defined in (6.21) can be rewritten as

$$\begin{aligned} &W_i\left(\boldsymbol{\eta}(r,k) \to (z_1, \ldots, z_5)|_{z_i=1}\right) \\ &\quad = \gamma P\Big(\Psi_B(\boldsymbol{\eta}(r,k)) + \cdots + \Psi_4(\boldsymbol{\eta}(r,k)) = 1\Big). \end{aligned} \tag{6.24}$$

Taking expectation values of (6.20) we get

$$\begin{aligned} f_i(r + mc_i, k+1) &= E\left(\mathcal{R}_i^{\mathrm{M}}\left(\boldsymbol{\eta}^{\mathrm{G}}(r,k)\right)\right) \\ &= \frac{1}{5}\sum_{l=1}^{5} E\left(\eta_l^{\mathrm{G}}(r,k)\right) \qquad \text{(isotropic random walk)} \\ &= \frac{1}{5}\sum_{l=1}^{5}\Big[f_l(r,k) + E\Big((1 - \eta_l(r,k))\zeta(\boldsymbol{\eta}(r,k))\Big)\Big] \\ &= \frac{1}{5}\sum_{l=1}^{5}\Big[f_l(r,k) + \gamma E\Big((1 - \eta_l(r,k)) \cdot \\ &\qquad\qquad (\Psi_B(\boldsymbol{\eta}(r,k)) + \cdots + \Psi_4(\boldsymbol{\eta}(r,k)))\Big)\Big], \end{aligned} \tag{6.25}$$

where we used (5.18) together with (6.18) and (6.24). Furthermore, as we show in appendix A, the following simplified LGCA growth rule leads to the same mean-field equation (6.25): *all* empty channels (r, c_i) simultaneously gain a particle with probability γ, if at least B particles are present at the node, i.e.,

$$\boxed{\mathrm{n}^{\mathrm{G}}(r,k) = \begin{cases} \tilde{b} & \text{with probability } \gamma, \text{ if } \mathrm{n}(r,k) \in [B, \tilde{b} - 1], \\ \mathrm{n}(r,k) & \text{else.} \end{cases}} \tag{6.26}$$

Here, we are interested in the *temporal* growth dynamics of particle density and for that reason we neglect the spatial dependence in (6.25), i.e., $f_i(k) \equiv f_i(r,k)$

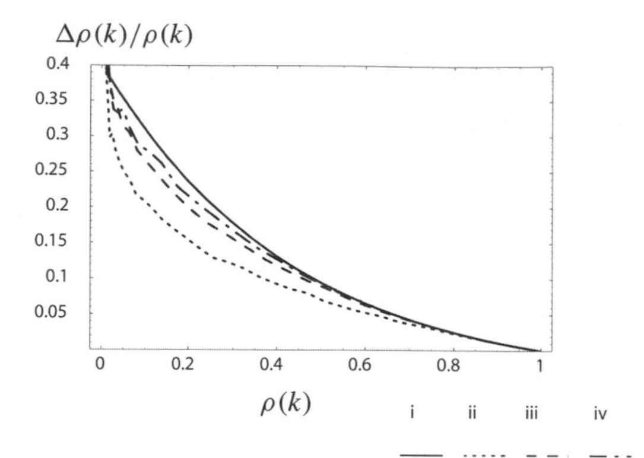

Figure 6.6. Comparison of relative growth rates: (i) mean-field equation (6.27), (ii) CA rule (6.8) with asynchronous movement ($\tau = 1$), (iii) LGCA rule (6.21) ($m = 1$), (iv) LGCA rule (6.26) ($m = 1$). Parameters: $L = 100 \times 100$, von Neumann interaction neighborhood, $\rho(0) = 0.01$, $B = 1$, and $\gamma = 0.1$.

for all $r \in \mathcal{L}$, and consider the spatially averaged local particle density $\rho(k) \in [0, 1]$ given by $\rho(k) := (1/\tilde{b}) \sum_{i=1}^{\tilde{b}} f_i(k)$. Thus, applying the mean-field assumption (4.19) to (6.25) and setting $B = 1$, we obtain rule (6.21)

$$\boxed{\rho(k+1) = \rho(k) + \gamma\,(1 - \rho(k))\left(1 - (1 - \rho(k))^4\right).} \tag{6.27}$$

Of course, this mean-field equation is identical to the one obtained from the CA growth rule (6.8) (eqn. (6.15)) due to the specific choice of parameters: for the LGCA we neglect the spatial dependence, assume $\tilde{b} = 5$, and choose the interaction neighborhood to be $\mathcal{N}_4^I(r) = \{r\}$, while for the probabilistic CA the size of the interaction neighborhood is $\nu = 5$. Accordingly, in both cases the creation of a new particle depends on the information encoded in a local configuration of "size 5." Note that the design of growth rules for LGCA, whose mean-field equations coincide with the mean-field equations of the probabilistic CA rules (6.9) and (6.10), is straightforward. As illustrated in fig. 6.6 mean-field approximation leads to a good prediction of the LGCA growth process combined with a synchronous random walk rule.

As a last example of modeling growth processes in LGCA, we change the point of view from an empty channel to an occupied channel: Each *particle* at a node r has an offspring with probability γ, but only a maximum number $\tilde{b} - \mathrm{n}(r, k)$ of these new particles are placed on the node. Then, the transition probability is given by

$$W_i\left(\boldsymbol{\eta}(r,k) \to (z_1, \ldots, z_5)|_{z_i=1}\right) \tag{6.28}$$
$$= \begin{cases} 0 & \text{if} \quad \mathrm{n}(r,k) \in \{0, 5\}, \\ \frac{1}{4}\gamma & \text{if} \quad \mathrm{n}(r,k) = 1, \\ \frac{1}{3}\gamma^2 + \frac{2}{3}\gamma(1-\gamma) & \text{if} \quad \mathrm{n}(r,k) = 2, \\ 1 - \left((1-\gamma)^3 + \frac{3}{2}\gamma(1-\gamma)^2\right) & \text{if} \quad \mathrm{n}(r,k) = 3, \\ 1 - (1-\gamma)^4 & \text{if} \quad \mathrm{n}(r,k) = 4, \end{cases}$$

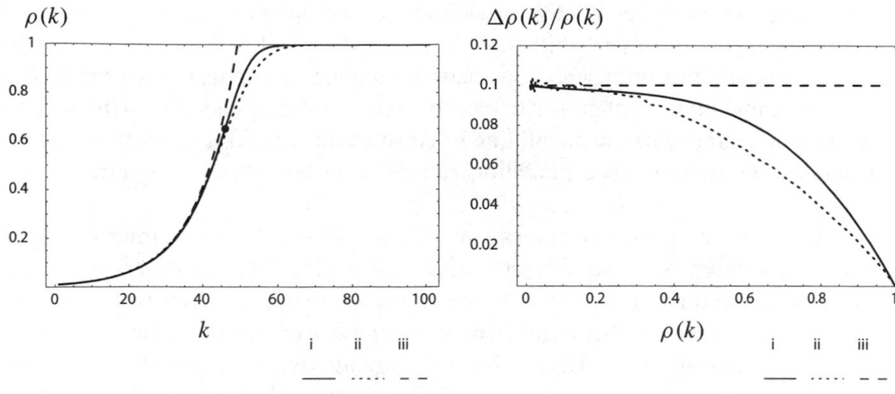

(a) Temporal dynamics; almost exponential growth for low densities. The dot indicates the inflection point of the mean-field curve.

(b) Relative growth rates

Figure 6.7. Lattice-gas growth model according to (6.28). Legend: (i) mean-field equation (6.29), (ii) simulated growth ($m = 1$), (iii) theoretical exponential growth. Parameters: $L = 100 \times 100$, $\rho(0) = 0.01$, and $\gamma = 0.1$.

$$= \frac{1}{4}\gamma P\left(\Psi_1(\boldsymbol{\eta}(r,k) = 1)\right) + \left(\frac{1}{3}\gamma^2 + \frac{2}{3}\gamma(1-\gamma)\right) P\left(\Psi_2(\boldsymbol{\eta}(r,k) = 1)\right)$$
$$+ \left(1 - \left((1-\gamma)^3 + \frac{3}{2}\gamma(1-\gamma)^2\right)\right) P\left(\Psi_3(\boldsymbol{\eta}(r,k) = 1)\right)$$
$$+ \left(1 - (1-\gamma)^4\right) P\left(\Psi_4(\boldsymbol{\eta}(r,k) = 1)\right).$$

From this, the mean-field equation for the spatially averaged particle density can be derived as

$$\rho(k+1) = (1+\gamma)\rho(k) + 2\gamma^2\rho(k)^2 + (6\gamma^2 - 2\gamma^3)\rho(k)^3 \qquad (6.29)$$
$$+ (-12\gamma^2 + 8\gamma^3 - \gamma^4)\rho(k)^4 + (-\gamma + 8\gamma^2 - 6\gamma^3 + \gamma^4)\rho(k)^5.$$

This growth process describes an almost exponential growth for low values of $\rho(k)$, as can be seen in fig. 6.7(a).

This behavior is qualitatively different from the growth models we introduced before: the relative growth rate (fig. 6.7(b)) is a convex curve, while all other relative growth rates are concave curves or decreasing lines (cp. figs. 6.5 and 6.6). In other words, in model (6.28) there is some threshold density below which the particles almost do not interfere with each other, while in the other models it is implicitly assumed that the relative growth rate is density-dependent even at the lowest densities.

Summary. In this chapter, we focused on spatial particle growth processes based on localized particle reproduction (see p. 131). With the help of various CA growth rules we demonstrated that the relative growth rate is sensitively dependent on the choice of the automaton rule. The relative growth rate cannot be adequately approximated by mean-field theory in probabilistic CA. But if the growth process is combined with an asynchronous or synchronous particle-moving mechanism, then the mean-field approximation is an appropriate tool for studying the extended growth process, since motion counteracts the upbuilding of correlations. In addition, particle motion accelerates the growth, since limitations imposed by local "crowding effects" are reduced.

Furthermore, we presented an example of a probabilistic CA with imposed asynchronous movement and two different LGCA, all of which can be described by the same mean-field equation (cp. fig. 6.6). This indicates that CA models including particle movement are less sensitive to the particular choice of the automaton interaction rule. We expect the mean-field approximation to qualitatively grasp the CA dynamics for extended particle interaction processes whenever particle motion is included. In the following chapters, examples of one-, two- and three-component interactions are provided to show that such expectations are indeed fulfilled. We put this in the framework of LGCA, because they are especially useful to model synchronous random cell movement (cp. sec. 5.4). Moreover, corresponding space- and time-dependent difference mean-field equations can be derived, which makes the automaton dynamics accessible not only to temporal but also to spatial pattern formation analysis.

6.4 Further Research Projects

1. **Analysis**
   (a) Define and analyze LGCA models for growth rules 6.7, 6.9 and 6.10. How can the different behaviors be quantitatively characterized?
   (b) Construct and analyze a LGCA that models particle birth **and** death processes.
2. **Growth and infection:** The propagation of an infectious disease has similarities to the growth dynamics discussed in this chapter. Modified growth rules can be chosen for the spread of an infectious disease. Construct a CA for the spread of an infection and consider susceptibles, infected and immune individuals.
3. **Differentiation:** Develop CA rules for age-dependent birth rates and density-dependent differentiation into a second cell type.
4. **Self-reproduction:** What is the difference of the growth models introduced in this chapter and models of self-reproduction (cp. ch. 4)?
5. **Modeling:** During tumor growth, it is often observed that tumor cells migrate into the surrounding tissue. Subsequently, a traveling wave of invasive tumor cells arises. Develop and analyze a CA model with a moving, non-proliferative and a non-moving, proliferative cell type. What kind of patterns emerge? In particular, do traveling waves occur and how could the wave speed be determined?

7

Adhesive Cell Interaction

> *"The questions which we shall pursue, then, are these: 'What may be the forces which unite cells into tissues,' and 'By what mechanism may cells exert preferences in their associations with other cells?'"*
>
> (Steinberg, 1958)

The *diffusive lattice-gas cellular automata (LGCA)* that we encountered in ch. 5 may serve as a model for random biological motion "without interaction." It is important to note that diffusive automata are not able to generate any patterns, as we demonstrated by Fourier analysis of the dominant modes that never become unstable. However, coupling of diffusion with an appropriate "reaction" may generate patterns—in sec. 5.5 we presented an example of (diffusion-limited) growth patterns generated as the result of an interplay of diffusion and "sticking." Historically, the discovery of diffusion-driven instabilities as a pattern-forming mechanism is due to Turing 1952 (cp. ch. 11). To what extent Turing instabilities account for biological pattern formation is not clear; they might influence the development of intracellular *prepatterns* (Hasslacher, Kapral, and Lawniczak 1993). An example of a "Turing LGCA" is introduced in ch. 11.

In this and the following chapters we address the implications of *direct cell–cell interactions*. Possibly, the role Turing played in the initiation of reaction-diffusion theories of biological morphogenesis (Turing 1952), Malcolm Steinberg played in the development of cell–cell interaction models (Steinberg 1963, 1970). From a historical perspective it is interesting that investigation of principles in morphogenesis started with *physical fields* (Thompson 1917), later realized the importance of *chemical dynamics* (Turing 1952), and only in the 1960s considered the importance of *biological*, namely cell–cell interactions (Steinberg 1963). Our main concern in the following chapters is the exploitation of the self-organizing potential of direct cell–cell interactions without assumption of external chemical or physical fields. In this chapter we analyze differential adhesion models without growth or loss of cells.

7.1 Cellular Patterns Originating from Adhesive Interaction

A variety of cell adhesion mechanisms underlie the way that cells are organized in tissues. Stable cell interactions are needed to maintain the structural integrity of tissues, and dynamic changes in cell adhesion are required in the morphogenesis of developing tissues. Stable interactions actually depend on active adhesion mechanisms that are very similar to those involved in tissue dynamics. Adhesion mechanisms are highly regulated during tissue morphogenesis and are intimately coupled to cell migration processes. In particular, molecules of the cadherin and integrin families are involved in the control of cell movement. Cadherin-mediated cell compaction and cellular rearrangements may be analogous to integrin-mediated cell spreading and motility on the extracellular matrix. Recently, the first in vivo example of cell sorting depending on differential adhesion mediated by cadherin has been demonstrated in *Drosophila* (Godt and Tepass 1998; Pfeifer 1998). Regulation of cell adhesion can occur at several levels, including affinity modulation, clustering, and coordinated interactions with the actin cytoskeleton. Structural studies have begun to provide a picture of how the binding properties of adhesion receptors themselves might be regulated. However, regulation of tissue morphogenesis requires complex interactions among the adhesion receptors, the cytoskeleton, and networks of signaling pathways. Signals generated locally by the adhesion receptors themselves are involved in the regulation of cell adhesion. These regulatory pathways are also influenced by extrinsic signals arising from the classic growth factor receptors. Furthermore, signals generated locally by adhesive junctions can interact with classic signal transduction pathways to help control cell growth and differentiation. This coupling between physical adhesion and developmental signaling provides a mechanism to tightly integrate physical aspects of tissue morphogenesis with cell growth and differentiation, a coordination that is essential to achieve the intricate patterns of cells in tissues.

An important role of adhesion in morphogenesis was already suggested by Holtfreter in the 1930s who found that fragments of young amphibian embryos showed marked preferences in their adhesive properties (Holtfreter 1939; fig. 7.1). "These preferences were correlated with their normal morphogenetic functions. For example, ectoderm and endoderm, isolated from a gastrula, would adhere to each other much as they do at the same stage in vivo" (quoted from Steinberg 1963).

Holtfreter also demonstrated that *cell affinity* is responsible for *tissue affinity*:

> He [Holtfreter] found that by subjecting a fragment of an amphibian gastrula to an environmental pH of about 10, he could cause the individual cells to separate and fall away from one another, ... Upon return to a more neutral pH, the amphibian cells would re-establish mutual adhesions, attaching themselves to any neighbors with which they came into contact, and building, in this manner, masses of tissue into which cells of the various germ layers were incorporated at random. ... Differences in the degree of pigmentation of the amphibian cells, together with their extraordinarily large size, allowed the investigator to follow the movements at least of the surface cells. Before his eyes the lightly pigmented mesoderm cells vanished into

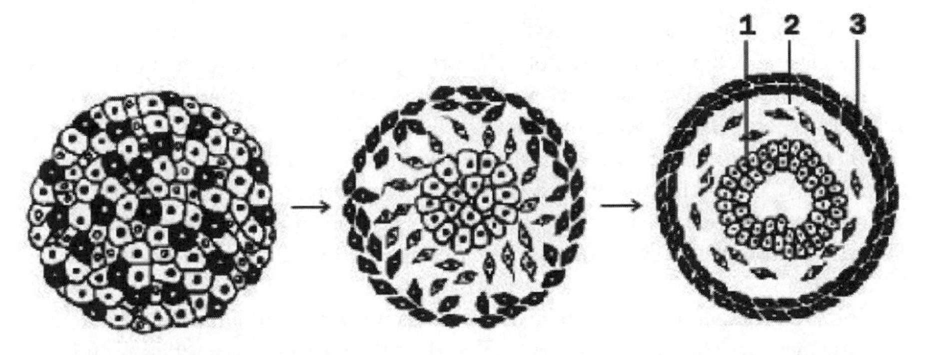

Figure 7.1. Cells from different parts of an early amphibian embryo will sort out according to their origins. In the classical experiment shown here, mesoderm cells, neural plate cells, and epidermal cells have been disaggregated and then reaggregated in a random mixture. They sort out in an arrangement reminiscent of a normal embryo, with a neural tube internally (1), epidermis externally (3), and mesenchyme in between (2). (Modified after Townes and Holtfreter 1955)

> the depths of the tissue mass, while darkly pigmented ectoderm and the almost pigment-free endoderm cells emerged to replace them at the periphery. Sorting out was a reality. (quoted from Steinberg 1963).

In the early 1960s the importance of cellular rearrangement due to differential adhesion seemed to be generally accepted, many experimental examples had been discovered. Moscona (Moscona 1961, 1962) and Steinberg were among the pioneers to recognize the importance of cellular reorganization in animal embryogenesis. The fundamental facts about *tissue reconstruction* were stated by Steinberg (Steinberg 1963):

- *Formation of aggregates.* After dissociation or mixing of different vertebrate embryonic cells, cells are capable of reestablishing "adhesions" with one another and constructing common aggregates.
- *Mixed aggregates.* Within cell mixtures containing cells from different tissues, the different types of cells regroup, to reconstruct the various original tissues.
- *Fixed relative positions.* Tissues are reconstructed in definite positions. For example, muscle is always built external to cartilage.
- *Geometry.* When the tissues employed are parts of a complex within the embryo, the geometry of the entire normal complex is reflected in the reestablished structures.

Steinberg already pointed out the basic morphogenetic mechanisms underlying the reconstruction scenario: mutual adhesivity of cells and cell motility. He suggested that the interaction between two cells involves an "adhesion surface energy," which varies according to the cell types. Steinberg interpreted cell sorting by the *differential adhesion hypothesis* (DAH), which states that cells can randomly explore various configurations and thereby reach the lowest energy configuration. Experimentally, the DAH has proven rather successful. There is experimental evidence that differential adhesion is the main source of cell sorting in *Hydra* cell aggregates (Technau and Holstein 1992). Type-selective surface adhesivity has turned out to be a nearly omnipresent property of cells; in morphogenesis, it is involved in cell recognition, gastrulation, cell shaping, and control of pattern formation (Armstrong 1985; Holtfreter 1943). Type-selective adhesivity may also play a role in the migration and invasiveness of cancer metastases (Takeichi 1991) and in immunological defenses (Schubert, Masters, and Beyreuther 1993; Schubert 1998) as well as wound healing (Drasdo 1993).

However, the particular mechanisms by which differential adhesion guides cell rearrangement are still unclear. In particular, the role of cell motion remains ambiguous. Glazier and Graner 1993 distinguished the following possibilities:

(A) Active motion: cells have an autonomous motility that allows for long-range migration.
(B) Stochastic cellular shape changes: random fluctuations of the cell surface permit a cell to locally explore its neighborhood.

It is in general not easy to distinguish these strategies experimentally. Glazier and Graner have shown that differential adhesion can cause various cellular rearrangement patterns *without* active cell motility (hypothesis B) with the help of an extended Potts model (Glazier and Graner 1993). The model has been extended to include chemotactic dynamics and been applied to various phases in the life cycle of the slime mold *Dictyostelium discoideum* (Marée and Hogeweg 2001; Savill and Hogeweg 1997).

Stochastic models of adhesively interacting cells producing cell sorting have been introduced before (Mochizuki et al. 1996, 1998). However, the implications of active motion (hypothesis A) have not been systematically investigated so far but will be analyzed here. We consider ς cell types characterized by their ability for active motion and adhesive interaction in a LGCA framework. Adhesion is modeled as a nearest-neighbor (attractive or repulsive) force without specifying the precise physico-chemical basis. In some detail, we describe the aggregation properties of the one-species model before demonstrating sorting behavior of the two-species model.

7.2 Adhesive Lattice-Gas Cellular Automaton

The ς-species LGCA (cp. the definition in sec. 4.3.4) is defined on a two-dimensional lattice $\mathcal{L}$ with coordination number $b = 4$ (square) or $b = 6$ (hexagonal) and no rest channels (i.e., $b = \tilde{b}$). We assume that $|\mathcal{L}| = L \cdot L = L^2$. Contrary to previous conventions we will commonly refer to "cells" instead of "particles" in this chapter. In the following, the **adhesive interaction rule**

$$\mathcal{R}^{\mathrm{I}} : \mathcal{E}^{\nu} \to \mathcal{E}, \qquad \text{where } \mathcal{E} = \{0, 1\}^{b\varsigma} \text{ and } \nu = |\mathcal{N}_b^I|$$

is defined, which transforms $\boldsymbol{\eta}_{\mathcal{N}(r)} \to \boldsymbol{\eta}^{\mathrm{I}}(r)$, where $\boldsymbol{\eta}(r), \boldsymbol{\eta}^{\mathrm{I}}(r)$ are the pre- and postinteraction state (cp. multicomponent state, p. 73), respectively, depending on the configuration in the interaction neighborhood $\mathcal{N}_b^I(r)$. The interaction neighborhood template $\mathcal{N}_b^I$ contains nearest-neighbor connections $\mathcal{N}_b$ (cp. p. 68) and $(0, 0)$, i.e., $\nu = b + 1$, $\nu_o = b$. Thus, for the square lattice $\mathcal{N}_4^I$ is the von Neumann neighborhood (cp. the definition on p. 70). The complete dynamics of the two-dimensional adhesive LGCA is defined by repeated application of operators $\mathtt{P} \circ \mathtt{I}$, where $\mathtt{P}$ is the propagation operator with speed $m = 1$ (cp. p. 77), i.e.,

$$\eta_{\sigma,i}(r + c_i, k + 1) = \eta^{\mathtt{I}}_{\sigma,i}(r, k) = \mathcal{R}^{\mathtt{I}}\left(\boldsymbol{s}_{\mathcal{N}(r)}(k)\right)$$

for each $\sigma = 1, \ldots, \varsigma$, $i = 1, \ldots, b$ and $c_i \in \mathcal{N}_b$.

Adhesive Interaction. Local interaction comprises cell reorientation according to adhesive interaction among cells of all types. In order to model the local adhesive interaction, we define

$$\boxed{\boldsymbol{G}_\sigma\left(\boldsymbol{\eta}_{\mathcal{N}(r)}\right) := \sum_{i=1}^{b} c_i \mathrm{n}_\sigma(r + c_i),} \tag{7.1}$$

which is the *gradient field* in the local density of cell type σ, $\sigma = 1, \ldots, \varsigma$. Thereby,

$$\mathrm{n}_\sigma(r) = \sum_{i=1}^{b} \eta_{\sigma,i}(r), \qquad \boldsymbol{\eta}(r) \in \mathcal{E},$$

denotes the number of cells of type σ at node r.

During interaction the number of cells of each type at each node remains constant (i.e., there is no creation, annihilation, or transformation between different cell types):

$$\mathrm{n}_\sigma(r) = \mathrm{n}^{\mathtt{I}}_\sigma(r) \qquad \text{for all } r \in \mathcal{L},$$

where n_σ and $\mathrm{n}^{\mathtt{I}}_\sigma$ are pre- and postinteraction cell numbers, respectively. Accordingly, the spatially averaged density of cells of type σ per node,

$$\bar{\rho}_\sigma = \left(\frac{1}{b|\mathcal{L}|}\right) \sum_{r \in \mathcal{L}} \mathrm{n}_\sigma(r) \in [0, 1],$$

is constant in time. Let

$$\boxed{\boldsymbol{J}\left(\boldsymbol{\eta}_\sigma(r)\right) := \sum_{i=1}^{b} c_i \eta_{\sigma,i}(r)}$$

denote the *cell flux* of type σ at node r.

The transition probability from $\boldsymbol{\eta}(r)$ to $\boldsymbol{\eta}^{\mathtt{I}}(r)$ in the presence of $\boldsymbol{G}\left(\boldsymbol{\eta}_{\mathcal{N}(r)}\right) = \{\boldsymbol{G}_\sigma(\boldsymbol{\eta}_{\mathcal{N}(r)})\}_{1 \le \sigma \le \varsigma}$ is given by

$$
\begin{aligned}
&W\left(\eta_{\mathcal{N}(r)} \to \eta^{\mathrm{I}} | \left(\alpha_{\sigma_1\sigma_2}\right)\right) \\
&= \frac{1}{Z} \exp\left(\sum_{\sigma_1=1}^{\varsigma} \sum_{\sigma_2=1}^{\varsigma} \alpha_{\sigma_1\sigma_2} \, \boldsymbol{G}_{\sigma_1}\left(\eta_{\mathcal{N}(r)}\right) \cdot \boldsymbol{J}\left(\eta^{\mathrm{I}}_{\sigma_2}\right)\right) \prod_{\sigma=1}^{\varsigma} \delta_{\mathrm{n}_\sigma, \mathrm{n}^{\mathrm{I}}_\sigma}, \qquad (7.2)
\end{aligned}
$$

where the normalization factor $Z = Z((\mathrm{n}_\sigma)_{1 \le \sigma \le \varsigma}, \boldsymbol{G})$ is chosen such that

$$
\sum_{\eta^{\mathrm{I}} \in \mathcal{E}} W\left(\eta_{\mathcal{N}(r)} \to \eta^{\mathrm{I}} | \left(\alpha_{\sigma_1\sigma_2}\right)\right) = 1 \qquad \forall \, \eta \in \mathcal{E}.
$$

The adhesion matrix $\left(\alpha_{\sigma_1\sigma_2}\right)_{1 \le \sigma_1, \sigma_2 \le \varsigma}$ ($\alpha_{\sigma_1\sigma_2} \in \mathbb{R}$) stores the adhesion coefficients of the corresponding cell types. The product over delta-functions assures conservation of cell numbers with respect to each individual cell type.

The key idea of the definition is that with large probability an output η^{I} results for which the argument of the exponential is maximized. The argument of the exponential is defined in such a way that cells of type σ_2 preferably move in the direction of increasing density of type σ_1 if $\alpha_{\sigma_1\sigma_2} > 0$, by that modeling adhesion (attraction) between cells. In contrast, if $\alpha_{\sigma_1\sigma_2} < 0$, then cells of type σ_2 predominantly migrate in the direction of decreasing density of type σ_1 mimicking repulsion. If $\alpha_{\sigma_1\sigma_2} = 0$, cells of types σ_1 do not influence cells of type σ_2. Note that not necessarily $\alpha_{\sigma_1\sigma_2} = \alpha_{\sigma_2\sigma_1}$. For example, even if $\alpha_{\sigma_1\sigma_2} = 0$ it may happen that $\alpha_{\sigma_2\sigma_1} > 0$; i.e., cell type σ_1 follows σ_2, while σ_2 is not influenced by σ_1.

This definition describes a rather general scenario of adhesive interaction. We will investigate two important special cases, namely $\varsigma = 1, 2$.

7.3 Analysis of Aggregation Dynamics in the Single-Cell-Type Adhesion Model

In the case of a single cell type ($\varsigma = 1$), the adhesive LGCA reduces to the "Alexander model" (Alexander et al. 1992), originally proposed as a model of antidiffusion and analyzed in Bussemaker (1996). Since it forms an important brick in the wall of our argument—it will turn out that it can be interpreted as a pattern-forming extension of the diffusive automaton—we briefly recapitulate the stability analysis here. The transition probability (7.2) simplifies to

$$
\boxed{W\left(\eta_{\mathcal{N}(r)} \to \eta^{\mathrm{I}} | \alpha\right) = \frac{1}{Z} \exp\left(\alpha \boldsymbol{G}\left(\eta_{\mathcal{N}(r)}\right) \cdot \boldsymbol{J}\left(\eta^{\mathrm{I}}\right)\right) \cdot \delta_{\eta, \eta^{\mathrm{I}}},} \qquad (7.3)
$$

where the normalization factor $Z = Z(\mathrm{n}, \boldsymbol{G})$ is chosen such that

$$
\sum_{\eta^{\mathrm{I}} \in \mathcal{E}} W\left(\eta_{\mathcal{N}(r)} \to \eta^{\mathrm{I}} | \alpha\right) = 1, \qquad \forall \, \eta \in \mathcal{E}.
$$

Note that superscripts have been neglected and $\alpha \in \mathbb{R}$ is the adhesiveness of the cells. Fig. 7.2 illustrates an example of the transition probability.

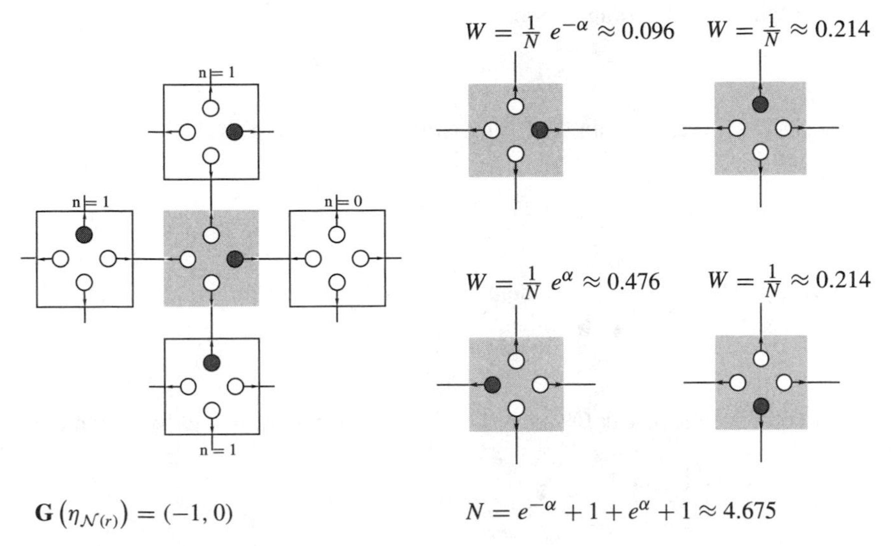

Figure 7.2. Example of adhesive interaction in the square lattice. Filled dots denote channels occupied by cells, while empty dots denote empty channels. With probability W each node configuration on the right side ($\boldsymbol{\eta}^{\mathrm{I}}$) is a possible result of the interaction step applied to the middle gray node ($\boldsymbol{\eta}$) of the initial configuration.

7.3.1 Linear Stability Analysis

To analyze the behavior of the adhesive LGCA model we consider the time evolution of a statistical ensemble of systems, i.e., the lattice-Boltzmann equation (cp. (4.23))

$$\begin{aligned}
& f_i(r + c_i, k + 1) - f_i(r, k) \\
&= \sum_{\boldsymbol{\eta}^{\mathrm{I}}(r) \in \mathcal{E}} \; \sum_{\boldsymbol{\eta}_{\mathcal{N}(r)} \in \mathcal{S}_{\mathcal{N}(r)}} \Bigg\{ \left(\eta_i^{\mathrm{I}}(r) - \eta_i(r)\right) W\left(\boldsymbol{\eta}_{\mathcal{N}(r)} \to \boldsymbol{\eta}^{\mathrm{I}}(r) | \alpha\right) \\
&\qquad \cdot \prod_{r_i \in \mathcal{N}_b^I(r)} \prod_{l=1}^{b} f_l(r_i, k)^{\eta_l(r_i)} \left(1 - f_l(r_i, k)\right)^{(1 - \eta_l(r_i))} \Bigg\} \\
&= \tilde{\mathcal{C}}_i\left(\boldsymbol{f}_{\mathcal{N}(r)}(k)\right), \qquad i = 1, \ldots, b,
\end{aligned} \tag{7.4}$$

where $\tilde{\mathcal{C}}_i\left(\boldsymbol{f}_{\mathcal{N}(r)}(k)\right)$ equals the average change in the occupation number of channel (r, c_i) during interaction. It can be shown that a possible solution to $\tilde{\mathcal{C}}_i\left(\boldsymbol{f}_{\mathcal{N}(r)}(k)\right) = 0$ is

$$f_i(r, k) = \bar{f} \qquad \text{for } i = 1, \ldots, b.$$

Note that in this case the stationary, spatially homogeneous single-channel density $\bar{f}$ coincides with the averaged density of cells at a node $\bar{\rho}$, since $\bar{\rho} = (1/b) \sum_{i=1}^{b} \bar{f}_i =$

$\bar{f}$. To assess the stability of this spatially homogeneous, isotropic, and stationary solution with respect to fluctuations

$$\delta f_i(r,k) = f_i(r,k) - \bar{f},$$

we linearize (7.4) and obtain the Boltzmann propagator (cp. (4.36))

$$\Gamma(q) = T\left\{\boldsymbol{I} + \boldsymbol{\Omega}^0 + \sum_{n=1}^{b} \boldsymbol{\Omega}^n\, e^{\mathbf{i}(2\pi/L)q\cdot c_n}\right\}, \qquad q = (q_1, q_2), \tag{7.5}$$

with $q_1, q_2 = 0, \ldots, L-1$, the identity matrix $\boldsymbol{I}$, and the "transport matrix"

$$T = \operatorname{diag}\left(e^{-\mathbf{i}(2\pi/L)(q\cdot c_1)}, \ldots, e^{-\mathbf{i}(2\pi/L)(q\cdot c_b)}\right).$$

The structure of the matrix $\boldsymbol{\Omega}^n$, $n = 0, \ldots, b$, is as follows: It can be shown that

$$\begin{aligned}
\Omega^0_{ij} &= \left.\frac{\partial\, \tilde{C}_i\left(\delta \boldsymbol{f}_{\mathcal{N}(r)}(k)\right)}{\partial \delta f_j(r,k)}\right|_{\bar{\boldsymbol{f}}} \\
&= \sum_{\boldsymbol{\eta}^{\mathrm{I}}(r)\in\mathcal{E}} \sum_{\boldsymbol{\eta}_{\mathcal{N}(r)}\in\mathcal{S}_{\mathcal{N}(r)}} \Bigg\{ \left(\eta^{\mathrm{I}}_i(r) - \eta_i(r)\right) W\left(\boldsymbol{\eta}_{\mathcal{N}(r)} \to \boldsymbol{\eta}^{\mathrm{I}}(r)|\alpha\right) \\
&\qquad\qquad \cdot \frac{\eta_j(r) - \bar{f}}{\bar{f}(1-\bar{f})} \cdot \prod_{r_m\in\mathcal{N}^I_b(r)} \prod_{l=1}^{b} \bar{f}^{\eta_l(r_m)} \left(1-\bar{f}\right)^{(1-\eta_l(r_m))} \Bigg\} \\
&= \frac{1}{b} - \delta_{ij},
\end{aligned} \tag{7.6}$$

where $\bar{\boldsymbol{f}} = \left(\bar{f}\right) \in [0,1]^b$, and that Ω^n_{ij} does not depend on its second index j, i.e., $\omega^n_i := \Omega^n_{ij}$. Furthermore, for $1 \le n \le b$ it is sufficient to give the structure of $\boldsymbol{\Omega}^1$ since all other matrices $\boldsymbol{\Omega}^n$ are related to the first by rotational symmetry[54] (Bussemaker 1996):

$$\omega^{n+l}_{i+l} = \omega^n_i \qquad \forall\, n > 1.$$

For the square lattice,

$$\boldsymbol{\Omega}^1 = \begin{pmatrix} a-b & a-b & a-b & a-b \\ b & b & b & b \\ -a-b & -a-b & -a-b & -a-b \\ b & b & b & b \end{pmatrix},$$

and for the hexagonal lattice,

[54] Indices and superscripts are taken modulo b.

$$\boldsymbol{\Omega}^1 = \begin{pmatrix} c-2e & c-2e & c-2e & c-2e & c-2e & c-2e \\ d+e & d+e & d+e & d+e & d+e & d+e \\ -d+e & -d+e & -d+e & -d+e & -d+e & -d+e \\ -c-2e & -c-2e & -c-2e & -c-2e & -c-2e & -c-2e \\ -d+e & -d+e & -d+e & -d+e & -d+e & -d+e \\ d+e & d+e & d+e & d+e & d+e & d+e \end{pmatrix}.$$

The coefficients can be evaluated as

$$\omega_i^n = \omega_i^n(\bar{f}, \alpha) = \left. \frac{\partial \tilde{\mathcal{C}}_i \left(\delta \boldsymbol{f}_{\mathcal{N}(r)}(k)\right)}{\partial \delta f_1(r_n, k)} \right|_{\bar{\boldsymbol{f}}}$$

$$= \sum_{\boldsymbol{\eta}^{\mathrm{I}}(r) \in \mathcal{E}} \sum_{\boldsymbol{\eta}_{\mathcal{N}(r)} \in \mathcal{S}_{\mathcal{N}(r)}} \Bigg\{ \left(\eta_i^{\mathrm{I}}(r) - \eta_i(r)\right) W\left(\boldsymbol{\eta}_{\mathcal{N}(r)} \to \boldsymbol{\eta}^{\mathrm{I}}(r)|\alpha\right)$$

$$\cdot \frac{\eta_1(r_n) - \bar{f}}{\bar{f}(1-\bar{f})} \cdot \prod_{r_m \in \mathcal{N}_b^I(r)} \prod_{l=1}^{b} \bar{f}^{\eta_l(r_m)} \left(1 - \bar{f}\right)^{(1-\eta_l(r_m))} \Bigg\}.$$

Note that ω_i^n does not depend on r since it represents a derivative evaluated in a spatially uniform state.

Together with (7.6), (7.5) becomes

$$\Gamma_{ij}(q) = e^{-\mathbf{i}(2\pi/L)q\cdot c_i} \left(\frac{1}{b} + \sum_{n=1}^{b} e^{\mathbf{i}2\pi/L)q\cdot c_n} \omega_i^n \right); \tag{7.7}$$

i.e., $\Gamma(q)$ is a $b \times b$ matrix with the following structure:

$$\Gamma(q) = (g(q), g(q), \ldots, g(q)),$$

where

$$g(q) := \begin{pmatrix} e^{-\mathbf{i}(2\pi/L)q\cdot c_1} \left(\frac{1}{b} + \sum_{n=1}^{b} e^{\mathbf{i}2\pi/L)q\cdot c_n} \omega_1^n \right) \\ \vdots \\ e^{-\mathbf{i}(2\pi/L)q\cdot c_b} \left(\frac{1}{b} + \sum_{n=1}^{b} e^{\mathbf{i}2\pi/L)q\cdot c_n} \omega_b^n \right) \end{pmatrix}.$$

The particular simple form of the propagator $\Gamma(q)$ implies that its eigenvalues are

$$\lambda_1(q) = \sum_{i=1}^{b} e^{-\mathbf{i}(2\pi/L)q\cdot c_i} \left(\frac{1}{b} + \sum_{n=1}^{b} e^{\mathbf{i}2\pi/L)q\cdot c_n} \omega_i^n \right), \tag{7.8}$$

$$\lambda_2(q) = \cdots = \lambda_b(q) = 0.$$

Note that $\mu(q) = |\lambda_1(q)|$ is the spectral radius of the propagator. For the square lattice, the symmetries[55] of ω_i^n can be used to obtain

$$\lambda_1(q) = \frac{1}{2}\,(\cos(q_1) + \cos(q_2)) + 4\left(\omega_1^1 + \omega_2^1\right)\left(\sin^2(q_1) + \sin^2(q_2)\right)$$
$$- 4\omega_2^1\,(\cos(q_1) - \cos(q_2))^2$$

or

$$\boxed{\lambda_1(q) \approx 1 - \left(\frac{1}{4} - 4\left(\omega_1^1 + \omega_2^1\right)\right)|q|^2 + O(|q|^4)} \tag{7.9}$$

with corresponding eigenvectors

$$\begin{aligned} v_1(q) &= g(q), \\ v_2(q) &= (1, 0, -1, 0) \quad =: v_{J_x}, \\ v_3(q) &= (0, 1, 0, -1) \quad =: v_{J_y}, \\ v_4(q) &= (1, -1, 1, -1) =: v_q, \end{aligned}$$

where v_{J_x} and v_{J_y} correspond to the x- and y-components of the total cell flux, while v_q corresponds to the difference between the number of horizontally and vertically moving cells. Let

$$\tilde{v}_1 = \begin{cases} v_1 & \text{if } \lambda_1(q) \neq 0 \\ (1, 1, 1, 1) := v_m & \text{if } \lambda_1(q) = 0, \end{cases}$$
$$\tilde{v}_j = v_j \quad \text{for } 2 \leq j \leq 4,$$

then $\{\tilde{v}^1, \tilde{v}^2, \tilde{v}^3, \tilde{v}^4\}$ are linearly independent vectors and form a basis of $\mathbb{C}^4$. A similar argument shows that for the hexagonal lattice there is an eigenvector basis of $\mathbb{C}^6$. In other words, $\Gamma(q)$ is diagonalizable, and therefore according to (4.41) the general solution of the lattice-Boltzmann equation (7.4) is given by

$$\delta f_i(r, k) = \frac{1}{L^2} \sum_q e^{-\mathbf{i}(2\pi/L)q\cdot r}\, F_i(q, k),$$

with

$$F_i(q, k) = d_1(q)\tilde{v}_{1i}(q)\lambda_1(q)^k,$$

where the constant $d_1(q) \in \mathbb{C}$ is specified by the initial condition

$$F_i(q, 0) = \sum_{l=1}^{b} d_l(q)\tilde{v}_{li}(q) = \sum_r e^{\mathbf{i}(2\pi/L)q\cdot r}\,\delta f_i(r, 0), \qquad i = 1, \ldots, b.$$

Therefore, the dominant eigenvalue $\lambda_1(q)$ fully determines the temporal growth of modes $F_i(q, k)$, i.e., the adhesive dynamics.

[55] That is, $\omega_i^n = -\omega_i^1$, $\omega_3^1 = -\omega_1^1$, and $\omega_4^1 = \omega_2^1$.

7.3.2 Spatial Pattern Formation

Recall that spatially inhomogeneous structures are determined by undamped modes according to wave numbers $q \in Q^c$ (sets of critical wave numbers $Q^c \supset Q^+ \cup Q^-$ were defined on p. 96). In two-dimensional systems, groups of unstable modes with identical absolute value of the wave number $|q| = \mathfrak{q}$ simultaneously start to grow; i.e., according to (4.42)

$$\delta f_i(r,k) \sim \sum_{\substack{q \in Q^c \\ |q| = \mathfrak{q}}} e^{-\mathrm{i}(2\pi/L)q \cdot r} F_i(q,k).$$

Therefore, according to linear theory, any superposition of these modes determines the dynamics of the system.[56]

Case 1: Growing modes with $q \in Q^+$

From (7.9) it follows that the adhesive LGCA for the square lattice may become unstable, i.e., $\lambda_1(q) > 1$, if $\left(\frac{1}{4} - 4\left(\omega_1^1 + \omega_2^1\right)\right) < 0$. The coefficients of the Boltzmann propagator ω_i^n depend on the sensitivity α and the averaged node density $\bar{\rho}$. Therefore, for a fixed averaged node density $\bar{\rho}$ a *critical sensitivity* α_c is defined by the condition

$$\lambda_1(q) \approx 1 \Leftrightarrow \left(\frac{1}{4} - 4\left(\omega_1^1(\bar{f}, \alpha_c) + \omega_2^1(\bar{f}, \alpha_c)\right)\right) = 0.$$

Fig. 7.3 shows the numerically determined phase diagram for the adhesive LGCA. For example, in the case of the half-filled square lattice ($\bar{\rho} = 0.5$) the critical sensitivity is $\alpha_c \approx 0.263$ (Bussemaker 1996).

The result of our analysis is that we expect a spatial pattern for parameters $\alpha > \alpha_c$ with a wavelength corresponding to the q_*-mode, $q_* \in Q^+$, for which $\lambda_1(q_*) = \max_{q \in Q^c} \lambda_1(q)$.

Simulations start from a spatially homogeneous initial condition, parameters are chosen from the unstable region, i.e., $\alpha > \alpha_c$. Typical density wave patterns on the square and on the hexagonal lattice are displayed in fig. 7.4. Furthermore, there is a clear preference for diagonal directions in the square lattice simulation, indicating an anisotropy of the square lattice. The anisotropy in the hexagonal lattice simulations is weaker. These simulation results can be explained by the Fourier analysis, as we will show in the following. In fig. 7.5 the dominant eigenvalue (cp. (7.9)) of the propagator is plotted which corresponds to the parameters of the square lattice simulation displayed in fig. 7.4. There is a nonzero maximum at $q_{1*} \approx 18$, i.e., $q_* \approx (18, 18)$, that belongs to a wavelength $L/|q_*| = L/\sqrt{18^2 + 18^2} \approx 3.93$. This value as well as the predicted periodicity of the pattern are in rather good agreement with the pattern observed in the simulation (fig. 7.4).

[56] For further reading see Mikhailov 1994.

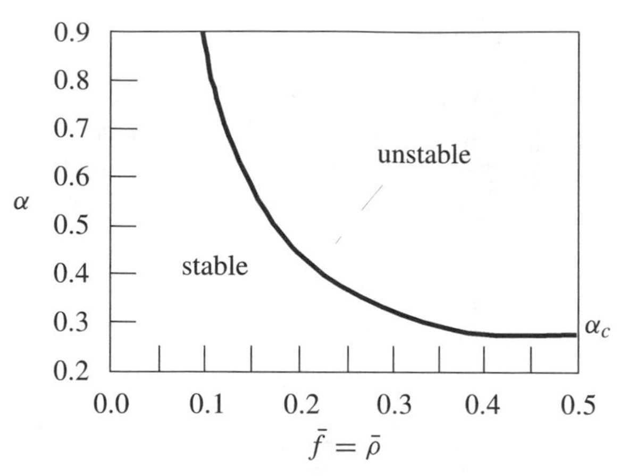

Figure 7.3. Phase diagram of the adhesive one-species LGCA model for the square lattice ($b = 4$). The regions of stable ($0 < \lambda_1(q) < 1$) and unstable ($\lambda_1(q) > 1$) behavior are shown as a function of sensitivity α and averaged single-channel density $\bar{f} = \bar{\rho}$. For any $\bar{\rho}$ the critical sensitivity α_c value can be determined.

The representation of the corresponding eigenvector $v_1(q_*)$ in the basis v_m, v_{J_x}, v_{J_y}, and v_q yields additional information about the qualitative behavior. It can be shown that the mass vector $v_m = (1, 1, 1, 1)$ is dominant in the expansion of $v_1(q_*)$ in this basis, indicating that the density is unstable, which explains the observed varying density wave pattern. Since $\text{Im}\,\lambda_i(q) = 0$ $(i = 1, \ldots, b)$ no traveling waves should occur, which agrees with the simulation.

However, linear Fourier analysis can not explain long-time behavior, i.e., the clustering in the further course of temporal development, which is caused by the growth of correlations and nonlinear influences for longer times (cp. fig. 7.4).

Anisotropy. In order to investigate the origin of the obvious anisotropy in the square LGCA simulation (cp. fig. 7.4), the spectral radius for square and hexagonal lattice in the unstable regime are compared (fig. 7.6).

On the square lattice the spectral radius is maximal for wave numbers q_* associated with the diagonal directions. Consequently, linear stability analysis predicts a spatial pattern with wave numbers q_* and strong anisotropies in diagonal directions. In contrast, the spectral radius of the hexagonal lattice model hardly depends on the direction of (q_1, q_2): all maxima are located along the same level line.

Case 2: Oscillating growing modes with $q \in Q^-$

Another type of pattern evolves if $\lambda_1(q)$ (cp. (7.8)) has a dominant instability at -1, i.e., $q_* \in Q^-$. This situation arises for example in the square LGCA model for $\alpha = -1$, $\bar{\rho} = 0.4$ and $L = 50$. Negative sensitivities $\alpha < 0$ imply a repulsive interaction since now cells try to arrange in directions opposite to the neighborhood momentum. (Compare the definition of transition probability in (7.3)) The dominant

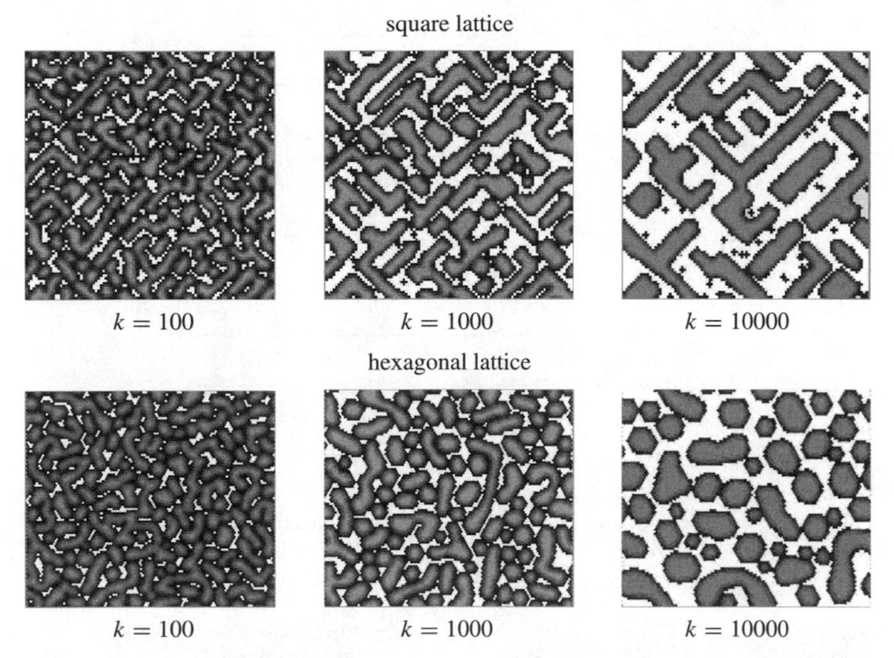

Figure 7.4. Adhesive pattern formation in the LGCA model on different lattices. Gray levels represent cell densities (dark: high cell density). On the square lattice, strong anisotropies are visible in diagonal directions which are absent on the hexagonal lattice. An explanation for the anisotropy is provided by the eigenvalue spectra which yield corresponding anisotropies for the square, but not the hexagonal lattice (cp. fig. 7.6); parameters: $\alpha = 0.8$ ($> \alpha_c$), $\bar{\rho} = 0.4$, $L = 100$.

mode of the corresponding eigenvalue $\lambda_1(q)$ (cp. fig. 7.7) is $q_* \approx 32$, which indicates small wavelengths. Since this mode grows with an oscillating sign of period 2, a *checkerboard-like structure* develops, as is shown in fig. 7.7.

In conclusion, Fourier analysis can deduce important features of the automaton patterns. In Bussemaker (1996), possible paths beyond mean-field analysis were described. As already noted, linear stability analysis does not provide any hints for long-time behavior, e.g., the "coagulation tendency" in the aggregation patterns (fig. 7.4).

7.4 Phase Separation and Engulfment in a Two-Cell-Type Adhesion Model

We discuss an extension of the LGCA with two cell types. It was Steinberg (1963) who considered the two-cell-type adhesion case for the first time in a "thought experiment." Taking into account an ensemble of adhesively interacting and moving cells, Steinberg assumed that cells would rearrange in order to minimize their free energy and examines a phase diagram as a function of "works of adhesion." This

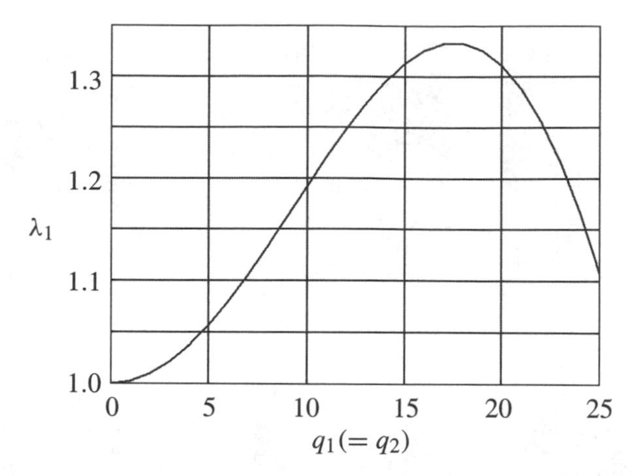

Figure 7.5. Dominant eigenvalue $\lambda_1(q)$, $q = (q_1, q_2)$, for diagonal waves ($q_1 = q_2$). The dominant unstable mode $q_* \approx (18, 18)$ corresponds to the maximum of the dominant eigenvalue. Its instability provides an indication of the initial wavelength in the simulations (cp. fig. 7.4, $k = 100$). Parameters: $\alpha = 0.8$ ($> \alpha_c$), $\bar{\rho} = 0.4$, $L = 100$, $b = 4$.

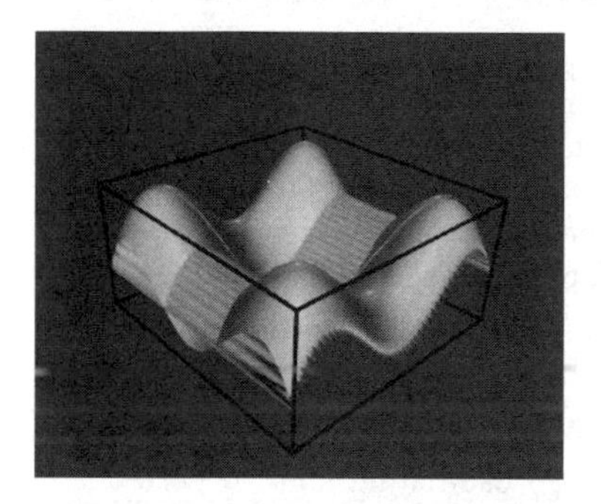

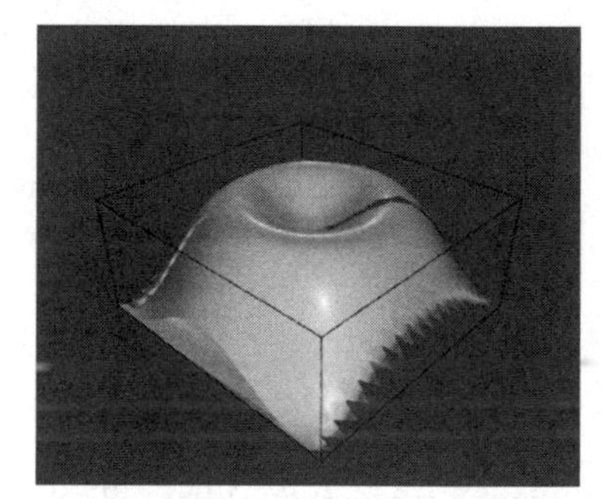

Figure 7.6. Comparison of spectra for the square (left) and hexagonal (right) lattice. Plot of the dominant eigenvalue $\lambda_1(q)$. The coordinate system for q_1- and q_2-values is centered in the middle of the cube, the vertical axis displays $\lambda_1(q)$ and allows visualization of dominant modes as elevations. Clearly, there are strong anisotropies on the square lattice which are absent on the hexagonal lattice. The maxima in the corners of the square lattice spectrum are responsible for the "diagonal anisotropies" in the square lattice simulations (cp. fig. 7.4). Parameters: $\alpha = 0.8$ ($> \alpha_c$), $\bar{\rho} = 0.4$, $L = 100$.

work may vary between equal and different cell types, denoted by w_a, w_b, and w_{ab}, respectively. According to Steinberg, just three scenarios are possible:

1. $w_{ab} \geq \frac{w_a+w_b}{2}$: this leads to *intermixing*,
2. $w_b \leq w_{ab} \leq \frac{w_a+w_b}{2}$: the corresponding behavior is *complete spreading*, or
3. $w_a \geq w_b > w_{ab}$: one should observe *self-isolation*.

Steinberg's accomplishments were to point to important instances of adhesion-based pattern formation, in modern notion checkerboard (intermixing), engulfment (com-

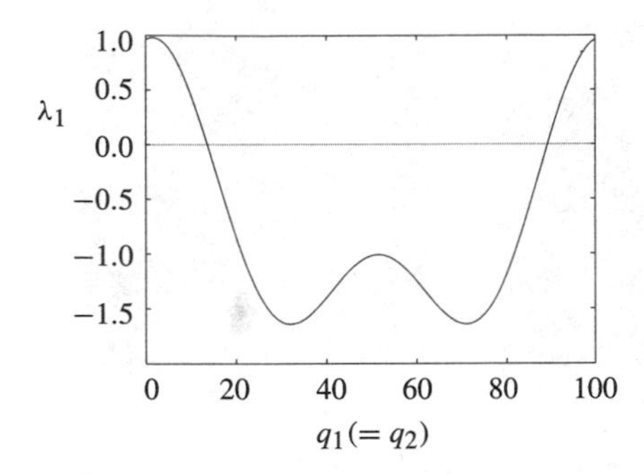

Figure 7.7. Checkerboard pattern formation in the adhesive LGCA model on the square lattice. A snapshot after 1000 time steps is shown (right). The corresponding spectrum (left) indicates small prevailing wave lengths $L/|q| \approx 50/\sqrt{32^2 + 32^2} \approx 1.1$ in diagonal directions ($q_1 = q_2$). Parameters: $\alpha = -1$, $\bar{\rho} = 0.4$, $L = 50$, $b = 4$.

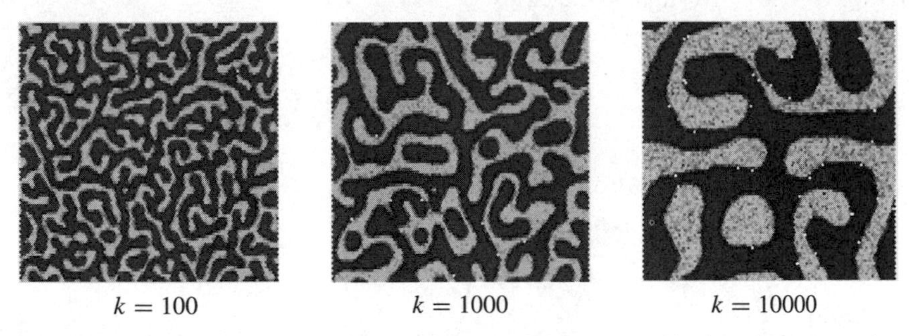

Figure 7.8. Sorting out in the two-species LGCA model (hexagonal lattice). Parameters (symmetric): $\alpha_{11} = \alpha_{22} = 0$, $\alpha_{12} = \alpha_{21} = -1$, $\bar{\rho}_1 = \bar{\rho}_2 = 0.4$, $L = 100$, $b = 6$. Light and dark gray levels represent the two cell types.

plete spreading), and sorting out patterns (self-isolation). Nevertheless, there are subtle weaknesses in his analysis. Besides that this parameter classification is rather artificial (it does not include repulsiveness, i.e., "negative works of adhesion"), the basic shortcoming is the missing cell motility, which Steinberg did not specify at all. Thirty years later, Glazier and Graner (1993) suggested a simulation model based on Steinberg's differential adhesion hypothesis. These authors distinguished between differential surface energies between cells (and the medium) and assumed an area constraint. There is no active motion of cells, explicitly. Glazier and Graner could show that even purely passively fluctuating cells can sort out partially or totally, engulf, disperse, and even form cavities.

Here, we address the implications of *active motion* and present typical simulations of the two-species LGCA model. There are six free parameters in the model: the averaged densities $\bar{\rho}_1$, $\bar{\rho}_2$ and the adhesion coefficients $(\alpha_{\sigma_1\sigma_2})_{1\leq\sigma_1,\sigma_2\leq 2}$. We always start out with a randomly mixed (homogeneous) initial configuration similar to biological rearrangement experiments.

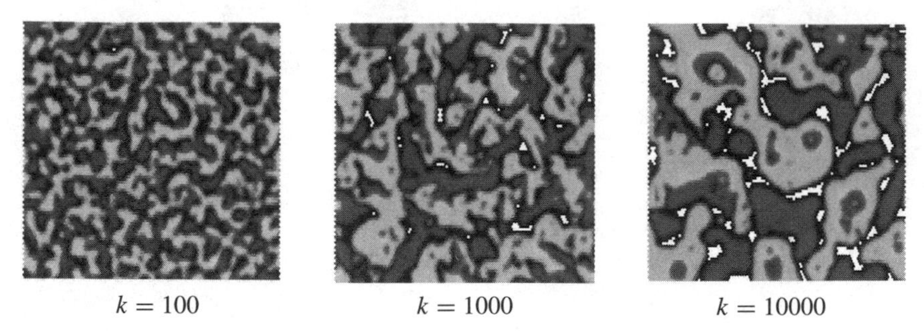

Figure 7.9. Engulfment in the two-species LGCA model (hexagonal lattice). Cell types 1 and 2 are marked by dark and light gray levels respectively, white denotes empty nodes. Parameters: $\alpha_{11} = 1.2$, $\alpha_{22} = 0.5$, $\alpha_{12} = -0.5$, $\alpha_{21} = 0.8$, $\bar{\rho}_1 = \bar{\rho}_2 = 0.4$, $L = 100$, $b = 6$.

Sorting Out. We investigated the case

$$(\alpha_{\sigma_1\sigma_2}) = \begin{pmatrix} > 0 & < 0 \\ < 0 & > 0 \end{pmatrix}$$

and observed sorting out. Also in the range

$$(\alpha_{\sigma_1\sigma_2}) = \begin{pmatrix} 0 & < 0 \\ < 0 & 0 \end{pmatrix}$$

sorting out is observed; i.e., the repulsiveness of different cell types is sufficient for sorting out. The initial prominent wavelength can be evaluated by linear stability analysis analogous to the single-cell-type case (cp. "Further research projects" at the end of the chapter). The long-time behavior implies phase separation that is also found in LGCA models with nonlocal interactions (Rothman 1989; Rothman and Zaleski 1997; fig. 7.8). Contrary to the adhesive LGCA such lattice gas models assume momentum conservation constraints.

Engulfment. Encapsulation is a morphogenetic process in which one cell type is *engulfed* by others, particularly implying compartmentalization (cp. gastrulation). Here, we ask for principles of encapsulating pattern formation within the framework of our adhesive LGCA model. An example of clustering is shown in fig. 7.9, which is based on asymmetric "adhesivities." One observes two types of patterns, either clusters of dark gray cells surrounded by free space or, alternatively, light gray cell compartments containing *nuclei* of dark gray cells, which themselves may have small light gray cell centers. Additionally, the light gray cell clusters are surrounded by a single membrane-like layer of dark gray cells again.

Summary. In this chapter we have analyzed spatial pattern formation in lattice-gas models for adhesive interaction. We have sketched the linear stability analysis of the single-cell-type case and have shown examples of pattern formation in the two-species model. Clustering of cells is the typical pattern characterizing adhesive interaction. In addition, sorting out and engulfment is observed in the two-species model. In later chapters we will show how the "adhesive model modul" can be used in models of pigment pattern formation and tumor growth (cp. ch. 9 and sec. 10.4).

7.5 Further Research Projects

1. **Analysis:** Perform a linear stability analysis for
   (a) the one-cell-type automaton model of adhesive interaction for the one-dimensional lattice, and
   (b) the two-cell-type automaton model of adhesive interaction for the one-, two-, and three-dimensional square lattice.
2. **Asynchronous automaton:** Develop a probabilistic asynchronous CA model for an adhesively interacting cell system.
3. **Pattern recognition:** How can the patterns observed in adhesive CA be described? Is it possible to develop a pattern recognition algorithm to distinguish the patterns?
4. **Cluster growth:** Cluster growth is typical of adhesively coupled cell systems. How can the cluster growth speed be determined?
5. **Modeling:** A necessary precondition of tumor invasion seems to be that tumor cells loose their affinity to the surrounding tissue. How would a simple CA model of tumor invasion that distinguishes healthy and tumor cells with different adhesivities look like? What are corresponding cell patterns and how can the invasion speed be calculated?

8

Alignment and Cellular Swarming

"A swarm is worth
A swarm in May is worth a load of hay,
A swarm in June is worth a silver spoon, or
A swarm in June is not worth a tune,
A swarm in July isn't worth a fly!"
(Honeykeeper's poem)

The LGCA introduced in this chapter presents a bottom-up view of swarming and may be characterized as an individual-based model defined by means of rules for motion and interaction of oriented entities (e.g., cells, molecules, organisms). The automaton is focused on a neighborhood-dependent interaction dynamics based on the concept of an *orientation field*. Mean-field analysis allows to view swarm initiation as a phase transition and indicates the formation of "orientational order."[57]

8.1 Orientation-Induced Pattern Formation

What is a Swarm? Swarming behavior may be characterized as the collective motion of oriented entities expressed by a large degree of local (and global) alignment. According to this interpretation, coupling of intrinsic individual propagation patterns is essential for swarm pattern formation. In addition to this aspect of pattern formation or initiation, the issue of pattern maintenance is also important since a swarm, essentially, is a nonstationary *moving pattern*. According to this interpretation, swarming might occur not just at organismic levels (birds and fishes exhibiting particularly striking examples), but even at cellular and intracellular scales whenever collective motion of biological, chemical, or physical entities is involved. Well-known examples at the cellular level are provided by streaming patterns of the cellular slime mold *Dictyostelium discoideum* and street formation of *Myxobacteria* (cp. fig. 8.1).

[57] The model described in this chapter was introduced and analyzed in Bussemaker, Deutsch, and Geigant 1997; Deutsch 1996.

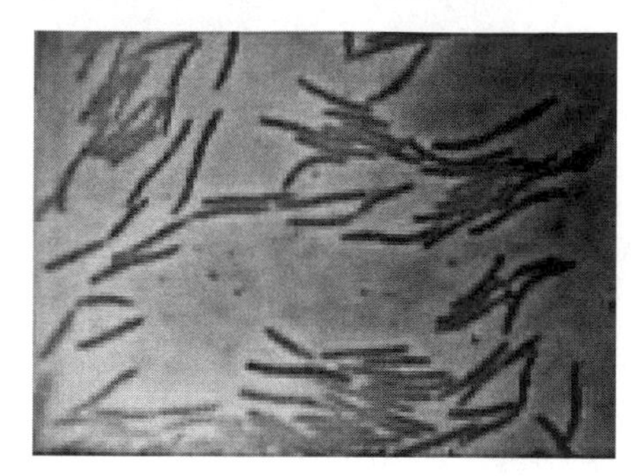

Figure 8.1. Street pattern formation of Myxobacteria (*Myxococcus xanthus*); individual bacteria (length $\approx$ 3–5μm) and aligned aggregates can be distinguished. Typically, the cells, which are about 7 μm long, move at a speed of about 1–2μm/min.

The transition to swarming patterns provides an indication of *social behavior* in these organisms.

Furthermore, the development of aggregation centers in cultures of fibroblasts (cells in connective tissue of mammals) can be interpreted as a swarming effect (fig. 8.2). At a particle level, one may consider the dynamics of charged particle groups, e.g., electron swarms (Kumar, Skullerud, and Robson 1980; Kumar 1984).

At least two different spatial and temporal scales have to be distinguished in swarms (Grünbaum 1994; Grünbaum and Okubo 1994). On a local (microscopic) level, characteristic local densities, interindividual distance, and orientational polarity of swarm members can be observed, while from a macroscopic perspective one can look at the distribution of group (patch) sizes within a swarm (Gueron and Levin 1995) or at long-range movement characteristics of the whole swarm, in particular transport properties such as motility and dispersivity, or the dynamics of typical front shapes (Gueron and Levin 1993). It is crucial to understand the conditions and

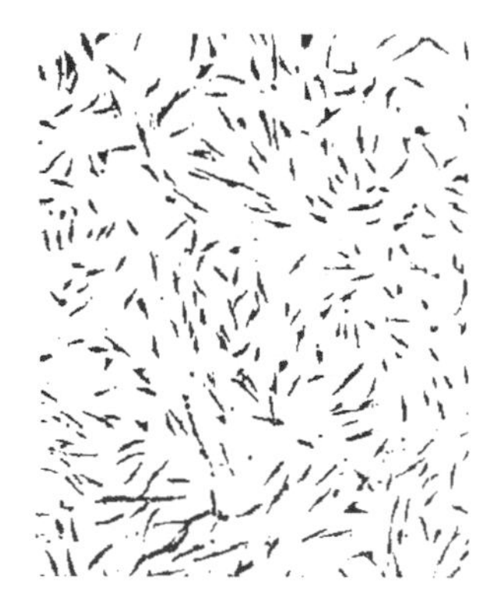

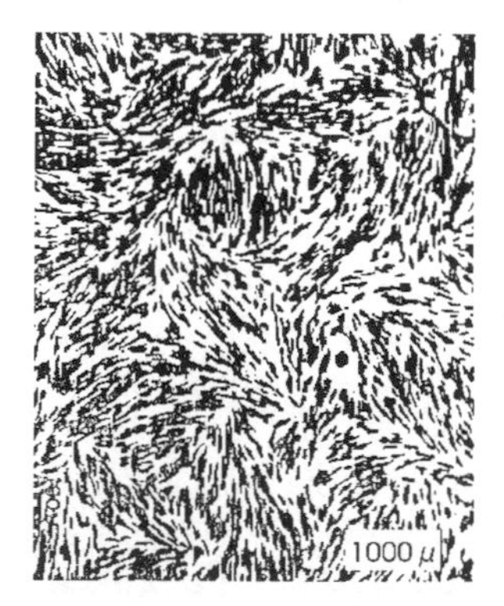

Figure 8.2. Fibroblast aggregate formation. Left: fibroblast culture one day after plating. The cells are randomly distributed. Right: fibroblast culture approaching confluence after several days. The monolayer is a patchwork of many small parallel associations.

mechanisms that govern pattern-forming processes of oriented entities. Hypothetical schemes include short-range repulsion (hereby maintenance of interindividual spacing), long-range attraction ("desire to be in a group"; Gueron et al. 1996), parallel alignment preference (Civelekoglu and Edelstein-Keshet 1994; Edelstein-Keshet and Ermentrout 1990; Mogilner and Edelstein-Keshet 1995, 1996b), differential adhesion (Steinberg 1970; cp. also ch. 7), or search for a "target density" of neighbors (Grünbaum 1994).

Models of Swarming. Even if the particular mechanisms might be rather different, the origin of structural similarities in swarming patterns can be addressed by means of mathematical models. A few models have already been designed that attack the problem from a top-down perspective; i.e., they handle (spatio-)temporal dynamics at the level of densities of oriented particles (Alt and Pfistner 1990; Civelekoglu and Edelstein-Keshet 1994; Edelstein-Keshet and Ermentrout 1990; Geigant 1999; Mogilner and Edelstein-Keshet 1996b; Okubo 1986; Pfistner 1995). However, the basis of such models is a stochastic process on a microscopic scale. Approximations for expectation values lead to the notion of macroscopic densities. These models are typically formulated as partial differential integral equations in which the integral term is due to the interaction property of the dynamics with regard to orientation changes within some "region of perception." A comparative study of microscopic and macroscopic perspectives in swarm models has already been carried out in a one-dimensional model that analyzes the relationship between a stochastic process defined in terms of microscopic variables and a partial integral differential equation for a particular interaction mechanism (Grünbaum 1994). Most often, some sort of linear analysis or perturbation theory is applied in order to characterize the nature of instabilities, but also tools from nonlinear analysis have been considered in "orientation models" (Geigant 1999; Mogilner and Edelstein-Keshet 1996a). An alternative way to handle the origin of orientational order is to use tensor expansion techniques (Cook 1995).

In the "swarm LGCA" to be introduced in this chapter, cells are assigned an orientation (and fixed absolute velocity) which determines their direction of motion. Cell orientations might change according to the "orientation field" in some neighborhood, i.e., by means of interaction with other members of the cell population (cp. the density-dependent interaction by means of a "density gradient field" in the "adhesive LGCA" introduced in ch. 7). A simple off-lattice model for collective motion of self-propelled particles on a plane has been introduced in Vicsek et al. 1995. Groups moving together can be observed in both the off-lattice and the cellular automaton (CA) model for a certain range of control parameters. While the off-lattice version gives a rather detailed picture of the alignment dynamics, the essential elements are also contained in the coarse-grained CA description, which, in addition allows a detailed analysis of the phase diagram on the level of the corresponding Boltzmann equation (Czirók, Deutsch, and Wurzel 2003).

8.2 A Swarm Lattice-Gas Cellular Automaton

The single-species[58] lattice-gas cellular automation (LGCA) is defined on a two-dimensional lattice $\mathcal{L}$ with coordination number $b = 4$ and no rest channels (i.e., $b = \tilde{b} = 4$) (cp. the LGCA definition in sec. 4.3.4). We assume that $\mathcal{L} = L \cdot L = L^2$. Here, we will commonly speak of cells instead of particles. We have to specify the *swarming interaction rule*

$$\mathcal{R}^{\mathtt{I}} : \mathcal{E}^{\nu} \to \mathcal{E}, \qquad \text{where } \mathcal{E} = \{0, 1\}^b \text{ and } \nu = |\mathcal{N}_b^I|,$$

which describes a transformation $\boldsymbol{\eta}_{\mathcal{N}(r)} \to \boldsymbol{\eta}^{\mathtt{I}}(r)$, where $\boldsymbol{\eta}(r)$, $\boldsymbol{\eta}^{\mathtt{I}}(r)$ are the pre- and postinteraction state, respectively, which depends on the configuration in the interaction neighborhood $\mathcal{N}_b^I(r)$ (the central node plus its nearest neighbor nodes). The repeated application of the swarming interaction operator ($\mathtt{I}$) together with the propagation operator ($\mathtt{P}$) defines the complete dynamics of the swarming LGCA.

Swarming Interaction. Recall that the number of cells at a node is denoted by

$$\mathrm{n}(r) = \sum_{i=1}^{b} \eta_i(r), \qquad \boldsymbol{\eta}(r) \in \mathcal{E}.$$

During interaction the number of cells at each node remains constant (i.e., there is no creation or annihilation). Consequently,

$$\mathrm{n}(r) = \mathrm{n}^{\mathtt{I}}(r) \qquad \text{for all } r \in \mathcal{L},$$

where n and $\mathrm{n}^{\mathtt{I}}$ are the pre- and postinteraction cell numbers, respectively. Accordingly, the spatially averaged density of cells per node, $\bar{\rho} = (1/b|\mathcal{L}|) \sum_{r \in \mathcal{L}} \mathrm{n}(r) \in [0, 1]$, is constant in time.

Local interaction comprises cell reorientation according to a local *alignment interaction*. In order to implement the local alignment dynamics, we define

$$\boxed{\boldsymbol{D}\left(\boldsymbol{\eta}_{\mathcal{N}(r)}\right) := \sum_{p=1}^{b} \sum_{i=1}^{b} c_i \eta_i(r + c_p) = \sum_{p=1}^{b} \boldsymbol{J}\left(\boldsymbol{\eta}(r + c_p)\right),} \tag{8.1}$$

which is the *director field* specifying the average flux of cells at the nearest neighbors of node r. Thereby, as in the adhesive LGCA,

$$\boldsymbol{J}\left(\boldsymbol{\eta}(r)\right) = \sum_{i=1}^{b} c_i \eta_i(r)$$

denotes the cell flux at node r.

[58] Extensions to more-species models are straightforward but will not be considered here.

Initial configuration: Result of interaction step ($\alpha = 1.5$):

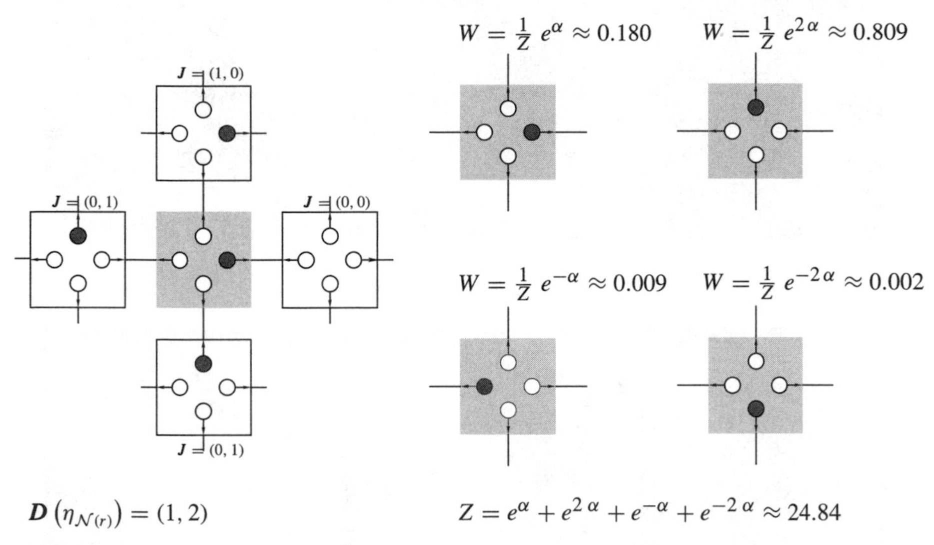

Figure 8.3. Example of a swarming interaction in the square lattice. Filled dots denote channels occupied by cells, while empty dots denote empty channels. With probability W each node configuration on the right side ($\boldsymbol{\eta}^{\mathrm{I}}$) is a possible result of the interaction step applied to the central gray node ($\boldsymbol{\eta}$) of the initial configuration.

The transition probability from $\boldsymbol{\eta}(r)$ to $\boldsymbol{\eta}^{\mathrm{I}}(r)$ in the presence of the director field $\boldsymbol{D}\left(\boldsymbol{\eta}_{\mathcal{N}(r)}\right)$ is given by

$$W\left(\boldsymbol{\eta}_{\mathcal{N}(r)} \to \boldsymbol{\eta}^{\mathrm{I}}|\alpha\right) = \frac{1}{Z}\exp\left(\alpha \boldsymbol{D}\left(\boldsymbol{\eta}_{\mathcal{N}(r)}\right) \cdot \boldsymbol{J}\left(\boldsymbol{\eta}^{\mathrm{I}}\right)\right) \cdot \delta_{\mathrm{n},\mathrm{n}^{\mathrm{I}}}, \tag{8.2}$$

where the normalization factor $Z = Z(\mathrm{n}, \boldsymbol{D})$ is chosen such that

$$\sum_{\boldsymbol{\eta}^{\mathrm{I}} \in \mathcal{E}} W\left(\boldsymbol{\eta}_{\mathcal{N}(r)} \to \boldsymbol{\eta}^{\mathrm{I}}|\alpha\right) = 1$$

for all $\boldsymbol{\eta}$; $\alpha \in \mathbb{R}$ is the alignment sensitivity of the cells. The idea is that an output $\boldsymbol{\eta}^{\mathrm{I}}$ results with large probability, which makes the argument of the exponential maximal. The interaction rule is designed to minimize the angle between the director field $\boldsymbol{D}$ and the postinteraction flux $\boldsymbol{J}$. Fig. 8.3 gives an example of the transition probability. The *alignment sensitivity* parameter α controls the degree of local alignment: for $\alpha = 0$ there is no alignment at all—the director field $\boldsymbol{D}$ does not have any effect on the outcome of the interaction: the outcome $\boldsymbol{\eta}^{\mathrm{I}}$ is chosen with equal probability among all states that have the same number of cells as $\boldsymbol{\eta}$, i.e., $\mathrm{n} = \mathrm{n}^{\mathrm{I}}$. The particular neighborhood configuration does not influence the dynamics. This "diffusive situation" is also observed in the adhesive LGCA model described in ch. 7. But for $\alpha \to \infty$ the particular neighborhood configuration becomes important since the inner product $\boldsymbol{D}\left(\boldsymbol{\eta}_{\mathcal{N}(r)}\right) \cdot \boldsymbol{J}\left(\boldsymbol{\eta}^{\mathrm{I}}\right)$—and therefore the local alignment—is maximized. It

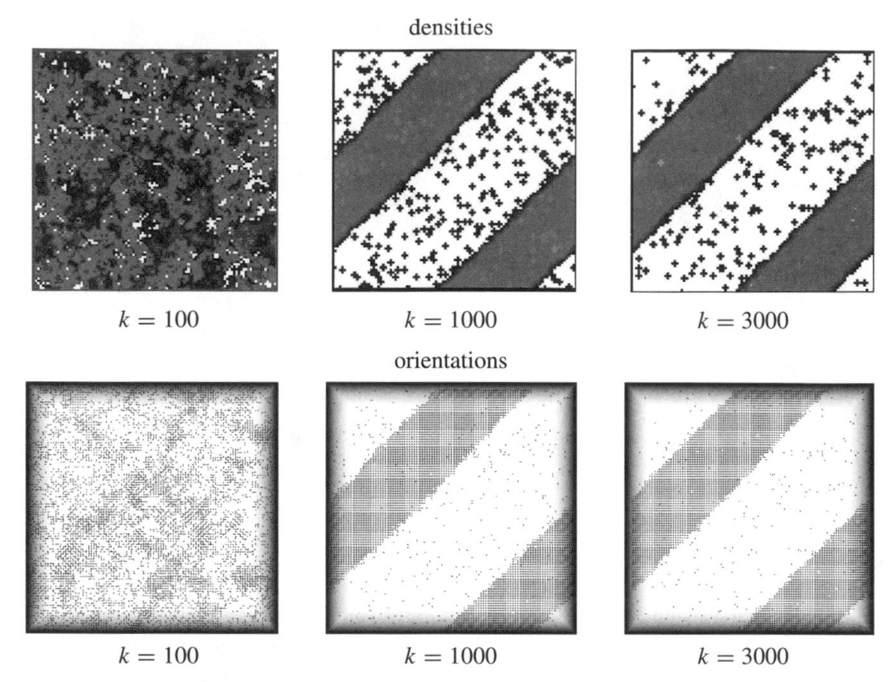

Figure 8.4. Swarming behavior in the swarm LGCA model (square lattice). Shown are configurations in two different representations. Top: densities are displayed as gray levels (white marks empty nodes); bottom: the cell orientations are shown. Clearly, formation of "streets" as aligned patches is visible. The formation of orientational order can be explained by the corresponding eigenvalue spectrum, in which the dominant mode has its maximum at $q_* = (0, 0)$ (cp. fig. 8.7). Parameters correspond to the unstable regime (cp. fig. 8.5): $\alpha = 1.5$, $L = 50$, $\bar{\rho} = 0.2$.

can be shown that a second-order dynamical phase transition occurs at a critical value α_c of the sensitivity (Bussemaker, Deutsch, and Geigant 1997). Fig. 8.4 shows the time evolution of an initially random distribution for $\alpha > \alpha_c$. In particular, formation of locally aligned patches ($k = 100$) can be observed. Furthermore, patch sizes grow in the course of temporal development. For long times ($k = 1000, 3000$) corresponding configurations can be explained by crossing of two large patches moving in rectangular directions. In the following, we focus on the *initialization of swarming*, i.e., the formation of oriented patches starting from a homogeneously distributed configuration. Note that our mean-field analysis of the corresponding nonlinear Boltzmann equation cannot explain long-time effects other than the initialization of swarming.

8.2.1 Linear Stability Analysis

To analyze the behavior of the swarm LGCA model we consider the time evolution of a statistical ensemble of systems, i.e., the lattice-Boltzmann equation (cp. (4.23))

$$f_i(r + c_i, k + 1) - f_i(r, k)$$

$$= \sum_{\boldsymbol{\eta}^{\mathrm{I}}(r)\in\mathcal{E}} \sum_{\boldsymbol{\eta}_{\mathcal{N}(r)}\in\mathcal{S}_{\mathcal{N}(r)}} \Bigg\{ \left(\eta_i^{\mathrm{I}}(r) - \eta_i(r)\right) W\left(\boldsymbol{\eta}_{\mathcal{N}(r)} \to \boldsymbol{\eta}^{\mathrm{I}}(r)|\alpha\right)$$

$$\cdot \prod_{r_i\in\mathcal{N}_b^I(r)} \prod_{l=1}^{b} f_l(r_i,k)^{\eta_l(r_i)} \left(1 - f_l(r_i,k)\right)^{(1-\eta_l(r_i))} \Bigg\}$$

$$= \tilde{\mathcal{C}}_i\left(\boldsymbol{f}_{\mathcal{N}(r)}(k)\right), \qquad i = 1, \ldots, b, \tag{8.3}$$

where $\tilde{\mathcal{C}}_i\left(\boldsymbol{f}_{\mathcal{N}(r)}(k)\right)$ equals the average change in the occupation number of channel (r, c_i) during interaction.

It can be shown that a possible isotropic solution to $\tilde{\mathcal{C}}_i\left(\boldsymbol{f}_{\mathcal{N}(r)}(k)\right) = 0$ is

$$f_i(r,k) = \bar{f} \qquad \text{for } i = 1, \ldots, b.$$

In this case, $\bar{\rho} = (1/b)\sum_{i=1}^{b} \bar{f}_i = \bar{f}$. To assess the stability of this spatially homogeneous, isotropic, and stationary solution with respect to fluctuations $\delta f(r,k) = f_i(r,k) - \bar{f}$, we linearize (8.3) and obtain the Boltzmann propagator (4.36)

$$\Gamma(q) = T\left\{\boldsymbol{I} + \boldsymbol{\Omega}^0 + \sum_{n=1}^{b} \boldsymbol{\Omega}^n e^{\mathbf{i}(2\pi/L)q\cdot c_n}\right\}, \qquad q = (q_1, q_2), \tag{8.4}$$

with $q_1, q_2 = 0, \ldots, L-1$, the identity matrix $\boldsymbol{I}$ and the "transport matrix"

$$T = \operatorname{diag}\left(e^{-\mathbf{i}(2\pi/L)(q\cdot c_1)}, \ldots, e^{-\mathbf{i}(2\pi/L)(q\cdot c_b)}\right).$$

As in the adhesive LGCA automaton (cp. ch. 7), it can be shown (Bussemaker 1996) that

$$\Omega_{ij}^0 = \frac{1}{b} - \delta_{ij}.$$

For $1 \le n \le b$, the elements

$$\Omega_{ij}^n = \Omega_{ij}^1$$

do not depend on n, as can be seen from the definition of the director field $\boldsymbol{D}$ (8.1).

Here, we concentrate on the LGCA model defined on a *square lattice*, i.e., $b = 4$. Matrix $\boldsymbol{\Omega}^1 = (\Omega_{ij}^1)$ has the following structure:

$$\boldsymbol{\Omega}^1 = \begin{pmatrix} c+d & -d & -c+d & -d \\ -d & c+d & -d & -c+d \\ -c+d & -d & c+d & -d \\ -c & -c+d & -d & c+d \end{pmatrix}.$$

To determine $c = c(\bar{f}, \alpha)$ and $d = d(\bar{f}, \alpha)$ for given values of the averaged density $\bar{\rho} = \bar{f}$ and given sensitivity α, we numerically evaluate the coefficients

$$\Omega^1_{ij}(\bar{f},\alpha)$$

$$= \left.\frac{\partial \tilde{C}_i\,(\delta f\,vNr(k))}{\partial \delta f_j(r_1,k)}\right|_{\bar{f}b}$$

$$= \sum_{\eta^{\mathrm{I}}(r)\in\mathcal{E}}\ \sum_{\eta_{\mathcal{N}(r)}\in\mathcal{S}_{\mathcal{N}(r)}} \Big\{ \left(\eta_i^{\mathrm{I}}(r) - \eta_i(r)\right) W\left(\boldsymbol{\eta}_{\mathcal{N}(r)} \to \boldsymbol{\eta}^{\mathrm{I}}(r)|\alpha\right)$$

$$\cdot \frac{\eta_j(r_1) - \bar{f}}{\bar{f}(1-\bar{f})} \cdot \prod_{r_i\in\mathcal{N}_b^I(r)} \prod_{l=1}^{b} \bar{f}^{\eta_l(r_i)}\left(1-\bar{f}\right)^{(1-\eta_l(r_i))}\Big\},$$

where $\bar{f}b = \left(\bar{f}\right) \in [0,1]^b$. The expression for Ω^1_{ij} does not depend on r since it represents a derivative evaluated in a spatially uniform state.

8.2.2 The Swarming Instability

Unfortunately, it is no longer possible to give a closed form of the eigenvalues of the Boltzmann propagator (8.4) as for the diffusive and the adhesive LGCA (cp. sec. 5.4 and ch. 7). We first investigate the stability of the spatially uniform state, i.e., $q = (0,0)$. Evaluating the eigenvalues and eigenvectors for $q = (0,0)$ *numerically*, we find

$$\begin{aligned} \lambda_1(0,0) &= 1, & v_m &= (1,1,1,1),\\ \lambda_2(0,0) &= 8c(\bar{\rho},\alpha), & v_{J_x} &= (1,0,-1,0),\\ & & v_{J_y} &= (0,1,0,-1),\\ \lambda_3(0,0) &= 16d(\bar{\rho},\alpha), & v_q &= (1,-1,1,-1), \end{aligned}$$

with $c(\bar{\rho},\alpha), c(\bar{\rho},\alpha) \in \mathbb{R}$, where $c > d$; d is about two orders of magnitude smaller than c. Note that the (α-independent) eigenvalue λ_1 with its corresponding eigenvector v_m reflects the fact that the total density is conserved (independent of the particular value of the sensitivity α). v_{J_x} and v_{J_y} correspond to the x- and y-components of the total cell flux, while v_q corresponds to the difference between the number of horizontally and vertically moving particles. Since all eigenvectors are linearly independent, the Boltzmann propagator $\Gamma(0,0)$ is diagonalizable and the temporal growth of modes $F_i((0,0),k)$ is determined by the dominant eigenvalue $\lambda_2(0,0)$ (cp. (4.39)). Hence, the onset of instability of the homogeneous state is determined by the "stability break" of $\lambda_2(0,0)$, i.e., the condition $\lambda_2(0,0) = 8c(\bar{\rho},\alpha) = 1$. The location of the critical line in the $(\bar{\rho},\alpha)$ parameter plane is shown in fig. 8.5, which was obtained by numerically solving $c(\bar{\rho},\alpha) = 1/8$ (cp. sec. 8.2.1).

Our mean-field stability analysis illuminates the nature of the observed phase transition, since the structure of the eigenspace corresponding to the dominant unstable eigenvalue $\lambda_2(0,0)$ indicates a global drift of cells: horizontal and vertical momentum both become unstable.

An appropriate order parameter comprising both horizontal and vertical momentum is the *spatially averaged velocity*,

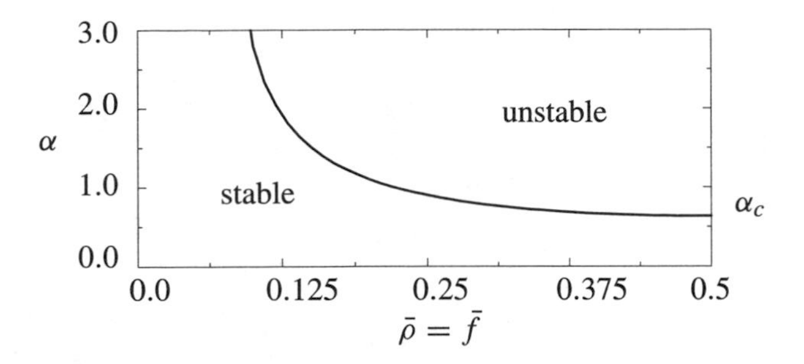

Figure 8.5. Phase diagram for the swarming model. Shown are the regions of stable ($0 < \lambda_2(0, 0) < 1$) and unstable ($\lambda_2(0, 0) > 1$) behavior, as a function of sensitivity α and averaged density $\bar{\rho} \in [0, 1]$. For given averaged density the critical sensitivity value α_c can be determined from the line.

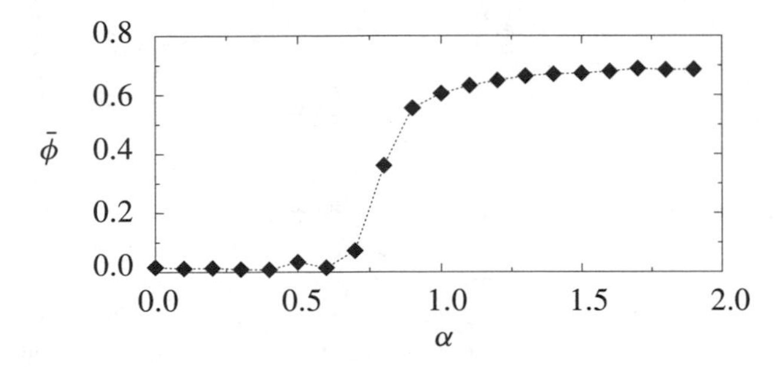

Figure 8.6. Mean velocity $\bar{\phi}$ versus sensitivity α. Obtained from simulation of the $L = 50$ system at averaged density $\bar{\rho} = 0.4$, after $k = 1000$ time steps. There is a phase transition for $\alpha_c \approx 0.7$ that can be interpreted as "swarming instability." The critical value is in good agreement with the value deduced from mean-field theory (cp. fig. 8.5).

$$\bar{\phi}(k) = \frac{1}{L^2} \left| \sum_{r \in \mathcal{L}} \boldsymbol{J}\left(\boldsymbol{\eta}(r, k)\right) \right| = \frac{1}{L^2} \left| \sum_{r \in \mathcal{L}} \sum_{i=1}^{b} c_i \eta_i(r, k) \right| ,$$

taking values between 0 and 1. For $\alpha < \alpha_c$ we have $\bar{\phi}(k \to \infty) \approx 0$. When the sensitivity parameter α reaches its critical value, this "rest" state becomes unstable, leading to a breaking of rotational symmetry, and a state where $\bar{\phi}(k \to \infty) \neq 0$.

We compare the results of the stability analysis with the results of computer simulations. Fig. 8.6 shows $\bar{\phi}(1000)$ plotted versus α for averaged density $\bar{\rho} = 0.4$. There is an abrupt change in $\bar{\phi}$ at $\alpha \approx 0.7$, which agrees very well with the prediction $\alpha_c = 0.67$ obtained from stability analysis (cp. fig. 8.5). In Bussemaker, Deutsch, and Geigant 1997 it was shown that the observed dynamic phase transition is of second order.

Spatial Patterns and Traveling Waves. In order to see if, in addition to the emergence of a global drift, we can explain the formation of spatial structure in terms of

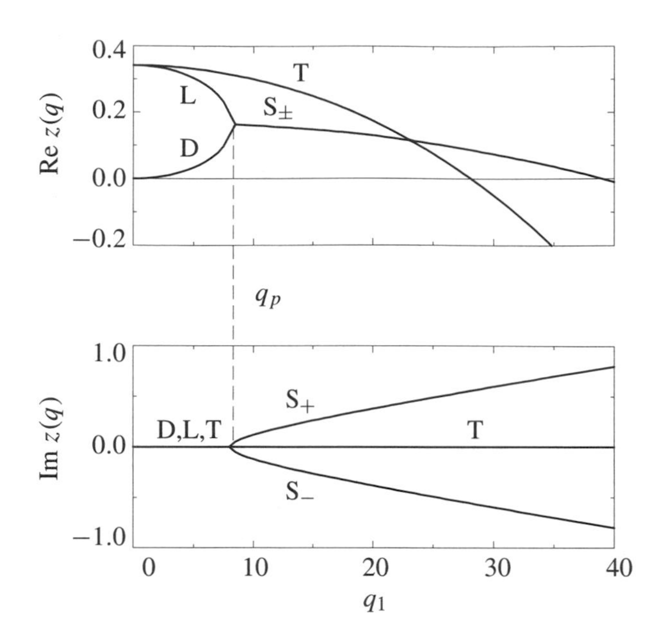

Figure 8.7. Numerically determined eigenvalue spectrum for $\bar{\rho} = 0.4$, $\alpha = 1.5$, and $q = (q_1, 0)$. D denotes the eigenvalue associated with $\lambda_1(0,0)$ (density momentum), T the eigenvalue associated with $\lambda_2(0,0)$ (transverse momentum), L an eigenvalue with longitudinal momentum, and $S_\pm$ the propagating modes with Im $z(q) \neq 0$. The stable eigenvalue associated with $\lambda_3(0,0)$ is not shown. The maximum Re $z(q)$ of the dominant mode T at $q_* = (0,0)$ indicates the formation of "orientational order" corresponding to street formation in the simulations (cp. fig. 8.4).

the eigenvalue spectrum, we study the case $q \in Q^+, q \neq (0,0)$ (cp. the definition of Q^+, p. 96). It proves useful to consider

$$z(q) := \ln \lambda(q) = \text{Re } z(q) + \mathbf{i} \text{ Im } z(q)$$

so that the dispersion relation for unstable modes becomes (cp. (4.41))

$$\begin{aligned} \delta f(r,k) \sim e^{-\mathbf{i}\frac{2\pi}{L} q \cdot r} \lambda(q)^k &= e^{-\mathbf{i}\frac{2\pi}{L} q \cdot r + z(q)k} \\ &= e^{\text{Re } z(q)\, k + \mathbf{i}\left(\text{Im } z(q)\, k - \frac{2\pi}{L} q \cdot r\right)}. \end{aligned} \tag{8.5}$$

Hence, unstable modes have Re $z(q) > 0$, while stable modes have Re $z(q) < 0$.

Fig. 8.7 shows the numerically determined eigenvalue spectrum of the Boltzmann propagator (8.4) for $\bar{\rho} = 0.4$, $\alpha = 1.5$, and $q = (q_1, 0)$. Although the fastest growth occurs at $q_* = (0,0)$, unstable modes with nonzero imaginary part (propagating mode) also grow for $|q| > q_p$ (see fig. 8.7 for q_p). The speed of propagation of these modes can be deduced from (8.5)

$$\text{Im } z(q)k - \frac{2\pi}{L} q \cdot r = \frac{2\pi}{L}|q| \left(\Delta k(q)^{-1} \frac{L}{|q|} k - \frac{q \cdot r}{|q|} \right),$$

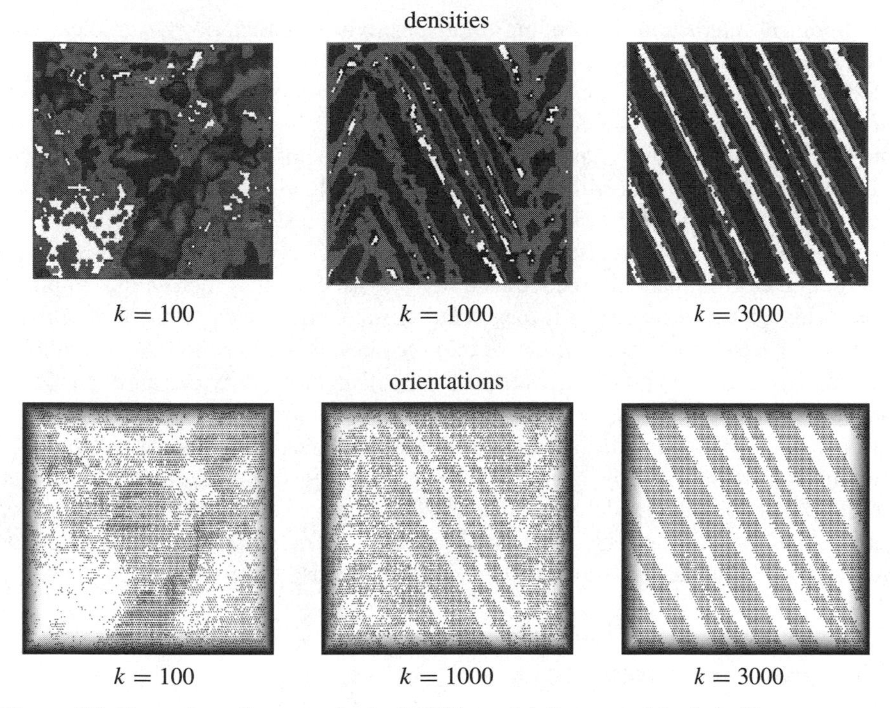

Figure 8.8. Formation of swarms in the LGCA model (hexagonal lattice). Shown are configurations in two different representations. Top: densities are displayed as gray levels (white marks: empty nodes); bottom: the cell orientations are shown. Formation of streets appears similar to the square lattice simulations (cp. fig. 8.4). Parameters: $\alpha = 1.5$, $L = 50$, $\bar{\rho} = 0.2$.

where $\Delta k = (2\pi/\text{Im } z(q))$ is the *period of oscillation* and $L/|q|$ the wave length. Therefore, soundlike modes propagate with *speed* $(L/|q|)/\Delta k(q) = (\text{Im } z(q)/2\pi) \cdot (L/|q|)$. Thus, traveling waves cannot occur on spatial scales larger than L/q_p. It is, however, questionable whether and how these traveling modes express themselves, since the fastest growth occurs at $q_* = (0, 0)$ and implies the "swarming instability." A corresponding analysis for the *hexagonal lattice* gives similar results: the fastest growing mode is also $q_* = (0, 0)$. Simulations are shown in fig. 8.8: analogous to the square lattice, formation of streets can be observed. Thus, it is sufficient to use a square lattice for the swarm interaction–based LGCA: Due to the properties of the spectrum ($q_* = (0, 0)$), spatial anisotropies cannot manifest themselves as strongly as in the adhesive interaction–based automaton in which nonzero maxima indicate prevailing spatial wave lengths in the patterns (cp. ch. 7). However, artefacts due to the lattice geometry are less pronounced in a hexagonal lattice since this allows a larger number of macroscopic directions (cp. figs. 8.4 and 8.8, $k \geq 1000$, as well as the discussion in secs. 12.1.2 and 12.5).

Summary. In this chapter a LGCA model for orientation-induced interaction was introduced. Interacting cells are equipped with an orientation that may change due to alignment interaction. Depending on a sensitivity parameter a dynamical phase transition is observed that can be analyzed in a mean-field description of the CA model. In conclusion, our analysis has demonstrated scenarios for the initialization of swarming viewed as the emergence of aligned patches. Statistical properties of the resulting patch patterns have been determined in a similar model (Deutsch 1996). The dynamical phase transition (Bussemaker, Deutsch, and Geigant 1997) suggests two possible scenarios for a change from random dispersal to cooperative behavior. On one hand, genetically caused minor microscopic effects on receptor properties of interacting cells influencing their sensitivity can have severe macroscopic implications with respect to swarming if they occur close to criticality (cp. fig. 8.5). On the other hand, a transition from the stable into the unstable region can also be achieved by simply increasing cell density (cp. fig. 8.5). This result provides an explanation for behavioral changes between cooperative and noncooperative stages in individual life cycles of Myxobacteria in which a reproductive feeding phase of individually moving cells is followed by social (coordinated) aggregation. Note that in Vicsek et al. 1995 an off-lattice model with a similar microscopic swarming interaction was introduced. It turns out that the coarse-grained CA perspective covers the essential aspects of cell interaction (Czirók, Deutsch, and Wurzel 2003).

8.3 Further Research Projects

1. **Asynchronous automaton:** Define a probabilistic CA that models cellular swarming.
2. **Initial/boundary conditions:** Analyze the influence of different initial and boundary conditions for swarm pattern formation.
3. **Predator-induced swarming:** Distinguish predator and prey individuals and test the following hypothesis by constructing a corresponding CA model: Prey swarms could arise simply as a consequence of "hunting pressure" imposed by the predator. Does this behavior depend on the choice of the boundary conditions?
4. **Modeling**
   (a) Define and analyze a lattice-gas cellular automaton model for contact guidance interaction; i.e., cells shall align to the orientation of the extracellular matrix, which can be assumed as stationary spatial structure in the model.
   (b) For the modeling of realistic swarms (in finite systems), an attracting (adhesive) interaction has to be considered. Construct a corresponding CA and analyze the differences from the CA model with solely alignment interaction.

9

Pigment Cell Pattern Formation

"If you see a whole thing—it seems that it's always beautiful. Planets, lives. . . . But close up a world's all dirt and rocks. And day to day, life's a hard job, you get tired, you lose the pattern. . . ."

(Ursula K. LeGuin)

In this chapter a LGCA is introduced as a model for complex pattern formation in salamander larvae. After providing the biological background the model is defined and tested by means of simulations. Model results demonstrate that the larval stripe pattern can arise solely as a consequence of direct cell interactions.

9.1 Principles of Pigment Cell Pattern Formation

Coat markings of zebras or other vertebrates are amongst the most striking biological patterns and are of both evolutionary and developmental interest (Gould and Morton 1983). Such patterns develop as characteristic pigment cell arrangements in the embryo. All pigment cells, except those in the retina, originate from the neural crest, a transient rod-like embryonic structure situated above the neural tube. Neural crest cells move actively *in vitro* and *in vivo* (Bonner-Fraser and Fraser 1988; Löfberg and Ahlfors 1980). In the course of development they migrate from the neural crest to various regions in the embryo, thereby initiating a variety of tissues, such as parts of the peripheral nervous system, the head, and the adrenal gland, besides giving rise to pigment cell patterns (LeDouarin 1982; Hall and Hörstadius 1988; Olsson and Löfberg 1993). Research on pigment pattern formation might not only offer new insights into an interesting and often esthetically appealing phenomenon but also help to improve understanding the principles underlying the development of cellular tissues.

Nature uses two different strategies for pigment pattern formation. In mammals and birds the primary process is cell distribution: undifferentiated cells leave the neural crest, distribute themselves homogeneously along the flanks of the embryo, and subsequently differentiate according to their final position. Mathematical models based on reaction-diffusion ideas try to explain such heterogeneous differentiation events as the result of an appropriate interplay of certain chemical "morphogens" (Meinhardt 1982; Murray 1981, 1988, Young 1984). Probably one of the first simulations of "morphogen-driven" pigment patterns (performed by hand!) was found in Turing's pioneering morphogenesis paper (Turing 1952; cp. also fig. 4.5).

In amphibia and fish the situation is the reverse, with differentiation being the primary process: here, fully differentiated pigment cells leave the neural crest and the final pattern results from interactions among the cells themselves and with the extracellular matrix (ECM). This is a three-dimensional fibrillous network through which the cells have to migrate in order to find their final destination (Löfberg et al. 1989). Several types of extracellular matrix molecules that influence cell migration have been localized, including collagens, glycoproteins as fibronectin, laminin and tenascin, and proteoglycans/glycosaminoglycans (cp. Olsson and Löfberg 1993). Cells in the neural crest are kept together by intercellular junctions and cell adhesion molecules (CAMs; Tucker et al. 1988). Initiation of cell migration is correlated with a loss of intercellular junctions, N-CAM and N-cadherin CAM. Cell migration is probably initiated when the crest cells shift from cell–cell adhesion to cell–substrate adhesion. There is experimental indication that fibronectin receptors are among the most important of the substrate adhesion molecules (see Olsson and Löfberg 1993 for a discussion). But how does a cell know where to move in the embryo?

Stripe Formation in the Salamander Family. A particularly well-studied biological example for the second type of pigment pattern formation based on active migration and cell–cell interactions is provided by salamanders. The axolotl *Ambystoma mexicanum* exhibits a periodic pigment pattern consisting of alternating "vertical stripes" of melanophores and xanthophores (Olsson and Löfberg 1993; cp. also fig. 1.6). Other species belonging to the same group (e.g., the alpine newt *Triturus alpestris*) possess two "horizontal melanophore stripes" with xanthophores in between. Furthermore there are salamanders with an "intermediate pattern" (fig. 9.1). What are the mechanisms responsible for such different patterns? There are extensive comparative experimental studies of the pattern-forming dynamics in salamanders (fig. 9.2). In the alpine newt, both pigment cell types leave the neural crest simultaneously and spread out evenly. Subsequently, the melanophores form two horizontal stripes along the larval body, and later the xanthophores migrate into the interstripe area. Experimental evidence indicates that melanophores recognize a prepattern of environmental cues laid down as a horizontal stripe prepattern in the extracellular matrix and cell–cell interaction is of minor importance for pattern formation (Epperlein and Löfberg 1990). In the following, we will focus on vertical stripe patterns as produced by the axolotl since cell migration is essential for this type of pattern formation. Various hypotheses regarding the obvious directionality of neural crest cell migration have been proposed, including long-range taxisms (chemo- and galvanotaxis; Erickson 1990). Nevertheless, experimental evidence indicates that the vertically directed pattern primarily arises as a result of adhesive cell–cell interactions. We will test this hypothesis with the help of an appropriately defined lattice-gas cellular automation (LGCA) based on adhesive/orientational interactions.

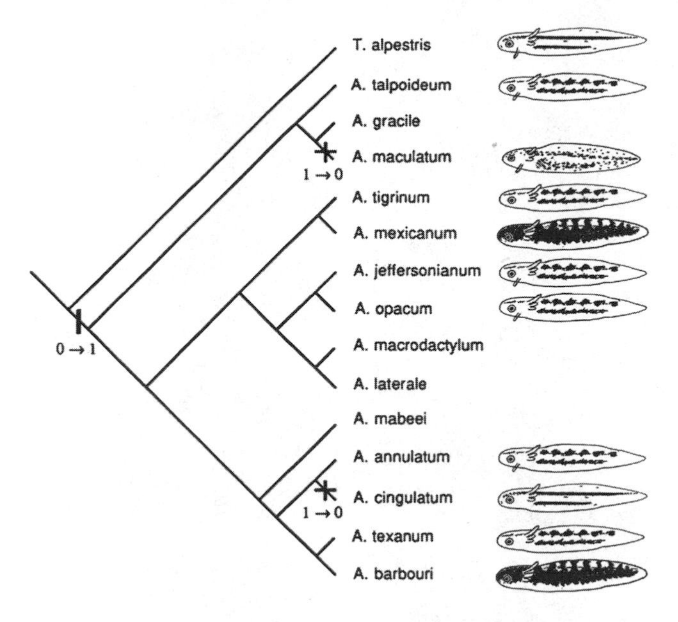

Figure 9.1. Phylogenetic tree of the salamander family (Ambystomatidae). (Courtesy of J. Löfberg and L. Olsson, University of Uppsala) The larval pigment patterns are shown if they are known. The aggregation mechanism that leads to vertical patterns has only evolved once and has been lost twice (in *Ambystoma maculatum* and *A. cingulatum*). 0: no aggregation patterns, 1: aggregation and stripes.

9.2 Automaton Model with Adhesive/Orientational Interaction[59]

A three-species LGCA is defined ($\varsigma = 3$; cp. sec. 4.3). In the LGCA, (biological) pigment cells and the ECM are represented by oriented (particles) cells of type σ, where

$$\sigma = 1\text{: melanophore,}$$

$$\sigma = 2\text{: xanthophore, and}$$

$$\sigma = 3\text{: ECM.}$$

We assume that all cells interact but only "pigment cells" (of types $\sigma = 1, 2$) move on a two-dimensional hexagonal lattice $\mathcal{L}$ ($b = 6$). We distinguish top and bottom regions of the lattice representing the neural crest and the ventral regions of the embryo, respectively. Thereby, we introduce a reference frame on the lattice. Various boundary conditions to be specified in the corresponding context will be considered.

The cells occupy lattice channels corresponding to nearest-neighbor vectors c_i (cp. definition on p. 68), associated with each node r. An exclusion principle prevents two (or more) cells of the same type from inhabiting the same channel. Accordingly, the state of node r is specified by $\boldsymbol{\eta}(r) = \left(\eta_{\sigma,i}(r)\right)_{1\leq\sigma\leq 3, 1\leq i\leq b}$, where $\eta_{\sigma,i}(r) \in$

[59] The simulations in this chapter have been performed by Michael Spielberg (Bonn).

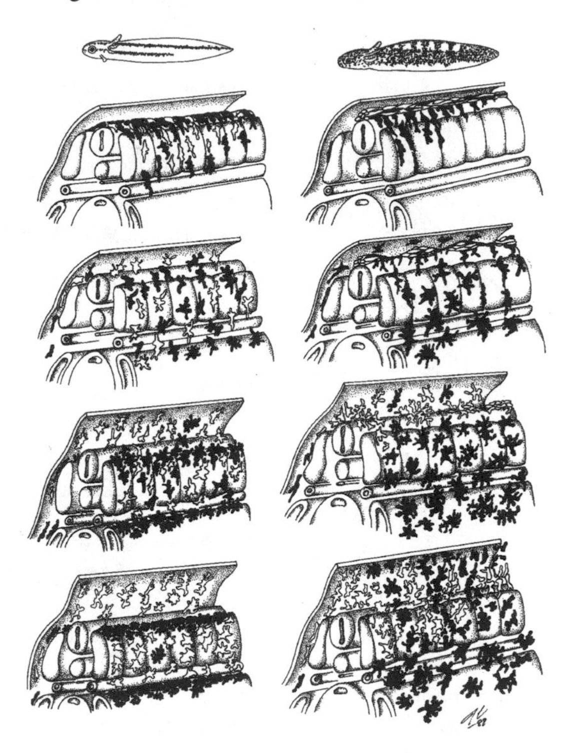

Figure 9.2. Sketch of horizontal and vertical pigment pattern formation in two salamander species (*Triturus alpestris* (left) and *Ambystoma mexicanum*); see explanations in the text. Courtesy of J. Löfberg and L. Olsson, University of Uppsala.

$\{0, 1\}$ denotes the absence (respectively, presence) of a cell of type σ in channel (r, c_i).

Cell–Cell Interactions. States may change according to local interactions. States of cell type $\sigma = 3$ (ECM) are assumed to maintain their initial value and not to change in time, thereby introducing an orientational anisotropy influencing the dynamics of the moving pigment cells.

Let the number of cells at node r of type σ be denoted by

$$\mathrm{n}_\sigma(r) = \sum_{i=1}^{b} \eta_{\sigma,i}(r).$$

During interaction the number of cells of each type remains constant (i.e., there is no creation, annihilation, or transformation of any kind). Consequently,

$$\mathrm{n}_\sigma(r) = \mathrm{n}_\sigma^{\mathrm{I}}(r) \qquad \forall\, r \in \mathcal{L},$$

where n_σ and $\mathrm{n}_\sigma^{\mathrm{I}}$ are the pre- and postinteraction cell numbers, respectively. Accordingly, the spatially averaged density of cells of type σ per node,

$$\bar{\rho}_\sigma = \left(\frac{1}{b|\mathcal{L}|}\right) \cdot \sum_{r \in \mathcal{L}} \mathrm{n}_\sigma(r) \in [0, 1],$$

is constant in time.

Local interaction comprises of cell reorientation according to local *adhesive interaction* among cells of all types and *local alignment* of cell types $\sigma = 1, 2$ to the orientation of cell type $\sigma = 3$ (ECM). In order to implement the local adhesive interaction we define

$$\boldsymbol{G}_\sigma\left(\boldsymbol{\eta}_{\mathcal{N}(r)}\right) := \sum_{p=1}^{b} c_p \mathrm{n}_\sigma\left(r + c_p\right),$$

which is the gradient field in the local density of cell type σ (cp. with the adhesive LGCA in sec. 7.2).

The information necessary for the alignment interaction is completely contained in the *director field*

$$\boldsymbol{D}_\sigma\left(\boldsymbol{\eta}_{\mathcal{N}(r)}\right) := \sum_{p=1}^{b} \sum_{i=1}^{b} c_i \eta_{\sigma,i}(r + c_p) = \sum_{p=1}^{b} \boldsymbol{J}\left(\boldsymbol{\eta}_\sigma(r + c_p)\right),$$

specifying the average flux of cells of type σ at the nearest neighbors of node r (cp. with the "swarm LGCA" in sec. 8.2). Here,

$$\boldsymbol{J}\left(\boldsymbol{\eta}_\sigma(r + c_p)\right) = \sum_{i=1}^{b} c_i \eta_{\sigma,i}(r + c_p)$$

denotes the cell flux of type σ at node r.

The probability for transition from $\boldsymbol{\eta}_\sigma(r)$ to $\boldsymbol{\eta}_\sigma^{\mathrm{I}}(r)$ in the presence of $\boldsymbol{G}(\boldsymbol{\eta}_{\mathcal{N}(r)}) = \{\boldsymbol{G}_\sigma\left(\boldsymbol{\eta}_{\mathcal{N}(r)}\right)\}_{1 \le \sigma \le \varsigma}$ and $\boldsymbol{D}(\boldsymbol{\eta}_{\mathcal{N}(r)}) = \{\boldsymbol{D}_\sigma(\boldsymbol{\eta}_{\mathcal{N}(r)})\}_{1 \le \sigma \le \varsigma}$ is given by

$$\begin{aligned} &W\left(\boldsymbol{\eta}_{\mathcal{N}(r)} \to \boldsymbol{\eta}^{\mathrm{I}} | \left(\alpha_{\sigma_1\sigma_2}\right), \gamma_\sigma\right) \\ &= \frac{1}{N} \exp\left(\left\{\sum_{\sigma_1=1}^{2} \sum_{\sigma_2=1}^{2} \alpha_{\sigma_1\sigma_2} \boldsymbol{G}_{\sigma_1}\left(\boldsymbol{\eta}_{\mathcal{N}(r)}\right) \cdot \boldsymbol{J}\left(\boldsymbol{\eta}_{\sigma_2}^{\mathrm{I}}\right)\right\}\right. \\ &\qquad \left. + \left\{\boldsymbol{D}_3\left(\boldsymbol{\eta}_{\mathcal{N}(r)}\right) \cdot \sum_{\sigma=1}^{2} \gamma_\sigma \boldsymbol{J}\left(\boldsymbol{\eta}_\sigma^{\mathrm{I}}\right)\right\}\right) \prod_{\sigma=1}^{3} \delta_{\mathrm{n}_\sigma, \mathrm{n}_\sigma^{\mathrm{I}}}, \end{aligned} \tag{9.1}$$

where the normalization factor $N = N((\mathrm{n}_\sigma)_{1 \le \sigma \le 3}, \boldsymbol{G}, \boldsymbol{D})$ is chosen such that

$$\sum_{\boldsymbol{\eta}^{\mathrm{I}} \in \mathcal{E}} W\left(\boldsymbol{\eta}_{\mathcal{N}(r)} \to \boldsymbol{\eta}^{\mathrm{I}} | \left(\alpha_{\sigma_1\sigma_2}\right), \gamma_\sigma\right) = 1 \qquad \forall\, \boldsymbol{\eta} \in \mathcal{E}.$$

The idea of this definition is that with large probability an output $\boldsymbol{\eta}^{\mathrm{I}}$ results which makes the argument of the exponential maximal. Thereby, the first term of the sum in the exponential is defined in such a way that cells of type σ_2 preferably move in the direction of increasing density of type σ_1 if $\alpha_{\sigma_1\sigma_2} > 0$, modeling adhesion (attraction) between cells (cp. sec. 7.2). Contrarily, if $\alpha_{\sigma_1\sigma_2} < 0$, then cells of type σ_2 predominantly migrate into the direction of decreasing density of type σ_1 mimicking contact inhibition. If $\alpha_{\sigma_1\sigma_2} = 0$, cells of types σ_1 and σ_2 do not interact with each other. The *adhesion matrix* $(\alpha_{\sigma_1\sigma_2})_{1\leq\sigma_1,\sigma_2\leq 3}$ $(\alpha_{\sigma_1\sigma_2} \in \mathbb{R})$ stores the adhesion coefficients of the corresponding cell types.

The second term of the sum models alignment of cell types $\sigma = 1, 2$ to the locally averaged direction of cell type $\sigma = 3$, the "ECM." The interaction rule is designed to minimize the angle between the given, temporarily fixed director field $\boldsymbol{D}_3$ and the postinteraction flux $\boldsymbol{J}\left(\eta_\sigma^{\mathrm{I}}\right)$, provided that $\gamma_\sigma > 0$ for $\sigma = 1, 2$, respectively.

Cell Division. In the biological system, cells differentiate at the edge of the neural crest from embryonic predecessor cells. In the automaton, this process is modeled by allowing in each time step for the creation of new cells along the top rows of the lattice which mimic the neural crest. Only at empty nodes in the top region, pigment cells (types $\sigma = 1, 2$) with no preferential orientation are introduced with equal rates into the system.

9.3 Simulation of Stripe Pattern Formation

In the simulations we neglect larval growth and use lattice sizes (40×20) and average pigment cell densities that correspond to a fully grown larva which is approximately 10 mm in length and which has a "mean settling density" of approximately 30 melanophores/mm^2 in the area between gills and cloaca (Epperlein and Löfberg 1990, p. 59). Accordingly, 4×4 lattice cells correspond to approximately 1 mm^2 of the real larva. Simulations always start with a distribution of the extracellular matrix (cell type $\sigma = 3$), which is associated with a certain density and mean orientation. The matrix distribution is maintained throughout the simulation. Furthermore, cells of types $\sigma = 1, 2$ are initially randomly distributed in the uppermost 10 rows of the lattice corresponding to the neural crest region. We distinguish hybrid open (for the top border, the "neural crest") and reflecting (for the bottom, left, and right borders) boundary conditions. In particular, reflecting boundary conditions are implemented by turning those cells that try to transcend the borderline around 180 degrees. If the corresponding channel is already occupied, the procedure leads to a loss of the cell that was not able to turn. Consequently, cell density at the corresponding node decreases. Note that the particular boundary conditions due to the cell flow induced by permanent creation of cells at one lattice side are an important precondition for pattern formation. If we start out from a homogeneous random initial condition, only patch, but no stripe formation is observed.

Adhesion and Contact Inhibition. We start our investigations by analyzing the adhesive impact. Simulations are performed with a fixed contact guidance (with respect to the ECM). Interestingly, "homotypic interactions" seem to be less important than "heterotypic interactions" (fig. 9.3).

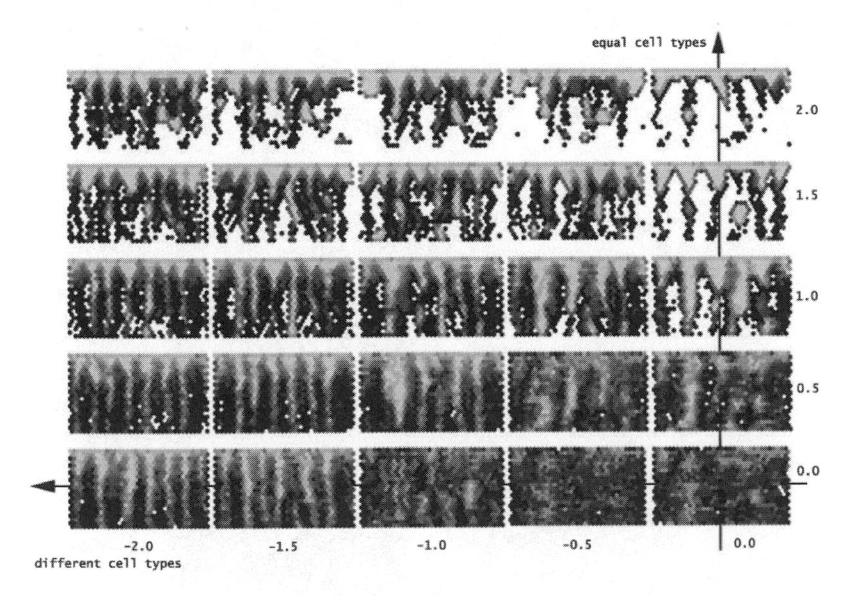

Figure 9.3. Phase diagram for the influence of the adhesive interaction. Melanophores and xanthophores are marked by different levels of gray. Adhesivity of equal cell types (y-axis) as well as repulsion between different types (x-axis) are varied. Parameters: contact guidance $\gamma_1 = \gamma_2 = 1.5$, density of ECM $\bar{\rho}_3 = 0.5$, configurations are shown after 50 time steps.

In particular, in the corresponding phase diagram we observe that contact inhibition of different cell types ($\alpha_{\sigma_1\sigma_2} < 0$) is a necessary condition for stripe pattern formation, while there can be zero adhesivity of equal cell types simultaneously. Thereby, the basic pattern formation always occurs in the first 5 to 10 time steps (fig. 9.4).

The width of the stripes seems to correlate with the interaction range (approximately 3–5 cells in the model), which is close to the stripe width of real salamanders (fig. 1.6). The stripe number (5–6) matches that of the real larvae. Lateral extension of the lattice corresponds to an extension of larval length and leads to an increase of the total number of stripes, while the lateral extension of individual stripes remains unchanged.

Contact Guidance and the Extracellular Matrix. Stripe pattern formation only occurs if, simultaneously, contact guidance is considered, i.e., $\gamma_k > 0, k = 1, 2$ (fig. 9.4). We examined the effect of the contact guidance by starting with a fixed adhesivity matrix (for which stripe patterns are observed) and varying the contact guidance specified by the parameters γ_k, $k = 1, 2$. For large values of $\gamma_1 = \gamma_2$ the ECM directionality dominates over the adhesive cell–cell interactions, which implies a suppression of stripe pattern formation (cp. the simulations in the phase diagram for $\gamma_1 = \gamma_2 > 3$; fig. 9.5). If, contrarily, interaction is restricted to adhesivity (i.e., $\gamma_\sigma = 0, \sigma = 1, 2$) there is still rudimentary stripe formation (cp. the simulations

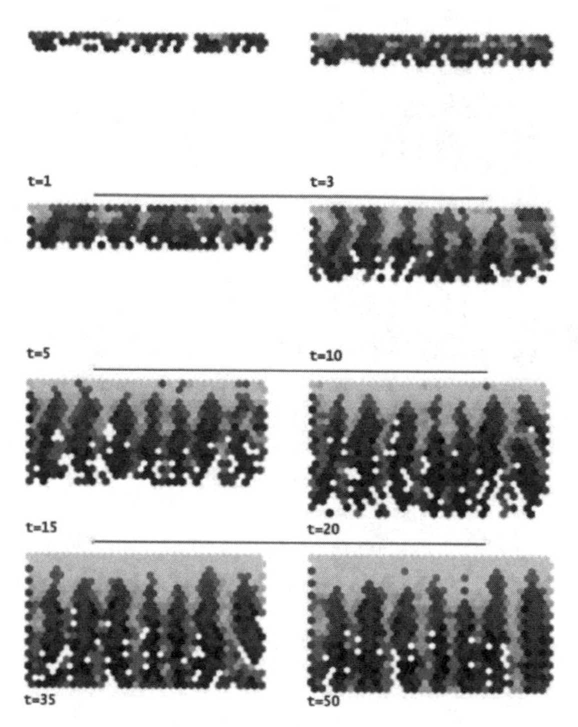

Figure 9.4. Stripe pattern formation in simulations of the LGCA with adhesive and orientational interaction. Melanophores and xanthophores are marked by different levels of gray and repell each other ($\alpha_{12} = \alpha_{21} = -2$). Parameters: $\alpha_{11} = \alpha_{22} = 0.5$, contact guidance $\gamma_1 = \gamma_2 = 1.5$, density of ECM $\bar{\rho}_3 = 0.25$.

in the phase diagram for $\gamma_\sigma = 0$; fig. 9.5). Because of the missing "orientation guidance," cell sorting occurs but, subsequently, cells clump together in the vicinity of their birth sites and large parts of the lattice will remain unoccupied. A similar result is obtained from simulations in which the concentration of the ECM has been varied. A "medium concentration" of ECM is required to obtain persistent stripe pattern formation (fig. 9.6). "Isotropic results" occur for average densities $\bar{\rho}_3 > 1/b$. For other values ($\bar{\rho}_3 \ll 1/b$ or $\bar{\rho}_3 \gg 1/b$) holes arise in the course of the temporal development. We conclude that a particular level of directionality of the ECM as a direction-giving cue and an intermediate ECM density are necessary for persistent stripe formation in the cellular automaton (CA) framework.

9.4 Development and Evolutionary Change

We have shown that a CA model based on cell–cell interactions and migration allows to analyze stripe pattern formation as observed, e.g., in larvae of the axolotl

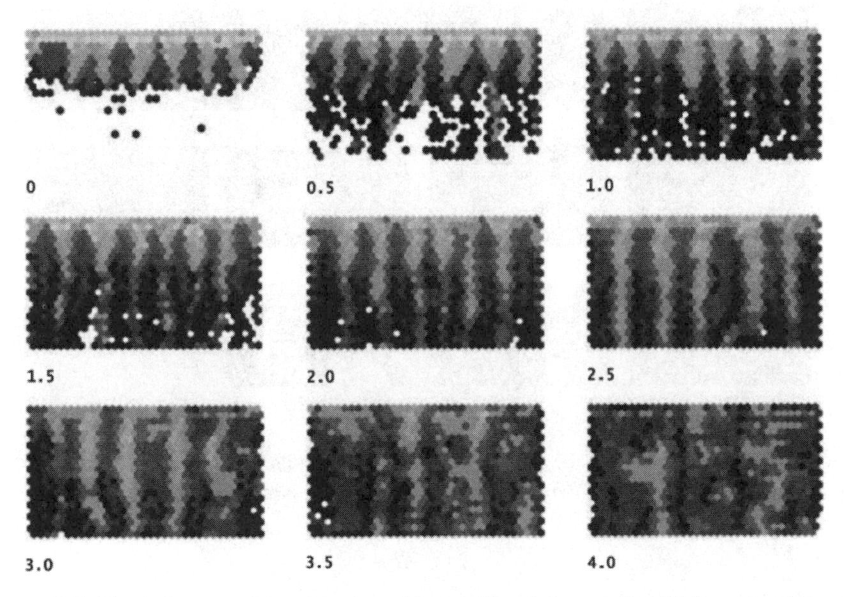

Figure 9.5. The influence of the contact guidance. Simulations (after 50 time steps) for different contact guidances ($\gamma_1 = \gamma_2$) are shown. Parameters: density of ECM $\bar{\rho}_3 = 0.25$, $\alpha_{12} = \alpha_{21} = -2, \alpha_{11} = \alpha_{22} = 0.5$.

Ambystoma mexicanum. In particular, stripe width and stripe number are remarkably similar in both the real larvae and the corresponding simulation pattern. Furthermore, we have demonstrated that the interplay of contact inhibition (of different pigment cell types) and contact guidance (along the direction imposed by the ECM) is essential for pattern formation. None of these processes alone is pattern-competent in itself. This has interesting (and testable) biological consequences: Do different pigment cell types repell each other in the biological system as well? And how strong is contact guidance imposed by the ECM in salamander larvae? Does it vary in different salamander species and/or different phases of the life cycle? Answers will provide evidence for the applicability of the proposed automaton model. Alternative model candidates accounting for phenomenologically similar patterns are reaction-diffusion drift models, but these are based on nonlocal cell–medium–cell interactions. It is the advantage of the CA introduced here that pattern formation can be deduced from local cell-cell and cell–ECM interactions, which agrees with experimental results (Olsson and Löfberg 1993).

Another interesting question is the connection of the observed ontogenetic patterns with their phylogenetic evolution. Is it possible to derive a pedigree based on morphogenetic bifurcation parameters (fig. 9.1)? Thereby, branching in the phylogenetic tree might be correlated with critical morphogenetic parameters inducing

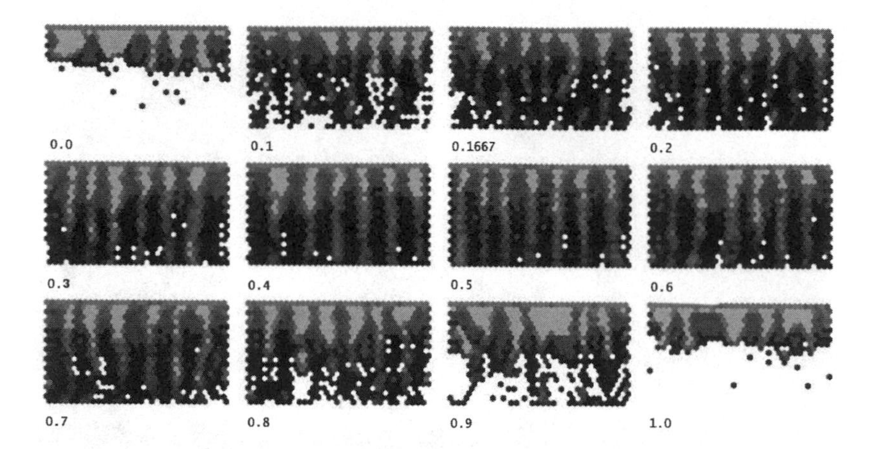

Figure 9.6. The influence of the extracellular matrix. Simulations (after 50 time steps) for different densities of the ECM ($\bar{\rho}_3$) are shown. Parameters: contact guidance $\gamma_1 = \gamma_2 = 1.5$, $\alpha_{12} = \alpha_{21} = -2, \alpha_{11} = \alpha_{22} = 0.5$.

certain morphogenetic phase transitions. Promising candidates for this kind of analysis are the contact guidance and adhesion parameters in the proposed model which have already proven their "critical potential" (fig. 9.5). So far all parameters have been considered constant in the simulations but might temporarily vary, mimicking heterochronic principles. Heterochrony assumes that changes in the relative timing of developmental processes could prove an important ontogenetic basis for phylogenetic change (Gould 1977).

Summary. A LGCA model for pigment cell pattern formation has been introduced in this chapter. Simulations provide hypotheses for stripe pattern formation in larval salamanders. An important motivation for using a CA model has been that it is feasible to analyze certain special cases analytically by restriction to approximative results (e.g., the mean-field assumption). It should be possible, e.g., to determine critical parameters along the lines of the analysis presented in preceding sections (7.2 and 8.2). In conclusion, the CA model based on the effects of adhesion and contact guidance for pattern formation can serve as a basic framework for future research focusing on the interplay of the influences of genetics, evolution, and development (cp. also the discussion in sec. 12.6).

9.5 Further Research Projects

1. **Asynchronous CA:** Define an asynchronous CA modeling the interaction of pigment cells with themselves and the ECM.

2. **Horizontal/intermediate patterns:** What are possible mechanisms that could explain the formation of horizontal and intermediate stripe patterns (cp. fig. 9.1)? Define corresponding automaton models.
3. **Adult pigment pattern formation:** Typically, the larvae significantly grow during the formation of adult pigment patterns. How could the growth of the larvae be considered in corresponding CA models?
4. **Pigment pattern formation in other organisms:** Discuss possible mechanisms of pigment pattern formation in other organisms, e.g., humans. What would corresponding CA rules look like?
5. **Melanoma:** Cutaneous melanoma are malignant degeneracies of melanocytes in the epidermis. Most frequently, superficial spreading melanoma occur that spread exclusively within the epidermis. Discuss possible mechanisms of melanoma formation. Might it be possible to construct a CA model for the spread of malignant melanocytes?
6. **Modeling:** During tumor invasion, it is often observed that tumor cells degrade surrounding ECM. Construct and analyze a simple model of tumor invasion in which tumor cells orient themselves according to the ECM and simultaneously degrade the matrix components. A traveling wave of tumor cells should appear. How can the speed of the invasion front be determined?

10

Tissue and Tumor Development

"All the extremely diverse structures of multicellular organisms may be traced back to the few modi operandi of cell growth, cell evanescence, cell division, cell migration, active cell formation, cell elimination and the quantitative metamorphosis of cells; certainly, in appearance at least, a very simple derivation. But the infinitely more difficult problem remains not only to ascertain the special role that each of these processes performs in the individual structure, but also to decompose these complex components themselves into more and more subordinate components. ..."

(W. Roux, quoted from Bard, 1990)

Tissues can be regarded as cell aggregates of one or more cell types developing in the course of embryonic development. Aggregates occur as different topological shapes, in particular layers, encapsulation, or globular patterns. What are the processes governing tissue pattern formation? One can distinguish "biological," chemical, and physical processes. Biological processes are, e.g., reproduction and cell death (apoptosis); chemical processes involve signaling and physical interactions are mediated predominantly by adhesive forces. All these processes are intertwined in producing particular tissue shapes.

It is important to analyze each of the individual interaction processes separately to get an understanding of their pattern-forming potential. In previous chapters we have already dealt with particular aspects of tissue formation as cell proliferation and adhesion (cp. chs. 6, 7, and 9). In the following, we introduce lattice-gas cellular automaton (LGCA) models, especially for contact inhibition and chemotactic interactions. Finally, we present a simple model of avascular tumor growth based on combinations of nearly all elementary interaction modules introduced in this book.

Besides the suggestion of cellular automaton (CA) models for tissue growth, the intention of the following chapter is to show how interaction modules for specific morphogenetic applications can be constructed. Development of further modules and combinations with respect to other morphogenetic questions as well as the mathematical analysis of the automata is open for further research (see also sec. 10.5 on "Further Research Projects").

10.1 Modeling Contact Inhibition of Movement in Lattice-Gas Cellular Automata

Contact inhibition of movement is one of the most important interactions between two cells and occurs when the leading edges of two migrating cells come into contact. In many cell types the encounter causes an immediate paralysis of the leading edge of each cell. For example, two fibroblasts that collide in culture cease generating mikrospikes and lamellipodia in the region of contact and begin to produce them elsewhere, so that eventually the cells move away from each other along new trajectories. The contact inhibition response involves rapid changes in the actin-based cortical cytoskeleton. Integrin receptors play a central role in cell migration through their roles as adhesive receptors for other cells and extracellular matrix (ECM) components. It is known that integrin and cadherin receptors coordinately regulate contact-mediated inhibition of cell migration (Huttenlocher et al. 1998). Contact inhibition of movement should not be confused with the contact inhibition of cell division observed in cultured cells, which have proliferated until they cover the entire surface of a culture dish. Contact inhibition of movement is an essential feature of wound healing in animals (and humans). Sheets of epithelial cells at the margin of a wound move rapidly out over the wounded area by extending lamellipodia; this movement ceases as soon as cells from different margins make contact across the gap created by the wound. For example, myoblasts ectopically expressing alpha5 integrin (alpha5 myoblasts) move normally when not in contact, but upon contact, they show inhibition of migration and motile activity (i.e., extension and retraction of membrane protrusions). As a consequence, these cells tend to grow in aggregates and do not migrate to close a wound (Huttenlocher et al. 1998). Contact inhibition of movement may also contribute to the selective bundling of axons in the developing nervous system (Oakley and Tosney 1993).

Here we define a LGCA and analyze the implications of contact inhibition. Due to contact inhibition cells tend to move to low-density neighboring nodes. A two-dimensional lattice with coordination number $b = 4$, one additional rest channel $\beta = 1$, i.e., $\tilde{b} = 5$, and von Neumann interaction neighborhood $\mathcal{N}_b^I(r)$ are assumed. The complete dynamics of the contact inhibition LGCA is defined by repeated application of the contact inhibition interaction (`I`) and propagation (`P`) operator, respectively.

Contact Inhibition Interaction Rule. In this model the number of cells at each node remains constant during interaction. The local contact inhibition interaction comprises cell reorientation according to a neighborhood-dependent rule:

(i) One cell always occupies the rest channel, if the outer interaction neighborhood contains at least one cell, i.e., $\mathcal{N}_{bo}^I \neq \emptyset$. Otherwise all cells are redistributed randomly on the channels;

(ii) the remaining cells are positioned at channels that point to low-density neighboring nodes (contact inhibition).

Finally, a propagation step follows the interaction step.

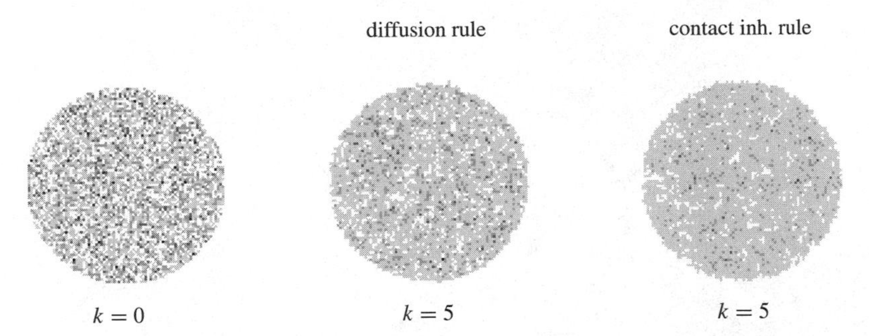

Figure 10.1. Comparison of the *contact inhibition* with a *diffusive* LGCA rules (cp. ch. 5). Snapshots of the initial spatial cell distribution and distributions after five time steps are shown. Parameters: average cell density $\bar{\rho} = 0.2$; initial circle with a radius of 20 (nodes) on a square lattice ($L = 100$). The gray level indicates cell number.

Simulations. We performed simulations on a square lattice with 100×100 lattice nodes, and started with an initial distribution of cells in a circle of radius 20 (nodes) in the lattice center. The averaged cell density at each node is $\bar{\rho} = 0.2$. Figs. 10.1 and 10.2 compare contact inhibition LGCA simulations with simulations of the diffusive LGCA (cp. sec. 5.4). From the snapshots shown in fig. 10.1 it can be seen that after five time steps, cells following the contact inhibition rule are more homogeneously distributed in space than cells following the diffusion rule.

The graphs in fig. 10.2 show that the initially present differences in cell densities (indicated by different gray levels) are reduced faster when applying the contact inhibition instead of the diffusive rule: the fraction of nodes with exactly one cell increases rapidly, while the fraction of nodes with more than one cell decreases. In total, the empty space, i.e., nodes with no cells ($\varrho = 0$), is filled faster using the contact inhibition rule than the diffusive rule. In this experiment, the differences between diffusive and contact inhibition LGCA rule are visible only at a short time scale. The effect of the contact inhibition rule is rather local. For long time scales or low initial differences in cell density, contact inhibition and diffusive LGCA are almost indistinguishable.

In the next section, we extend the LGCA model of contact inhibition to a simple model of tissue growth, in which the number of cells, which are now tissue cells, is no longer conserved. Thus, local fluctuations in the cell density emerge. The application of the contact inhibition rule implies a decrease of local density.

10.2 Tissue Growth

So far, we studied growth models without cell loss (cp. ch. 6). Including cell loss processes may lead to the following question: Does the relationship between growth and loss rates play an important role in the LGCA growth dynamics? For example, what effect does this relationship have on the spread of the aggregate and the relative growth rate? In order to illustrate this point, in this section we introduce a LGCA

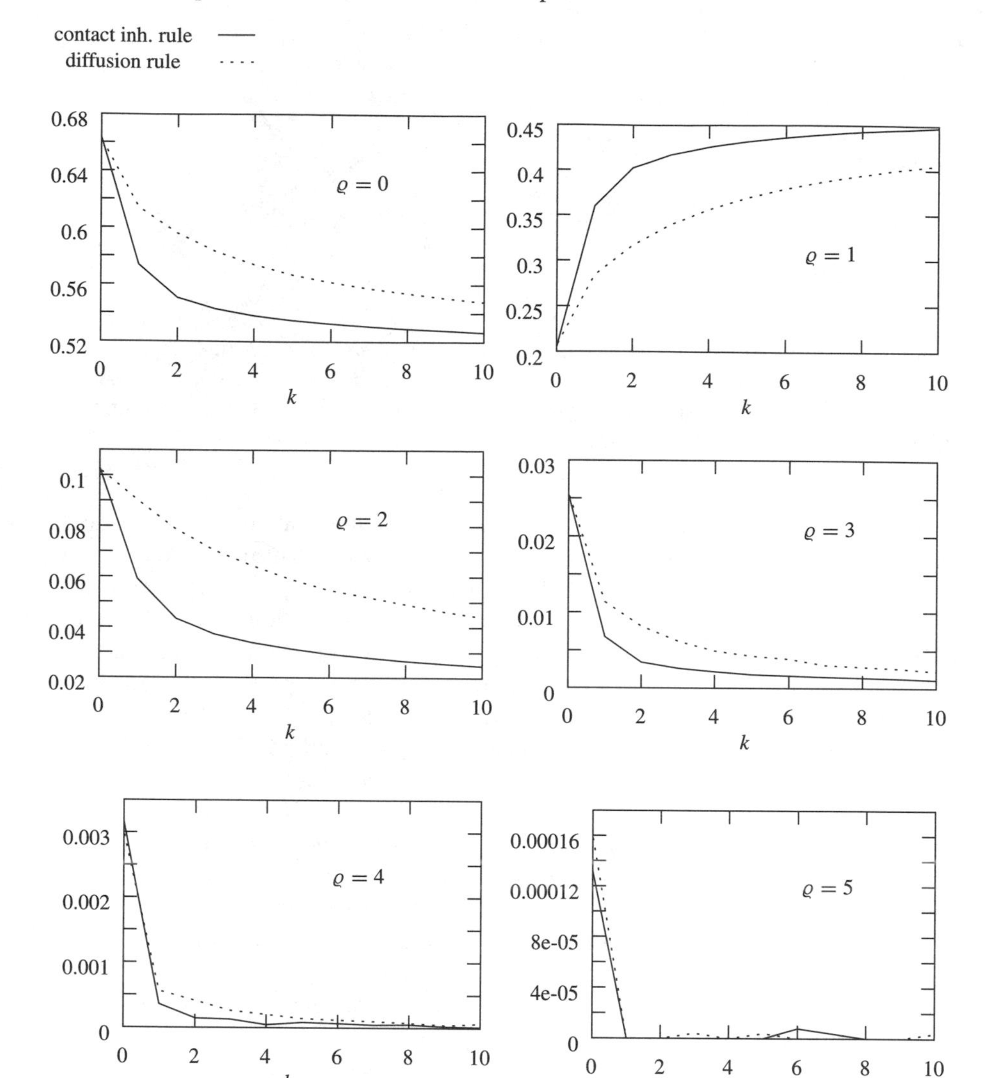

Figure 10.2. Comparison of the *contact inhibition* LGCA rule with a *diffusive* LGCA rule (cp. ch. 5). ϱ denotes the number of cells at a node. The vertical axes represent the average fraction of nodes that are occupied by ϱ cells. Parameters: average cell density $\bar{\rho} = 0.2$; initial circle with a radius of 20 (nodes) on a square lattice ($L = 100$).

model for *in vitro* tissue growth. Cells now represent tissue cells that either proliferate, stay quiescent, or die (programmed cell death = apoptosis).

As in the contact inhibition LGCA, we take a two-dimensional lattice with $b = 4$ and $\tilde{b} = 5$ and the von Neumann interaction neighborhood into account. The interaction step is composed of two parts.

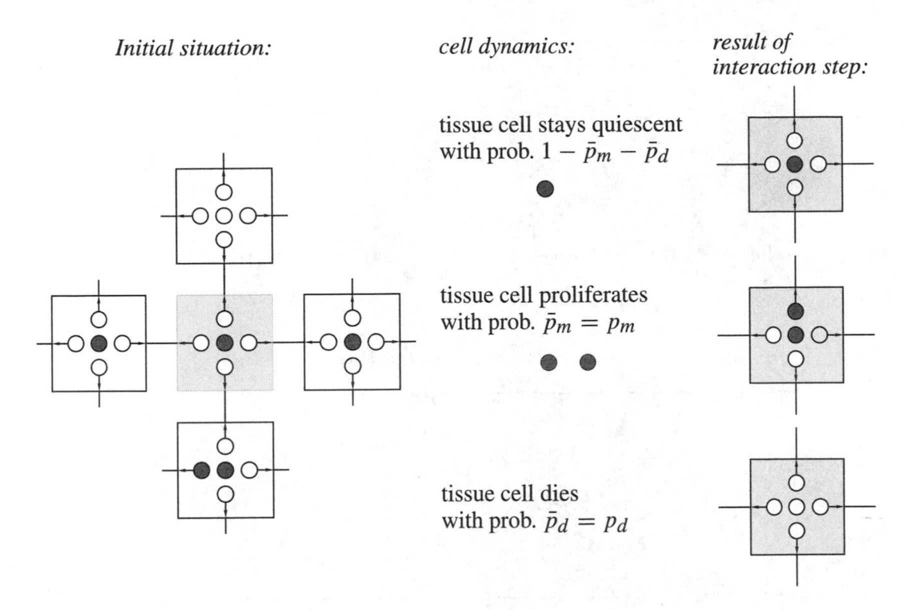

Figure 10.3. Example of an interaction step for tissue growth. Gray dots denote tissue cells, while white dots denote empty channels. Each configuration on the right side is a possible outcome of the interaction step applied to the central gray node (on the left side).

Tissue Growth Interaction Rule. Tissue cell dynamics is modeled in a probabilistic way. For each node, rates for mitosis, quiescence, and apoptosis are determined. Afterwards, each tissue cell at the node either divides (if unoccupied channels exist), remains quiescent, or dies. New cells are introduced randomly at empty nodes. We assume that all rates are linearly density-dependent, such that the rate of cell division, $\bar{p}_m(r)$, decreases with the number of tissue cells present at a node, i.e., $\mathrm{n}(r)$, while the rate of cell death, $\bar{p}_d(r)$, increases. Hence, $\bar{p}_m(r) := p_m/\mathrm{n}(r)$ and $\bar{p}_d(r) := p_d\,\mathrm{n}(r)$.

In the second part all tissue cells at a node are redistributed on the channels according to the contact inhibition interaction rule (cp. p. 186).

Fig. 10.3 shows an example for this interaction step. Since, according to the contact inhibition rule, the rest channel always gains a tissue cell if the outer interaction neighborhood is not empty, a compound of tissue cells can grow; i.e., cells "stick together" (adhesion). Furthermore, the spatial scale of the lattice (i.e., the area of a lattice node) is chosen such that contact inhibition of movement is induced if more than one tissue cell is present at a node. Therefore, cells are moving towards neighborhoods with low cell density; i.e., cells take a line of least resistance. Finally, a propagation step follows the interaction step.

Simulations. We performed simulations with $p_m - p_d = 0.04$ [1/h], starting from an initial aggregate disc of 45 tissue cells. Within this disc each lattice node is occupied by one tissue cell residing on the rest channel. The time scale for k is set to *hour*. It is observed that the growth process sensitively depends on the magnitudes of mitosis

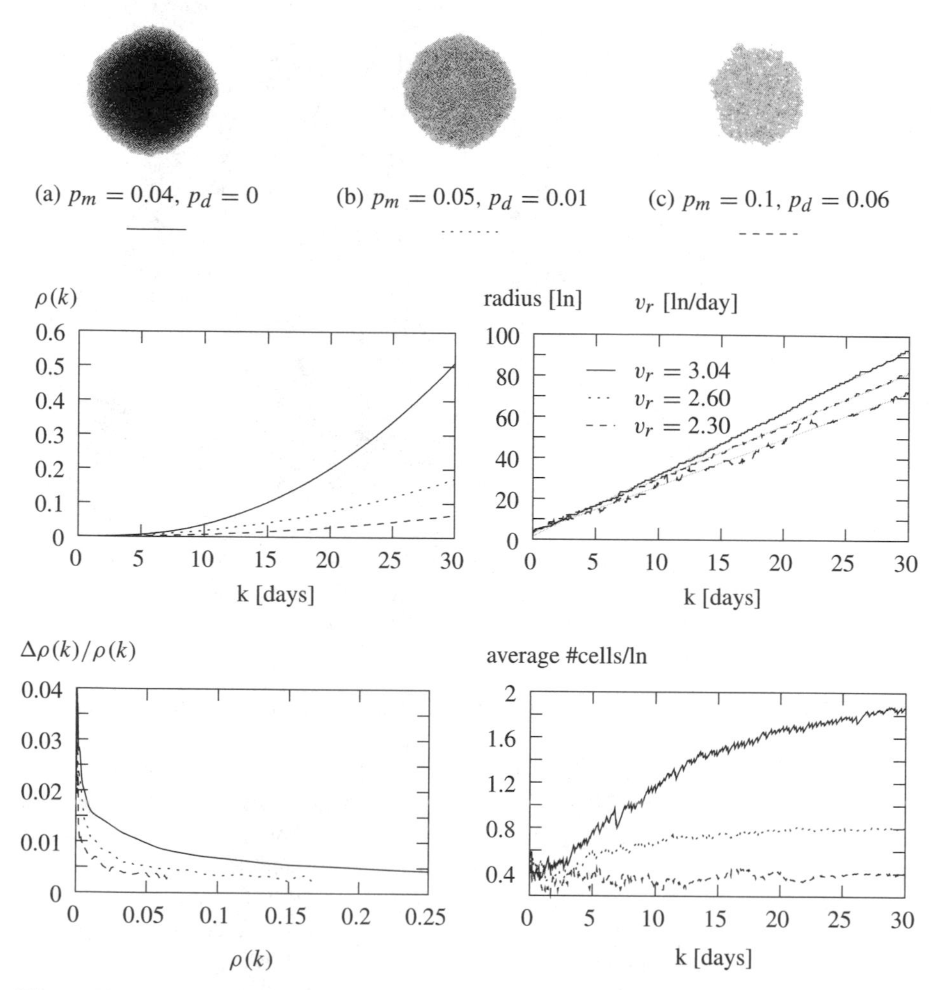

Figure 10.4. Simulations of tissue growth with $p_m - p_d = 0.04$ [1/h]. This corresponds to a density-independent doubling time of ca. 16 hours for each individual cell. The data on the relative growth rates are smoothed (interpolated); here, "ln" denotes lattice node. Parameters: $L = 200 \times 200$ and $\rho(0) = 0.000225$.

and apoptosis rates (fig. 10.4). While the average number of tissue cells per lattice node[60] for the parameter set (a) is approximately 2, it is approximately 0.2 for parameter set (c). The higher density in compound (a) results in a faster spread due to contact inhibition ($v_r = 3.04$ [ln/day]; ln: lattice node) than the spread of compound

[60] The average number of cells per lattice node is calculated as average #cells/ln $= (\rho(k)\, 200^2 \cdot 5)/(2\pi \,\text{radius}(k)^2)$. Note, that the *radius* is determined as the maximal distance from the center of the lattice, for which tissue cells exist, ln: lattice node).

(c) ($v_r = 2.3$ [ln/day]). Clearly, in compound (c) local regions of low density are filled with tissue cells moving into this area.

10.3 A Cellular Automaton Model for Chemotaxis

Chemotaxis describes the dependence of individual cell movement on a chemical signal gradient field (see the glossary by Alt/Tranquillo in Alt and Hoffmann 1990). Accordingly, spatial patterns at the level of cells and chemical signals can be distinguished. Chemotactic patterns result from the coupling of two different spatio-temporal scales at the cell and the molecular level, respectively. Morphogenesis of the cellular slime mold *Dictyostelium discoideum* has developed as an experimental model system for chemotactic dynamics, particularly governing the formation of aggregation centers and the formation of slugs and stalks (Marée and Hogeweg 2001). In the case of the slime mold, the major signaling molecule has been identified as cyclic monophosphate (cAMP, see Chang 1968; Chen, Insall, and Devreotes 1996).

The first mathematical model of chemotactic pattern formation was proposed in the early 1970s and was formulated as partial differential equation system (Keller and Segel 1970, 1971a, 1971b). Later, hybrid models distinguishing discrete cells and continuous signal concentrations were suggested (e.g., Dallon, Othmer, v. Oss, Panfilov, Hogeweg, Höfer, and Maini 1997; Savill and Hogeweg 1997). So far, hybrid models have been solely studied by means of simulations (Weimar 2001). Here, we present a hybrid LGCA that allows us to explicitly analyze spatial pattern formation depending particularly on the chemotactic sensitivity of cells to the gradient field (see suggestions for future research projects)[61]. The analysis is similar to the adhesive interaction model (cp. ch. 7) since in both cases gradients—in cellular density and chemical concentration, respectively—are governing the interaction (see sec. 10.5, "Further Research Projects").

Our model is based on motile cells (e.g., myxobacteria) moving on a two-dimensional lattice ($b = 4$) with one rest channel ($\tilde{b} = 5$). We assume that $\mathcal{L} = L_1 \cdot L_2 = L^2$. The cells are able to produce and secrete a substance. In the absence of this substance, the cells perform a random walk in the lattice; otherwise they are sensitive to the substance such that they move towards regions of higher concentrations of the signal (c_{sig}) in the outer von Neumann interaction neighborhood. The signal decays continuously and diffuses on the lattice.

Chemotactic Interaction Rule. The interaction rule of the LGCA for chemotaxis comprises the following processes in sequential order:

1. the decay of chemical substance;
2. the secretion, with a given probablity, of a fixed amount of chemical substance;
3. the diffusion of chemical substance; and

[61] No confusion should arise with hybrid CA defined by nonuniform rules; i.e., different rules are applied to the cells of the lattice (Chaudhuri et al. 1997).

4. the redistribution of cells on node channels depending on the local signal concentration.

The interaction step is followed by a propagation step.

Diffusion of the signal (ad 3). In order to model diffusion of the signal, a method of "moving averages" is applied as a discrete approximation of the continuous diffusion equation. This method allows an efficient implementation of the diffusion process even for large lattices (Weimar, Tyson, and Watson 1992).

Redistribution of cells (ad 4). In order to determine the probability for transition from the preinteraction state $\boldsymbol{\eta}(r) = (\eta_1, \ldots, \eta_5)$ to the post-interaction state $\boldsymbol{\eta}^{\mathrm{I}}(r) = \left(\eta_1^{\mathrm{I}}, \ldots, \eta_5^{\mathrm{I}}\right)$, we define rules for the occupation of the rest channel (c_5) and the velocity channels ($c_1, \ldots, c_4$) separately. The rest channel of node r gains a cell with probability $\mathrm{n}(r)/5$, where $\mathrm{n}(r) = \sum_{i=1}^{5} \eta_i(r)$ is the number of cells at node r. Let $\hat{\mathrm{n}}(r) \leq \mathrm{n}(r)$ be the number of the remaining cells at r.

For the chemotactic response to the local signal concentration of the remaining cells, we define the **normalized signal gradient field**

$$\bar{\boldsymbol{G}}_{\mathrm{sig}}\left(\boldsymbol{\eta}_{\mathcal{N}(r)}\right) := \sum_{p=1}^{4} c_p \quad \mathrm{on}\left(r + c_p\right), \tag{10.1}$$

where $\mathrm{on}(r + c_p)$ is the *order number* of the signal concentration $c_{\mathrm{sig}}(r + c_p)$. It is determined by ordering the four neighboring nodes of r according to their signal concentration,[62] such that node r' with maximal signal concentration receives $\mathrm{on}(r') = 4$, and node r'' with the lowest concentration $\mathrm{on}(r'') = 1$. Note that we consider order numbers rather than signal concentrations in the computation of the gradient field, in order to normalize absolute differences of neighboring signal concentrations (which are usually very small) to relative differences (which are at least 1 after normalization).

As in the adhesive LGCA (cp. sec. 7.2) the probability for transition from $\hat{\boldsymbol{\eta}}(r) := (\eta_1, \ldots, \eta_4)$ to $\hat{\boldsymbol{\eta}}^{\mathrm{I}}(r) := \left(\eta_1^{\mathrm{I}}, \ldots, \eta_4^{\mathrm{I}}\right)$ in the presence of $\bar{\boldsymbol{G}}_{\mathrm{sig}}\left(\boldsymbol{\eta}_{\mathcal{N}(r)}\right)$ is given by

$$\boxed{W\left(\hat{\boldsymbol{\eta}}_{\mathcal{N}(r)} \to \hat{\boldsymbol{\eta}}^{\mathrm{I}} | \alpha\right) = \frac{1}{Z} \exp\left(\alpha\, \bar{\boldsymbol{G}}_{\mathrm{sig}}\left(\hat{\boldsymbol{\eta}}_{\mathcal{N}(r)}\right) \cdot \boldsymbol{J}\left(\hat{\boldsymbol{\eta}}^{\mathrm{I}}\right)\right) \delta_{\hat{\mathrm{n}}, \hat{\mathrm{n}}^{\mathrm{I}}},} \tag{10.2}$$

where the normalization factor $Z = Z(\hat{\mathrm{n}}, \bar{\boldsymbol{G}}_{\mathrm{sig}})$ is chosen such that

$$\sum_{\hat{\boldsymbol{\eta}}^{\mathrm{I}} \in \{0,1\}^4} W\left(\hat{\boldsymbol{\eta}}_{\mathcal{N}(r)} \to \hat{\boldsymbol{\eta}}^{\mathrm{I}} | \alpha\right) = 1 \qquad \forall\, \hat{\boldsymbol{\eta}} \in \{0, 1\}^4$$

and the cell flux is defined as

$$\boldsymbol{J}\left(\hat{\boldsymbol{\eta}}(r)\right) = \sum_{i=1}^{4} c_i \eta_i(r);$$

α is the *chemotactic sensitivity* of the cells. Fig. 10.5 gives an example of the transition probability.

[62] If, e.g., $c_{\mathrm{sig}}(r + c_3) > c_{\mathrm{sig}}(r + c_4) > c_{\mathrm{sig}}(r + c_2) > c_{\mathrm{sig}}(r + c_1)$, then $\mathrm{on}(r + c_3) = 4$, $\mathrm{on}(r + c_4) = 3$, $\mathrm{on}(r + c_2) = 2$, $\mathrm{on}(r + c_1) = 1$.

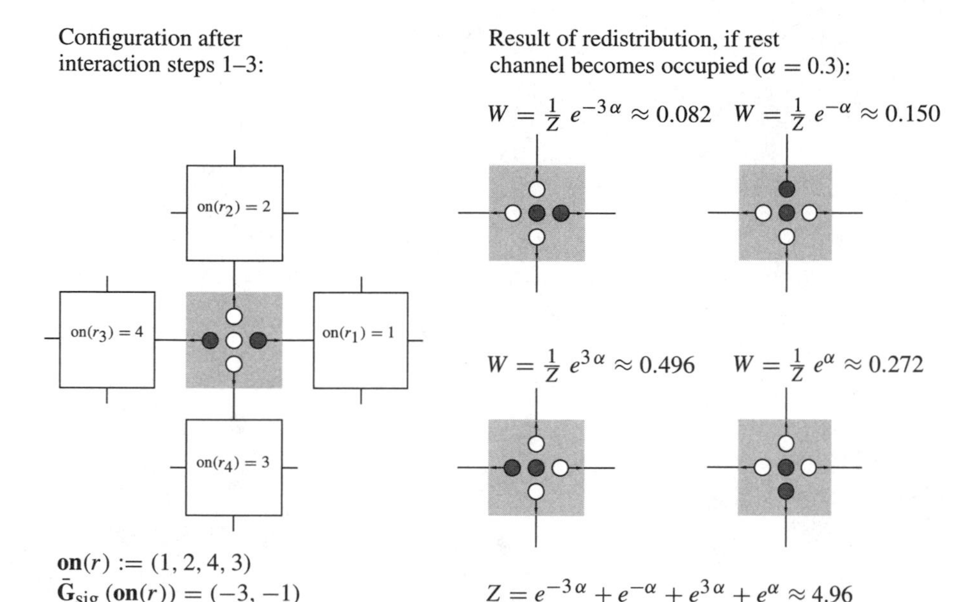

Figure 10.5. Example of cell redistribution on a lattice node. The rest channel becomes occupied with probability $2/5$. The signal concentration in the interaction neighborhood can be ordered such that $c_{\text{sig}}(r_3) > c_{\text{sig}}(r_4) > c_{\text{sig}}(r_2) > c_{\text{sig}}(r_1)$ (cp. *chemotactic interaction rule*).

Simulations. We performed simulations on a 100×100 square lattice with a random spatial distribution of cells of density 0.05. Fig. 10.6 shows resulting spatial cell distributions for chemotactic sensitivities $\alpha = 0.1$ and $\alpha = 0.3$ and different signal diffusion constants.

Spatial patterns are observed for both sensitivities, but on different time scales. Clearly, a high sensitivity enhances the formation of aggregation clusters (cp. fig. 10.6, $k = 100$). If simultaneously the diffusion constant of the chemical signal is increased, fewer but larger clusters appear.

Note that the presented chemotaxis LGCA is an *extension* of the adhesive LGCA (introduced in ch. 7). With a particular choice of parameters, the chemotactic interaction rule reduces to the adhesive interaction rule: namely in this case, that each cell secretes equal amounts of chemical substance, and that the substance *decays instantaneously* (no diffusion!), the order number is proportional to the number of cells in the outer interaction neighborhood (i.e., $\bar{\boldsymbol{G}}_{\text{sig}}\left(\boldsymbol{\eta}_{\mathcal{N}(r)}\right) \propto \boldsymbol{G}\left(\boldsymbol{\eta}_{\mathcal{N}(r)}\right)$, cp. (7.1), and (10.1)), and hence both models are determined by identical transition probabilities.

An *incomplete decay* of the diffusing signal introduces on one hand a memory effect (the information about local cell densities is "stored" in the signal concentration for longer times) and on the other hand an indirect interaction between cells of a larger spatial range than the interaction neighborhood (transmission of the information about local cell densities by the diffusion of the signal).

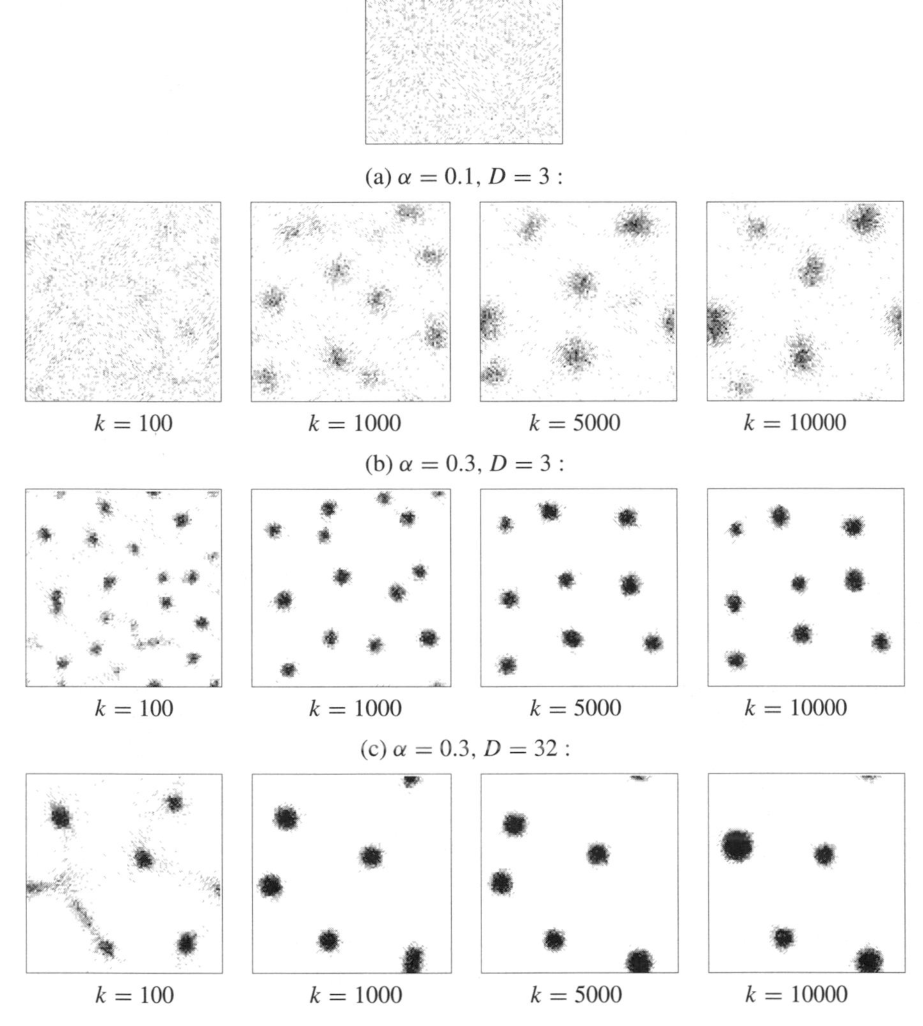

Figure 10.6. Chemotactic pattern formation in the LGCA model. Gray levels represent cell densities (dark: high cell density). Parameters: cell density: 0.05, a cell produces a signal of amount 1 with probability 0.01, continuous signal decay of 70%, diffusion coefficient of signal: $D[\Delta l^2/\Delta k]$ (Δl: unscaled unit for length of lattice node, Δk unscaled unit for time step), chemotactic sensitivity: α, lattice size: $L = 100$.

It is possible to extract the patterns in the chemotactic CA from a stability analysis of the automaton Boltzmann equation (see "Further Research Projects" at the end of this chapter).

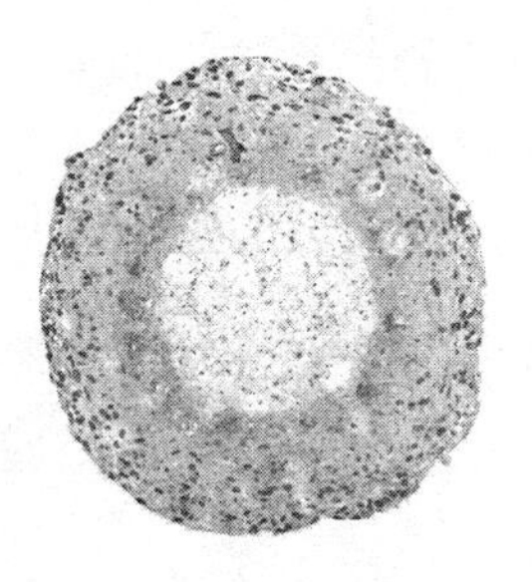

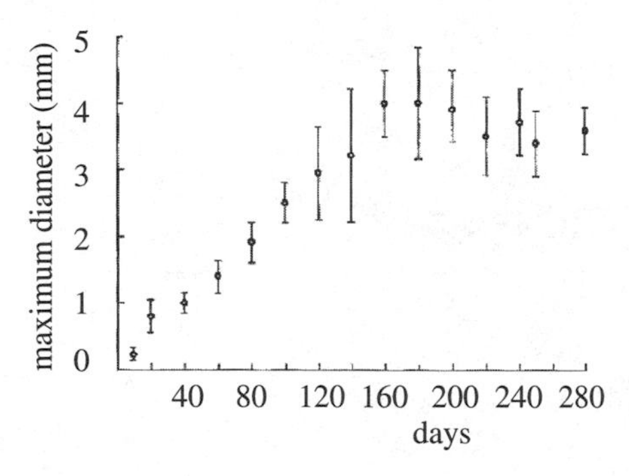

Figure 10.7. Folkman and Hochberg (1973) studied the growth of isolated spheroids from V-79 Chinese hamster lung cells, repeatedly transferred to new medium. Left: a cross section of a V-79 spheroid is shown, 1.0 mm in diameter and 20 days old. Viable cells are labeled with [^{3}H]thymidine. Right: mean diameter and standard deviation of 70 isolated spheroids of V-79 cells.

"Growth for the sake of growth is the ideology of the cancer cell."
(Edward Abbey)

10.4 Avascular Tumor Growth[4]

Tumor growth always starts from a small number of malignantly proliferating cells, the tumor cells. The initial avascular growth phase can be studied *in vitro* by means of multicellular spheroids. In a typical experiment, tumor cells are grown in culture and are repeatedly exposed to fresh nutrient solution. Interestingly, after an initial exponential growth phase which implies tumor expansion, growth saturation is observed even in the presence of a periodically applied nutrient supply (Folkman and Hochberg 1973). A section of the tumor spheroid shows a layered structure: a core zone composed mainly of necrotic material is surrounded by a thin layer of quiescent tumor cells and an outer ring of proliferating tumor cells (fig. 10.7). A better understanding of the processes which are responsible for the growth of a layered and saturating tumor is crucial. It has been realized that mathematical modeling can contribute to a better understanding of tumor growth (Gatenby and Maini 2003). In particular, various models have been suggested for the avascular growth phase (Drasdo et al. 2004). We show here with a hybrid CA model that the layered pattern can be explained solely by the self-organized growth of an initially small number of tumor cells. A better knowledge of the spatio-temporal tumor dynamics should allow one to design treatments that transfer a growing tumor into a saturated (nongrowing and undangerous) regime by means of experimentally tractable parameter shifts.

A realistic model of avascular solid tumor growth should encompass mitosis, apoptosis, and necrosis, processes which are particularly dependent on growth factors and nutrient concentrations (cp. fig. 10.8). Growth inhibitors play an impor-

[63] Parts of this section were published in Dormann and Deutsch (2002).

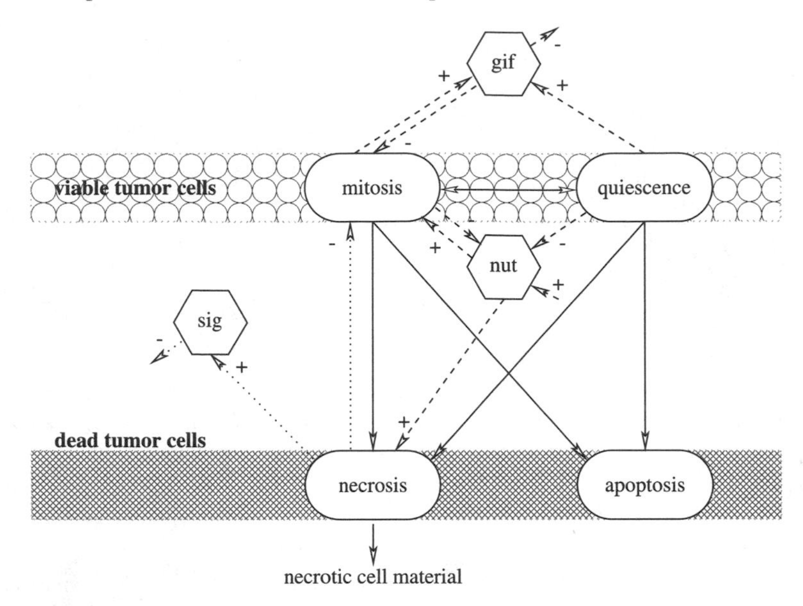

Figure 10.8. Cell dynamics for solid *in vitro* tumor growth. nut: nutrient dependency, gif: growth inhibitor factor dependency, sig: necrotic signal dependency.

tant regulative role during tumor growth. Several models suggest that diffusible inhibitors are produced internally (e.g., metabolic waste products) and that mitosis is completely inhibited if the concentrations are too large. With increasing size and cell number, the spheroid requires more energy (nutrient). Since the nutrient concentration is lowest in the center of the avascular nodule, cells will starve here at first and may eventually die (necrosis). Under necrosis, cells swell and burst, forming a necrotic site. There is experimental evidence that toxic factors are released or activated in this region and alter the microenvironment of the viable cells (Freyer 1988). Contrarily, cells, which exceed their natural lifespan (apoptosis), shrink and are rapidly digested by their neighbors or by other specialized cells (macrophages) (Arends and Wyllie 1991).

Traditional mathematical models of avascular solid tumor growth are formulated as deterministic integro-differential equations incorporating mitosis, apoptosis and necrosis inside the tumor (e.g., Adam and Bellomo 1996; Chaplain 1996; Greenspan 1972; Preziosi 2003). These models are based on the assumptions (i) that the tumor is spherically symmetric at all times and (ii) that the tumor sphere comprises a multilayered structure, particularly a central necrotic core surrounded by an outer ring of proliferating tumor cells. Tumor growth is modeled by following the translocation of the outer radii of these layers. A cell-based Monte-Carlo approach has been introduced as a model of the initial exponential growth phase (Drasdo 2000).

Here, we ask how the saturation of growth can be explained and how the layered tumor structure can form. We present a two-dimensional hybrid LGCA model for the avascular growth phase defined in terms of lattice-gas terminology (Dormann,

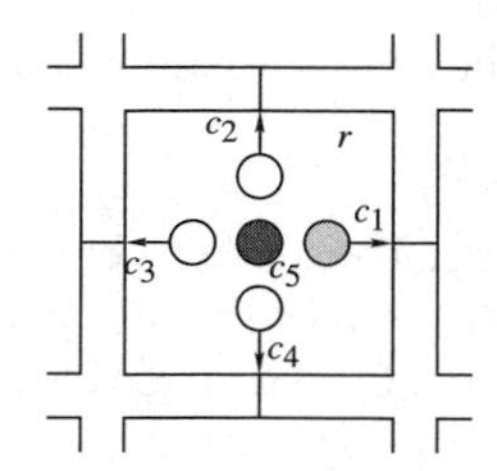

Figure 10.9. Example of a cell configuration at a lattice node r. The dark-gray dot and the light-gray dot denote the presence of a tumor and a necrotic cell, respectively.

Deutsch, and Lawniczak 2001). As we show in this book CA allow for the systematic analysis of cooperative effects in interacting cell systems. In contrast to differential equation models, it is possible to follow the fate of individual cells. All cells are subject to identical interaction rules. Every cell can proliferate, be quiescent, or die due to apoptosis and necrosis depending on its microenvironment.

Experimental work indicates that not only are cells moving towards the periphery, but a significant number of proliferative and quiescent tumor cells are moving from the periphery towards the core area (Dorie et al. 1982, 1986). This inward cell flow is a necessary condition for the growth saturation characterizing multicellular spheroids. If there would be no cell flow towards the center but only resting cells and cells moving in the direction of the periphery, constant nutrient delivery would imply unbounded tumor growth. Accordingly, two oppositely moving cell populations have to be considered. In the model, it is assumed that migration of cells depends on a chemical signal emitted by cells when they become necrotic. The chemotactic motion induces an antagonistic process to tumor expansion since some cells will migrate into the opposite direction, namely into the direction of the necrotic center. Based purely on local cell dynamics, formation of a two-dimensional multilayered tumor can be observed. We will also present results of statistical analysis of simulation runs. A different type of hybrid model has been previously introduced as a model of angiogenetic pattern formation which can follow the avascular growth phase (Andrecut 1998). Note that in Alarcon, Byrne, and Maini 2003, a hybrid cellular automaton has been suggested for the vascular growth phase.

10.4.1 A Hybrid Lattice-Gas Cellular Automaton Model for Tumor Growth

The tumor LGCA model is an extension of the "tissue growth model," which was introduced in sec. 10.2. Here, cells represent tumor cells (C) and necrotic cells (N), which reside on the same two-dimensional square lattice ($b = 4$). With each lattice node ($r = (r_1, r_2)$) four velocity channels $c_1 = (1, 0), c_2 = (0, 1), c_3 = (-1, 0), c_4 = (0, -1)$ and one resting channel $c_5 = (0, 0)$ are associated, i.e., $\tilde{b} = 5$. Each channel can be occupied by at most one tumor (C) or necrotic (N) cell (fig. 10.9). The von Neumann interaction neighborhood is considered.

Furthermore, **diffusion of chemicals** (nutrient and necrotic signal) is modeled explicitly. Nutrient is consumed by proliferating and quiescent tumor cells. When tumor cells become necrotic, they burst, leaking cell contents and necrotic signal into the surrounding tissue.

Mitosis, Apoptosis, and Necrosis. The model is based on conventional cell kinetics. Mitosis ($\bar{p}_m(r)$), apoptosis ($\bar{p}_d(r)$), and necrosis ($\bar{p}_n(r)$) rates depend on the local nutrient concentration ($c_{nut}(r)$) and local cell density (node configuration). They are defined as[64]

$$\bar{p}_m(r) := \begin{cases} \frac{p_m}{\mathrm{n}_C(r)} \left(\frac{c_{nut}(r) - t_{nut}}{1 - t_{nut}} \right) & \text{if} \quad \mathrm{n}_N(r) = 0 \ \wedge \ c_{nut}(r) > \mathrm{n}_C(r)\, t_{nut}, \\ 0 & \text{else,} \end{cases}$$

$$\bar{p}_d(r) := \begin{cases} p_d \, \mathrm{n}_C(r) & \text{if} \quad c_{nut}(r) > \mathrm{n}_C(r)\, t_{nut}, \\ 0 & \text{else,} \end{cases}$$

$$\bar{p}_n(r) := \begin{cases} 0 & \text{if} \quad \mathrm{n}_N(r) = 0 \ \wedge \ c_{nut}(r) > \mathrm{n}_C(r)\, t_{nut}, \\ p_n & \text{if} \quad \mathrm{n}_N(r) > 0 \ \wedge \ c_{nut}(r) > \mathrm{n}_C(r)\, t_{nut}, \\ 1 & \text{else,} \end{cases}$$

where $p_m, p_d, p_n < 1$, $\mathrm{n}(r) = \mathrm{n}_C(r) + \mathrm{n}_N(r)$ denotes the number of all cells at r, and $t_{nut} \leq 1$ is the critical nutrient concentration for necrosis. According to these rates each tumor cell at a node either proliferates (i.e., divides, if unoccupied channels exist), remains quiescent, dies or becomes necrotic. Nutrient is consumed by proliferating and quiescent tumor cells at a constant rate ($\bar{c}_{nut}$). Note that the presence of necrotic material at a node leads to a complete inhibition of mitosis and might even act toxic for all tumor cells present at that node. This assumption is based on evidence that cell quiescence is due to factors other than nutrients, such as cell contact effects (Casciari, Sotirchos, and Sutherland 1992).

All cells propagate simultaneously according to their orientation; only cells residing in "rest channels" do not move. Redistribution of cells at each lattice node, which is similar to the rules defined in the "tissue growth model" (cp. p. 186), is defined by rules specifying

(i) adhesion;
(ii) contact inhibition: cells are moving towards neighborhoods with low cell density; and
(iii) chemotactic motility: tumor cells move into the direction of the maximal signal gradient.

The following two node configurations can be distinguished:

1. *Presence of tumor cells but no necrotic cells.* One tumor cell always occupies the rest channel, if the outer interaction neighborhood contains at least one tumor cell; the remaining tumor cells are placed at channels which point to low-density

[64] Mostly, $\bar{p}_m + \bar{p}_d + \bar{p}_n \leq 1$; if this is not the case, the parameters are normalized.

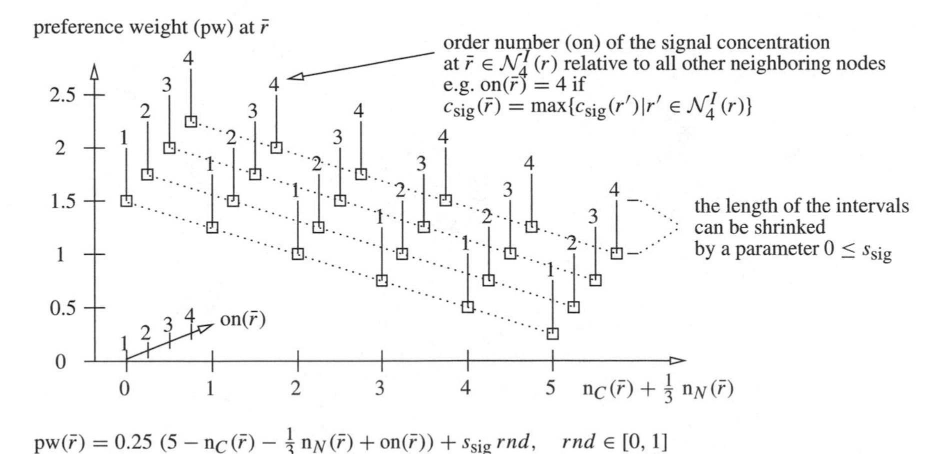

Figure 10.10. Interval of preference weights (pw) for each neighborhood configuration of a node r, concerning number of cells and signal concentration (c_{sig}). For example, if $s_{\text{sig}} = 0$, then a neighboring node $\bar{r}$ with no cells and maximal signal concentration ($\text{on}(\bar{r}) = 4$) always receives the highest preference weight ($\text{pw}(\bar{r}) = 2.25$), while a neighboring node r' with no cells and an order number of 1 obtains a smaller weight ($\text{pw}(r') = 1.5$) than a node r'' with either one tumor cell or three necrotic cells and order number 3 ($\text{pw}(r'') = 1.75$). Compare also fig. 10.11.

neighboring nodes mimicking the influence of contact inhibition. Thus, cells follow a track of least resistance (*passive motion*). Note that in this model, the *density* of a node is assumed to be the number of tumor cells (n_C) plus a third of the number of necrotic cells (n_C). This models the smaller volume of necrotic cells viewed as burst tumor cells. The spatial scale of the lattice (i.e., the area of a lattice node) is chosen such that contact inhibition movement is induced whenever more than one cell is present at a node.

In addition, we assume that the chemotactic response to the chemical signal contributes to the motility of tumor cells (*active motion*). This assumption is inspired by the experimental observation that there is a significant number of cells which drift from the viable rim of spheroids to the necrotic core (Dorie et al. 1982, 1986).

In order to specify the impact of active and passive motion the following rules for the successive occupation of velocity channels are defined: First, the four neighboring nodes are ordered according to the chemical signal concentration (cp. sec. 10.3). The density of cells together with the order number of a neighboring node $\bar{r}$ define an interval from which a *preference weight* (pw) is randomly chosen (cp. fig. 10.10). Finally, the velocity channels of node r are ordered according to the magnitude of preference weights (the highest value is first) of the neighboring nodes to which they point, and the remaining tumor cells are sequentially placed on the channels. Fig. 10.11 shows an example of this process. A special situation occurs if no signal and no tumor cells are present in the near-

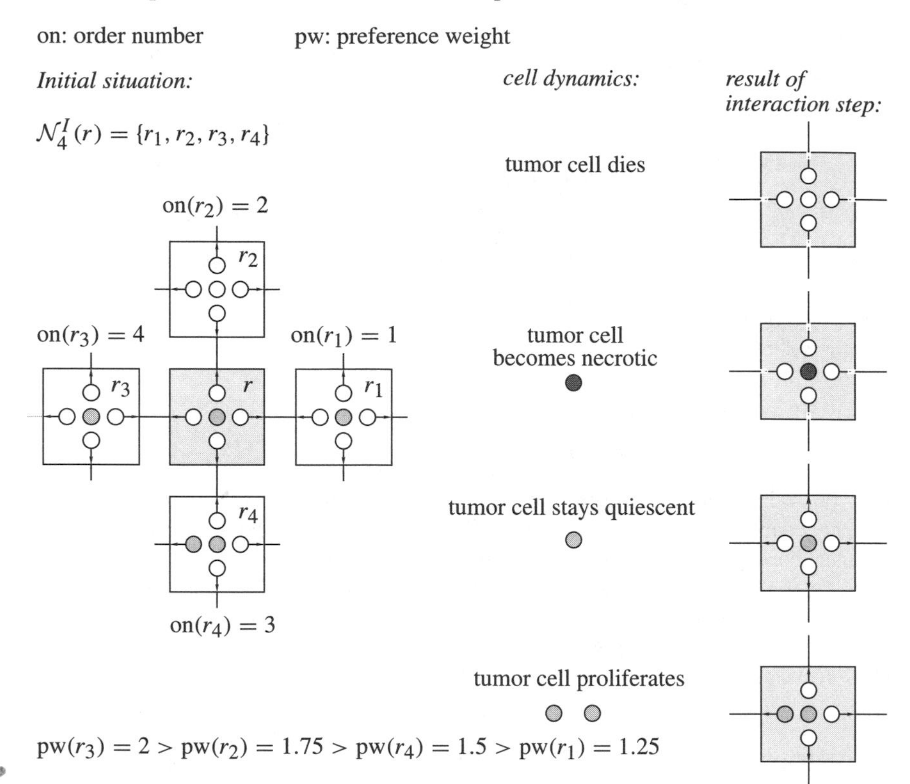

Figure 10.11. Example of the redistribution of cells at a lattice node. Since the focal (gray) node possesses more than one tumor cell in its outer interaction neighborhood, the rest channel always gains a tumor cell. With $s_{\text{sig}} = 0$ the preference weights are uniquely determined (cp. fig. 10.10). Channel c_3 is associated with the maximal weight.

est neighborhood of r. Then, all cells are redistributed randomly on the channels. Accordingly, the cells perform a random walk.

2. *No tumor cells but presence of necrotic cells.* Necrotic cells are always distributed at first among the channels. If necrotic cells reside at a lattice node, then the rest channel receives a necrotic cell and the remaining necrotic cells are distributed at the velocity channels according to a line of least resistance with respect to the densities of the corresponding neighbor nodes.
   If tumor cells are simultaneously present at the node, they are placed at the remaining channels according to their preference weights. This rule mimics the fact that necrotic cell material that is in contact with tumor cells decreases the adhesivity of the cells.

The model dynamics is summarized in fig. 10.12. The automaton is *scaled* as follows:

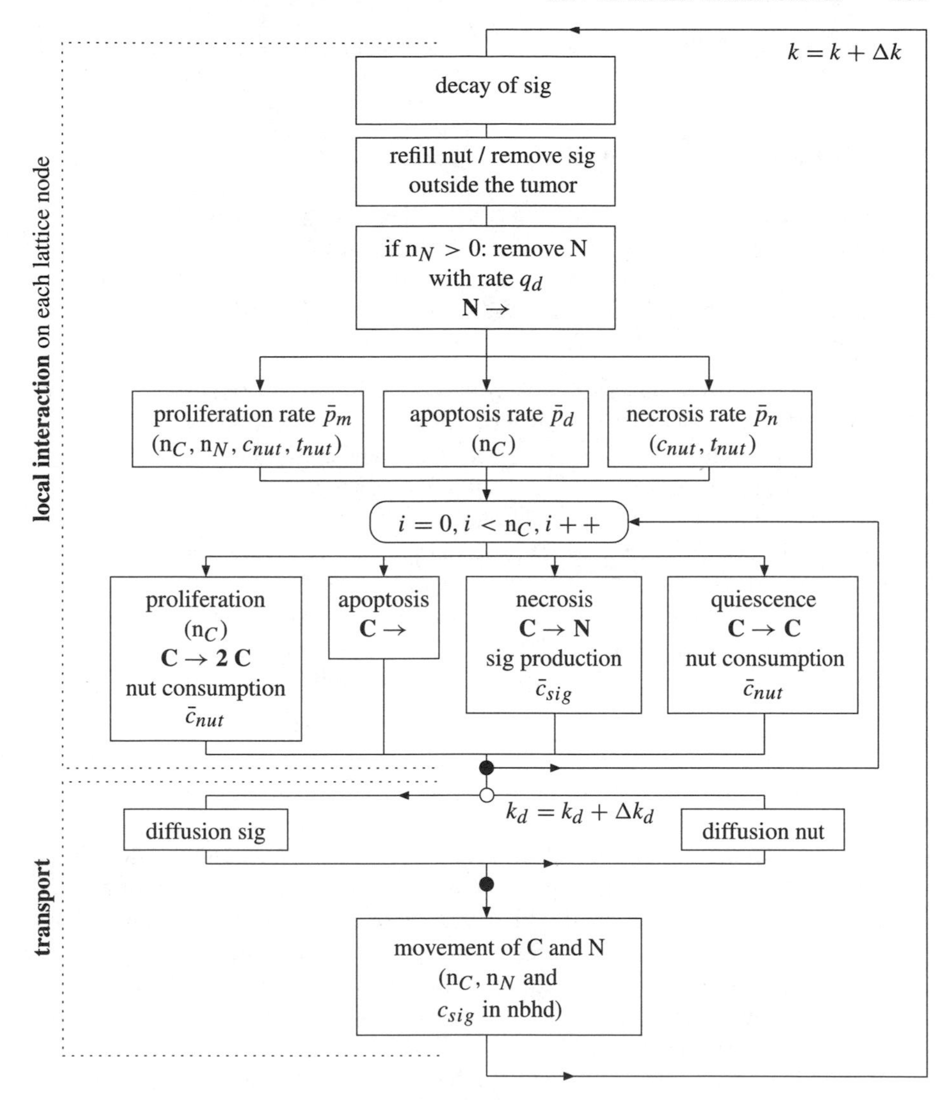

Figure 10.12. Schematic representation of the model dynamics. Parameters: n_C, number of tumor cells and n_N, number of necrotic cells at a node; c_{nut}, nutrient and c_{sig}, signal concentration at a node; t_{nut}: critical nutrient concentration, $\bar{c}_{nut}$: nutrient consumption of a tumor cell; $\bar{c}_{\text{sig}}$: signal production during necrosis; nbhd: neighborhood.

- *Tumor cell size.* Tumor cells have a volume of about 3.36×10^{-5} mm^3 (V-79 cells; Folkman and Hochberg 1973); necrotic cells are assumed to occupy one-third of this volume. It is supposed that cells are "packed" in the volume of a cubic lattice node, which is chosen as 2 × volume of one tumor cell (6.7×10^{-5} mm^3). Accordingly, the length of a square lattice area is $\Delta l := 0.04$ mm.
- *Time steps.* For cell dynamics $\Delta k = 1$ h, for chemical diffusion $\Delta k_d = 1$ min.

- *Diffusion coefficients of nutrient and necrotic signal.* $D = 10^{-6}$ cm^2/s $= 3.64$ Δl^2/min.

10.4.2 Simulations

We have performed simulations starting from a small number of active tumor cells[65] and applying realistic parameter sets (fig. 10.13). Parameters taken from the literature and incorporated in the automaton rules are glucose uptake rate, critical glucose concentration and doubling times for V-79 cells (Freyer 1988; Hlatky, Sachs, and Alpen 1988; Ward and King 1997):

- *Glucose uptake rate.* Investigations with V-79 cell cultures (Hlatky, Sachs, and Alpen 1988) indicated that, if the external glucose concentration is approximately $1.15\,10^{-5}$ mg/mm^3 ($\rightarrow 7.7 \cdot 10^{-8}$ mg/Δl^3), then the consumption rate of glucose can be determined as $7.2 \cdot 10^{-8}$ mg/cell h. Hence, during one hour, all available nutrient is consumed, i.e., $\bar{c}_{nut} = 1$ 1/cell h,
- *Critical glucose concentration.* The critical glucose concentration is about $1.4 \cdot 10^{-4}$ mg/mm^3 ($\rightarrow 9.38 \cdot 10^{-9}$ mg/Δl^3), hence $t_{nut} = 0.12$,
- *Doubling times.* Doubling times for V-79 cells are measured to be 10–19 hours (Freyer 1988; Ward and King 1997). Assuming an initial doubling time of 16 hours, the growth rate of the initial exponential growth period is $p_m - p_d = \ln(2)/16\ \mathrm{h}^{-1}$, hence $p_m - p_d = 0.04\ \mathrm{h}^{-1}$.

In the simulations nutrient is regularly applied and the chemical signal regularly removed (every hour) outside of the tumor, i.e., at nodes that have no tumor or necrotic cell material in the Moore neighborhood with range 3 (48 empty neighboring nodes). Furthermore, the size of the lattice (200×200 nodes $\approx 8 \times 8$ mm) is chosen sufficiently large such that the boundaries do not influence the tumor growth within the considered time interval. The formation of a layered pattern comprised of a central necrotic core, a rim of quiescent tumor cells, and an outer thin ring of proliferating cells can be observed. After an initial exponential growth phase, growth significantly slows down (fig. 10.14). This is accompanied by the increase of the necrotic cell population and simultaneous decrease of the tumor cell number.

The nutrient concentration in the tumor decreases until the onset of necrosis and increases afterwards since the necrotic core does not consume nutrient (fig. 10.15). For comparison, we performed simulations without considering a necrotic signal (i.e., no chemotactic influence on the tumor cells). The result is an unlimited growth of the spheroid (fig. 10.16).

The CA introduced here reproduces experimental results, particularly the formation of a layered structure and growth saturation observed in multicellular spheroids (Folkman and Hochberg 1973). Purely local rules (cell–cell interactions) allow for the transition from an initially small number of tumor cells to the final structured tumor. There are other CA models of avascular tumor growth, but these are based on nonlocal rules (Kansal et al. 2000; Qi et al. 1993). Kansal, Torquato, Harsh, Chiocca,

[65] The initial number of tumor cells is always 44.

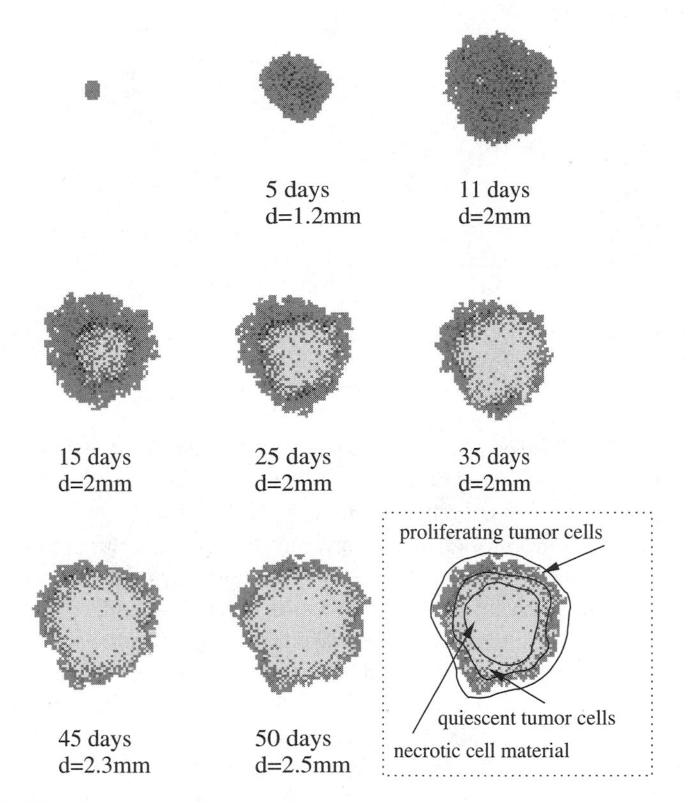

Figure 10.13. Simulation of tumor growth with a CA. A layered tumor forms, comprised of necrotic cell material and quiescent and proliferating tumor cells. Parameters: mitosis rate $p_m = 0.05$, apoptosis rate $p_d = 0.01$, necrosis rate $p_n = 0.008$, rate for necrotic cell dissolution $q_d = 0.0005$, production rate for chemical signal $\bar{c}_{\text{sig}} = 1$, decay rate for chemical signal 0.8, strength of chemical signal $s_{\text{sig}} = 0.4$, lattice size $|\mathcal{L}| = 200 \times 200$.

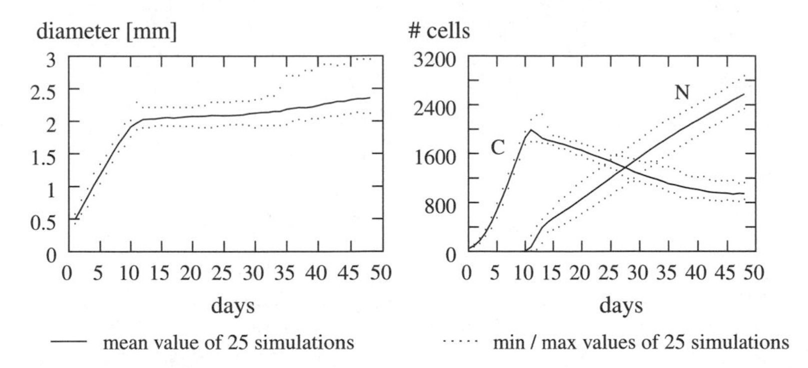

Figure 10.14. Simulation of diameter and cell number of 25 tumor growth simulations with "necrotic signaling" (C: tumor cells, N: necrotic cells).

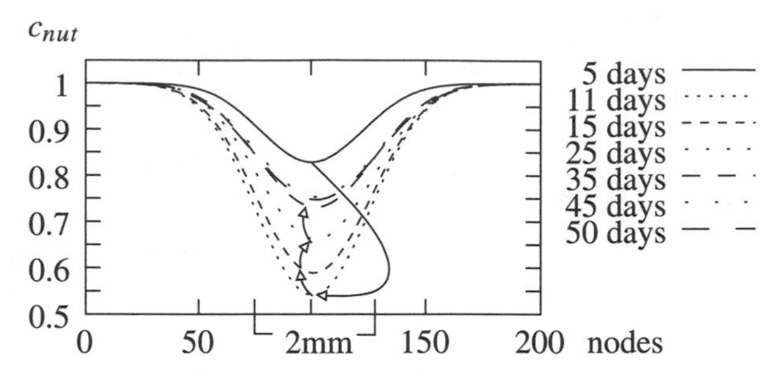

Figure 10.15. Simulation of nutrient concentration in a system of 200 × 200 lattice nodes (cp. fig. 10.13). The figure shows a section of the lattice at row 100.

and Deisboeck (2000) used a Delaunay triangulation instead of a regular lattice. The hybrid CA approach presented here incorporates both the dynamics of discrete cells and the dynamics of chemical concentrations.

A sufficient condition for growth saturation during avascular growth even in the case of periodic nutrient supply is to guarantee a tendency of tumor cell motion into the direction of the necrotic core. Otherwise the tumor would continue to expand until finally the tumor compound would break up as a result of necrotic material dissolution. The "antagonistic growth direction" is established in the simulations by the chemotactic migration of tumor cells into the direction of the maximum necrotic signal gradient. Accordingly, in the model it is assumed that a diffusible signal emitted by bursting tumor cells is attracting living tumor cells. This mechanism produces a cell flow towards the center. Initially, the inward flow is small since the necrotic core doesn't exist or is small. Accordingly, the outmoving cell population dominates; i.e., the tumor expands. But, later in development, if the necrotic core has reached a critical size the inward flow takes over, which limits further growth.

Our cellular model in principle allows to manipulate single cells or the microenvironment and to simulate the consequences of various treatments. For example,

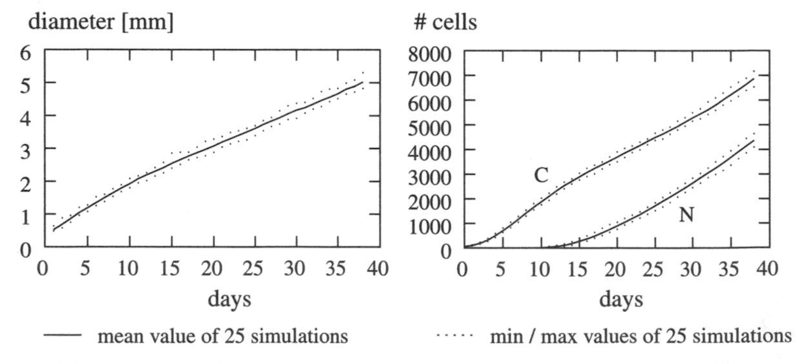

Figure 10.16. Simulation of diameter and cell number of 25 tumor growth simulations without a necrotic signal (C: tumor cells, N: necrotic cells).

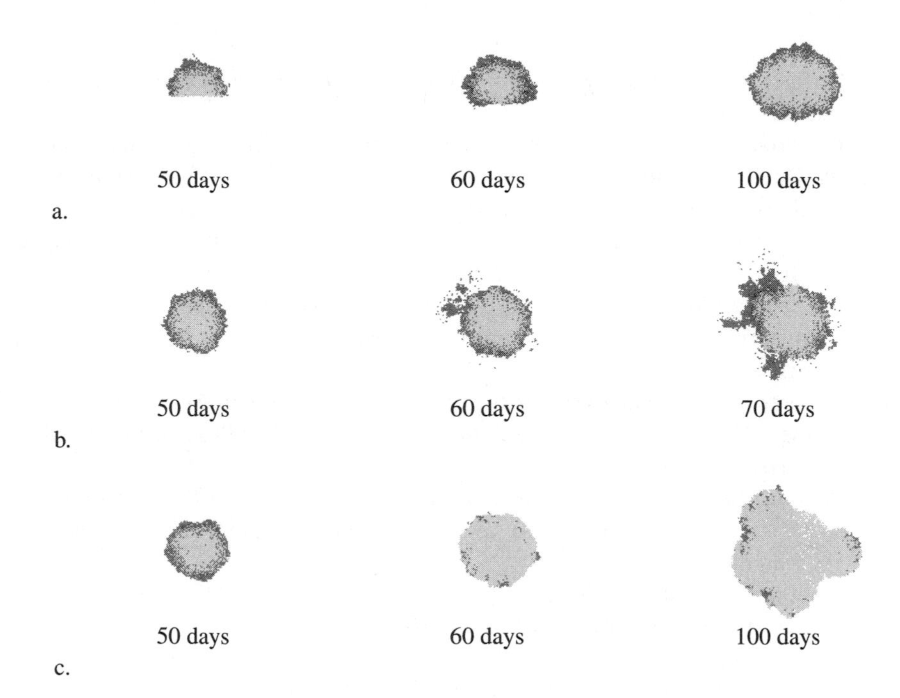

Figure 10.17. Simulation of tumor growth as in fig. 10.13. Various "treatments" are simulated: (a) After 50 days one-half of the tumor is removed. The tumor recovers from this surgery. (b) After 50 days the cell–cell adhesion is lowered. (c) After 50 days the necrosis rate is magnified by a factor 10 ($p_n = 0.08$). However, tumor cells still survive.

tumor growth can be followed after parts of the tumor have been "surgically" removed (fig. 10.17a). Tumor spread is observed if the cell–cell adhesion is lowered by some substance (fig. 10.17b). Finally, even if the cells have been manipulated such that they become necrotic (i.e., burst), survival of some tumor cells might occur (cp. fig. 10.17c). Particularly, fig. 10.17b demonstrates that a lowering of cellular adhesivity might have important consequences for the onset of tumor invasion. It is also straightforward to incorporate interactions with the immune system in the model.

Summary. In this chapter we have focused on elementary interactions in tissue and tumor growth. In particular, CA models for contact inhibition and chemotactic interaction have been presented. Later we showed how different modeling modules can be combined in order to produce a model for the avascular tumor growth phase that can be analyzed in multicellular spheroids. The LGCA model is hybrid: it represents biological cells as discrete entities and molecular concentrations as continuous states.

10.5 Further Research Projects

1. **Analysis**
   (a) Perform a linear stability analysis of the *tissue growth model* defined in this chapter (sec. 10.2).
   (b) Develop and analyze three-dimensional versions of the tissue growth model.
2. **Lattice-Boltzmann model:** Develop, simulate, and analyze lattice-Boltzmann models for chemotactic interaction, for the tissue and for the tumor growth model. In which situations is a coupling of CA and lattice-Boltzmann methods useful? Discuss corresponding hybrid models.
3. **Pattern recognition:** Is it possible to design algorithms for recognizing patterns that characterize the avascular tumor growth phase?
4. **Chemotaxis modeling**
   (a) Perform a stability analysis for the *chemotaxis model* defined in this chapter (sec. 10.3) and discuss similarities and differences with respect to local adhesive interactions (cp. ch. 7).
   (b) Provide examples of quantities that can be obtained from the mean-field analysis such as clustering rate, average cluster size, and average distance between clusters.
   (c) What are qualitative differences between the adhesive and the chemotactic LGCA model? (Compare figs. 7.4 and 10.6.)
   (d) Develop and study a LGCA model of chemotaxis for *two cell types*, which produce and interact via cell-type-specific chemicals.
5. **Tissue modeling**
   (a) Is it possible to determine mechanic properties (macroscopic quantities) of the simulated tissue?
   (b) Biological tissues are often viewed as "porous media." Discuss and develop corresponding CA and lattice-Boltzmann models.
6. **Tumor modeling**
   (a) Design CA models that incorporate interactions with the immune system and the extracellular matrix as well as the effects of angiogenesis.
   (b) What are "minimal models" for the avascular, the vascular, and the invasive tumor growth phases?
   (c) How could the results of these models be used to design alternative therapies?
   (d) Which cancer-specific problems could be analyzed with CA models?

11

Turing Patterns and Excitable Media

"It is suggested that a system of chemical substances, called morphogens, reacting together and diffusing through a tissue, is adequate to account for the main phenomena of morphogenesis. ..."

(from Turing, 1952)

In this chapter we demonstrate how cellular automata (CA) can be used to analyze Turing-type interactions and excitable media. In particular, we show that mean-field analysis of the CA models allows us to deduce the important pattern characteristics observed in simulations.

11.1 Turing Patterns[1]

A concept of pattern formation in biological systems was suggested by Alan Turing in his paper "The Chemical Basis of Morphogenesis" (Turing 1952). He demonstrated that a spatially homogeneous stable steady state of a reactive system can lose its stability when diffusive transport is included. The diffusive instability is able to enhance local random fluctuations, and as a result a spatially heterogeneous pattern of chemical (morphogen) concentrations may arise from initially homogeneous conditions. The wavelengths of these patterns are functions solely of the values of the diffusion coefficients and the kinetic parameters and not of domain size. In the following we refer to this type of pattern as a "Turing pattern." It was a revolutionary concept that diffusion, usually considered a *stabilizing* (spatially homogenizing) process (Landau and Lifshitz 1979), can actually cause instability. An experimental verification of this kind of pattern formation was possible in the chlorite-iodide-malonic acid[67] (CIMA) and in the polyacrylamide-methylene blue-sulfide-oxygen (PA-MBO) reactions (Castets et al. 1990; Ouyang and Swinney 1991). Inhowfar diffusive instability accounts for biological pattern formation is not clear. A review of its role in other biological contexts can be found in Murray 2002.

[66] Parts of this section are based on Dormann 2000.

[67] $ClO_2^- + 4\,I^- + 4\,H^+ \to Cl^- + 2\,I_2 + 2\,H_2O, \quad MA + I_2 \to IMA + I^-.$

Later, Turing pattern formation in activator–inhibitor systems was proposed as a model of biological morphogenesis (Gierer and Meinhardt 1972). Typically, the activator is characterized by local self-enhancement, while the inhibitor has a long-range antagonistic effect. Such two-component systems are traditionally analyzed with the help of appropriately constructed partial differential equations (Gierer and Meinhardt 1972; Murray 2002). Lately, the idea of local self-enhancement combined with a long-range antagonisic effect has been suggested as a mechanism for the determination of the division site in *Escherichia coli* (Meinhardt and de Boer 2001; cp. also an alternative model in Kruse 2002).

Note that the analysis of Turing pattern formation with partial differential equations does not provide insight into the microscopic basis. This problem inspired the development and application of LGCA models for reactive/interactive systems (Hasslacher, Kapral, and Lawniczak 1993; Kapral, Lawniczak, and Masiar 1991; Lawniczak, Dab, Kapral, and Boon 1991). In particular, Turing patterns observed in LGCA simulations can be analyzed by means of linear stability analysis of corresponding partial differential equations (Hasslacher, Kapral, and Lawniczak 1993; Lawniczak, Dab, Kapral, and Boon 1991). In addition, CA models mimicking activator–inhibitor interactions were suggested, e.g., as models of vertebrate coat markings or shell pattern formation (Young 1984; Markus and Schepers 1993). Investigation of those automata is based on computer simulations.

In the following, we focus on the analysis of a CA with activator–inhibitor–like interactions. In particular, we study pattern formation in a two-component LGCA. Particles (which can symbolize cells or organisms) are created or destroyed at the nodes of the lattice. In addition, particle movement resembles a random walk. This property distinguishes the LGCA from deterministic and probabilistic CA models. Furthermore, the LGCA rules have been inspired by local interactions of the components in systems capable of Turing pattern formation. The considered rules can be viewed as caricatures of local activator–inhibitor interactions and could e.g., serve as a gross simplification of predator–prey interactions.

Here, our emphasis is on linear stability analysis of discrete space and time automaton mean-field (Boltzmann) equations. Particularly, a critical wavelength and a *Turing condition* for the onset of pattern formation are derived. Although the mean-field description is not exact it is important to understand how well it can predict and characterize Turing patterns observed in LGCA simulations. The LGCA model allows us to analyze the influence of fluctuations and initial conditions on pattern formation.

11.1.1 Turing Pattern Formation in Macroscopic Reaction–Diffusion Systems

The objective of this subsection is to outline some important mathematical aspects of the appearance of Turing structures in *macroscopic* reaction-diffusion systems of two components.[68] Originally, Turing studied a "*mathematically convenient,*

[68] There exists an extensive literature on the problem of Turing pattern formation; for example Engelhardt 1994; Maini 1999; Murray 2002 give good reviews.

though biologically unusual system."[69] He considered a one-dimensional ring of cells ($r = 1, \dots, L$), each of which contains various "morphogens." In particular, he studied two morphogens whose concentrations are given by a_r and b_r, $r = 1, \dots, L$, and whose dynamics within each identical cell is described by a system of coupled differential equations of the form

$$\begin{aligned} \partial_t a_r &= F(a_r, b_r) + D_a\left((a_{r+1} - a_r) - (a_r - a_{r-1})\right), \\ \partial_t b_r &= G(a_r, b_r) + D_b\left((b_{r+1} - b_r) - (b_r - b_{r-1})\right), \end{aligned} \tag{11.1}$$

where D_a and D_b are "cell–cell" diffusion constants. Later, systems defined in continuous space that are mathematically described by partial differential equations have been examined. A one-dimensional reaction-diffusion model[70] defined on a line of length l, i.e., $x \in [0, l]$, is

$$\begin{aligned} \partial_t a &= F(a, b) + D_a \partial_{xx} a, \\ \partial_t b &= G(a, b) + D_b \partial_{xx} b. \end{aligned} \tag{11.2}$$

In order to develop a pattern or structure due to an instability of the homogeneous equilibrium, which is triggered by random disturbances, both systems (11.1) and (11.2) have to fulfill the following two **Turing conditions** (e.g., Engelhardt 1994; Murray 2002):

1. The system has a *spatially uniform stationary state* $(\bar{a}, \bar{b})$, i.e.,

   $$F(\bar{a}, \bar{b}) = G(\bar{a}, \bar{b}) = 0,$$

   which is *linearly stable* against spatially homogeneous perturbations. For this the necessary conditions are

   $$\operatorname{tr} \boldsymbol{J} = \partial_a F + \partial_b G < 0 \quad \text{and} \quad \det \boldsymbol{J} = \partial_a F \partial_b G - \partial_b F \partial_a G,$$

   where

   $$\boldsymbol{J} = \left.\begin{pmatrix} \partial_a F & \partial_b F \\ \partial_a G & \partial_b G \end{pmatrix}\right|_{(\bar{a}, \bar{b})}$$

   is the *Jacobian matrix* of system (11.2).
2. In addition, Turing instabilities can occur only if the diffusion coefficients of the two components differ significantly, i.e., $|D_a - D_b| \gg 0$.
   To be more precise, the *critical diffusion ratio* of system (11.2) is given by (Engelhardt 1994)

   $$\frac{D_b}{D_a} > \frac{(\det \boldsymbol{J} - \partial_b F \partial_a G) + 2\sqrt{-\partial_b F \partial_a G \det \boldsymbol{J}}}{(\partial_a F)^2} \tag{11.3}$$

[69] From Turing 1952.

[70] Note that this system is not the result of a formal derivation of (11.1), in particular D_a and D_b in (11.1) and (11.2) are not identical.

or

$$\frac{D_a}{D_b} > \frac{(\det \boldsymbol{J} - \partial_b F \partial_a G) + 2\sqrt{-\partial_b F \partial_a G \det \boldsymbol{J}}}{(\partial_b G)^2}, \tag{11.4}$$

where $D_a \neq D_b$. These inequalities hold for Jacobian matrices $\boldsymbol{J}$ with signs

$$\begin{pmatrix} + & - \\ + & - \end{pmatrix} \text{ or } \begin{pmatrix} - & + \\ - & + \end{pmatrix} \quad \text{"real activator–inhibitor models"}^{71} \text{ or}$$

$$\begin{pmatrix} + & + \\ - & - \end{pmatrix} \text{ or } \begin{pmatrix} - & - \\ + & + \end{pmatrix} \quad \text{"activator-substrate depleted models"}^{72} \tag{11.5}$$

In other words, for two-component systems, one component has to be autocatalytic and the other one has to be self-inhibiting. Furthermore, the cross-activations or -inhibitions need to be of opposite sign, i.e., if one activates the other, the other component has to inhibit the first (or vice-versa). The notion of "activator-inhibitor model" refers to systems in which one component activates itself and the other, while the second component inhibits itself and the first. "Activator-substrate depleted models" are characterized by either component promoting increase in the second component and decrease in the first. The two alternatives in the structure of the Jacobian for each category result from the identification of the self-activating and self-inhibiting components.

From the linearized reaction-diffusion system (11.2) around $(\bar{a}, \bar{b})$, the so-called *dispersion relation* can be derived with the help of Fourier transformation. The dispersion relation is given by

$$\lambda^2 - \lambda\left(\operatorname{tr} \boldsymbol{J} - \tilde{q}^2 (D_a + D_b)\right) + \det \boldsymbol{J} - \tilde{q}^2 (D_b \partial_a F + D_a \partial_b G) + \tilde{q}^4 D_a, D_b = 0, \tag{11.6}$$

where those $\tilde{q}$ are relevant for which the largest root $\lambda(\tilde{q}) > 0$.

The *fastest growing mode* is found to have the wave number $\tilde{q}_*$ given by

$$\tilde{q}_*^4 = \frac{\det \boldsymbol{J}}{D_a D_b}. \tag{11.7}$$

If the system (11.2) is defined in one spatial dimension with system length l, then

$$q_* = \frac{l}{2\pi} \tilde{q}_* .$$

---

[71] Such as the Lengyel-Epstein and the Gierer-Meinhardt models (Engelhardt 1994).

[72] Such as the Selkov, Brusselator, and Schnakenberg model (Engelhardt 1994); another terminology is "positive-feedback system."

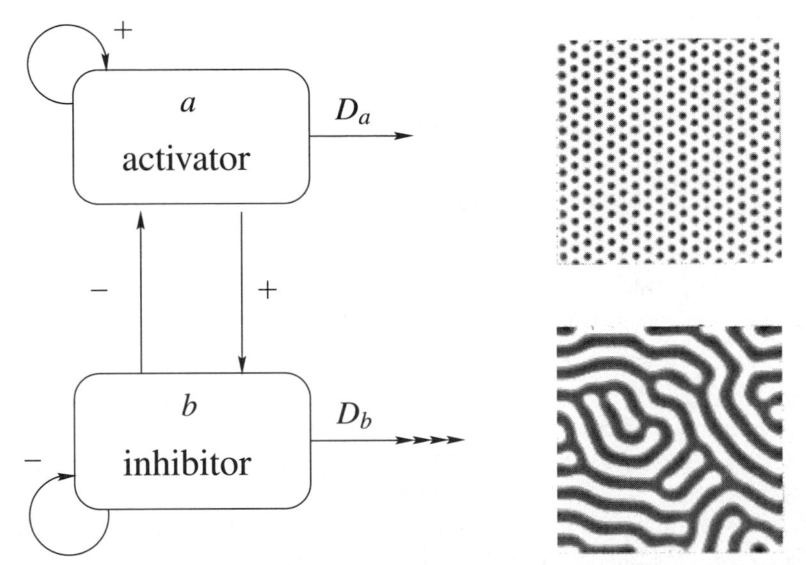

(a) Schematic representation of the "real activator–inhibitor" interaction according to (11.5).

(b) Basic types of two-dimensional Turing patterns in activator concentration depending on different parameter values of a continuous "real activator–inhibitor model."

Figure 11.1. Turing pattern formation in a "real activator-inhibitor model."

Note that in a finite domain situation wave numbers $q = (l/2\pi)\tilde{q}$ are discrete and so q_* may not be an allowed wave number. In this case the integer number closest to the analytically determined q_* characterizes the resulting patterns (Murray 2002).

Under these restrictions a and b concentrations develop *steady state* heterogeneous spatial patterns at the onset of instability with an *intrinsic wavelength* $2\pi/\tilde{q}_* = l/q_*$; i.e., the wavelength depends only on the kinetic parameters and diffusion coefficients but not on the geometrical length of the system domain (cp. (11.7)). In "real activator–inhibitor models" (cp. fig. 11.1) high concentrations of both components are found in the same spatial region (they are *in phase*), while patterns resulting from "activator-substrate depleted models" are *out of phase*; i.e., the concentration of one component is high where the concentration of the other is low.

11.1.2 A Lattice-Gas Cellular Automaton Model for Activator–Inhibitor Interaction

The model system consists of two moving and interacting "species" $\sigma, \sigma \in \{A, I\}$, the activator species A and the inhibitor species I. Particles X_σ of each species move on its own one-dimensional periodic lattice $\mathcal{L}_\sigma \subset \mathbb{Z}$. Later, we will extend the model to two-dimensional lattices. We assume that $\mathcal{L}_A$ and $\mathcal{L}_I$ possess an identical labeling of nodes and the number of nodes is equal to L. Then, for convenience, we

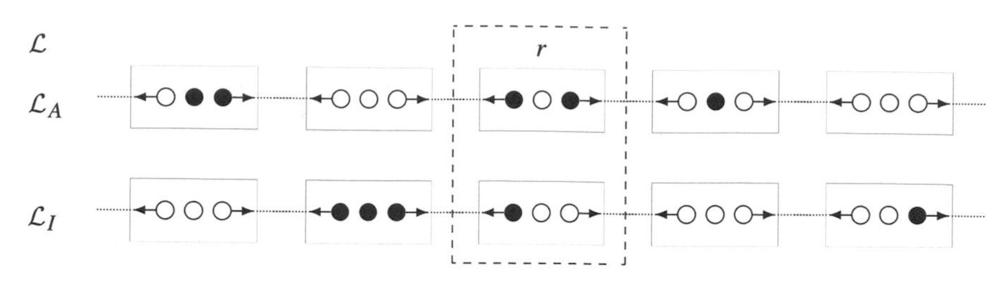

Figure 11.2. Possible one-dimensional lattice ($\tilde{b} = 3$) configuration for a two-component model; local node configurations: $\boldsymbol{\eta}_A(r) = (1, 1, 0)$, $\boldsymbol{\eta}_I(r) = (0, 1, 0)$, and hence $\boldsymbol{\eta}(r) = (1, 1, 0, 0, 1, 0)$.

identify both lattices and denote them by $\mathcal{L}$. Each node $r \in \mathcal{L}$ can host up to three particles, i.e., $\tilde{b} = 3$, of each species σ which are distributed in different velocity channels $(r, c_i)_\sigma$, $1 \leq i \leq 3$, with at most one particle of a given species per channel (fig. 11.2). Two velocity channels correspond to nearest-neighbor directions, i.e., $c_1 = 1$ and $c_2 = -1$, the third is a rest channel, $c_3 = 0$. Hence, the global automaton configuration $\boldsymbol{s}(k) \in \mathcal{S}$ at discrete time k is described locally at each node r by

$$\begin{aligned}\boldsymbol{\eta}(r, k) &= \Big(\eta_{A,1}(r, k), \eta_{A,2}(r, k), \eta_{A,3}(r, k), \eta_{I,1}(r, k), \eta_{I,2}(r, k), \eta_{I,3}(r, k)\Big) \\ &= (\boldsymbol{\eta}_A(r, k), \boldsymbol{\eta}_I(r, k)) \in \{0, 1\}^6,\end{aligned}$$

with Boolean components $\eta_{\sigma,i}(r, k)$ (cp. fig. 11.2). $\eta_{\sigma,i}(r, k) = 1$ represents the presence and $\eta_{\sigma,i}(r, k) = 0$ the absence of a particle of species σ at time k in channel $(r, c_i)_\sigma$. The number of particles of species σ at node r at time k is given by

$$\mathrm{n}_\sigma(r, k) = \sum_{i=1}^{3} \eta_{\sigma,i}(r, k).$$

Automaton dynamics arises from repetitive applications of reactive and diffusive "interactions" applied simultaneously at all lattice nodes at each discrete time step. First a reactive interaction step is performed during which particles X_σ of each species σ are destroyed or created according to a stochastic rule. Next, in a diffusion step particles X_σ perform a random walk at their individual lattices, independently from the other species (cp. sec. 5.4).

Activator–Inhibitor Interaction. The reactive interaction rule (R) is designed to capture the main characteristics of activator–inhibitor system dynamics, as is shown in fig. 11.1a. The model also contains a particle-motion process, and for the sake of simplicity, the interaction neighborhood will be restricted to the node itself, i.e., $\mathcal{N}_2^I(r) = \{r\}$. Next, we define sequences of node- and time-independent identically distributed Bernoulli-type random variables $\{\zeta_c(r, k) : r \in \mathcal{L}, k \in \mathbb{N}\}$ and $\{\zeta_d(r, k) : r \in \mathcal{L}, k \in \mathbb{N}\}$, which govern the creation and destruction of particles, such that

$$p_c := P\left(\zeta_c(r, k) = 1\right) \quad \text{and} \quad p_d := P\left(\zeta_d(r, k) = 1\right). \tag{11.8}$$

Then, the number of particles X_σ at a node r at time k after the reactive interaction step R took place, $\mathrm{n}_\sigma^{\mathrm{R}}(r,k)$, is defined as

$$\mathrm{n}_\sigma^{\mathrm{R}}(r,k) = \begin{cases} \tilde{b} & \text{with prob. } p_c \text{ if } \mathrm{n}_A(r,k) > \mathrm{n}_I(r,k) \geq 0, \quad \text{(i)} \\ 0 & \text{with prob. } p_d \text{ if } 0 \leq \mathrm{n}_A(r,k) < \mathrm{n}_I(r,k), \quad \text{(ii)} \\ \mathrm{n}_\sigma(r,k) & \text{otherwise} \end{cases} \tag{11.9}$$

for each $\sigma \in \{A, I\}$. Rule (11.9) (i) states that the activator A autocatalytically activates its own production and that of the inhibitor I. Note that the activator A in the absence of the inhibitor I performs a "growth process" that corresponds to the growth rule (6.26) studied earlier with the corresponding parameters $B = 1$ and $\gamma = p_c$. In turn, the dynamics of the inhibitor I is determined by its own degradation and by suppression of activator growth A (11.9)(ii). Fig. 11.3 illustrates this rule. Since rule (11.9) depends only on the number of particles of each species at a node, we use the previously defined indicator functions $\Psi_a(\boldsymbol{\eta}(r,k))$ (cp. p. 139) to derive a microscopic description of the model. Then, with $\Psi := (\Psi_0, \Psi_1, \Psi_2, \Psi_3)$ the action of the reactive interaction operator R can be written as

$$\begin{aligned} \eta_{\sigma,i}^{\mathrm{R}}(r,k) &= \mathcal{R}_{\sigma,i}^{\mathrm{R}}(\boldsymbol{\eta}(r,k)) \\ &= \Psi(\boldsymbol{\eta}_A(r,k))\, M_\sigma\left(\eta_{\sigma,i}(r,k)\right) \Psi^T(\boldsymbol{\eta}_I(r,k)), \end{aligned} \tag{11.10}$$

with the 4×4 "interaction matrices" M_σ. Each matrix M_σ has a nonzero entry if an interaction takes place, depending on the $\mathrm{n}_A(r,k)$ and $\mathrm{n}_I(r,k)$ relationship. They are defined as

$$M_A\left(\eta_{\sigma,i}(r,k)\right) := \begin{pmatrix} 0 & 0 & 0 & 0 \\ \alpha_c & 0 & \alpha_d & \alpha_d \\ \alpha_c & \alpha_c & 0 & \alpha_d \\ 0 & 0 & 0 & 0 \end{pmatrix}, \quad M_I\left(\eta_{\sigma,i}(r,k)\right) := \begin{pmatrix} 0 & \alpha_d & \alpha_d & \alpha_d \\ \alpha_c & 0 & \alpha_d & \alpha_d \\ \alpha_c & \alpha_c & 0 & \alpha_d \\ \alpha_c & \alpha_c & \alpha_c & 0 \end{pmatrix},$$

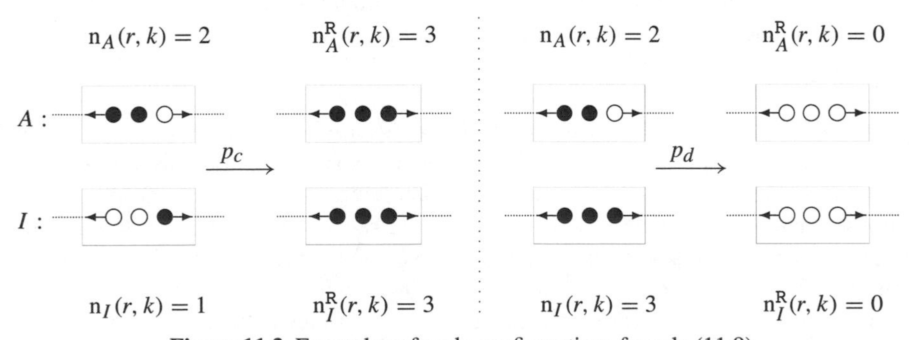

Figure 11.3. Examples of node configurations for rule (11.9).

with

$$\alpha_c := \zeta_c(r,k)\,\big(1-\eta_{\sigma,i}(r,k)\big) \quad \text{and} \quad \alpha_d := -\zeta_d(r,k)\,\eta_{\sigma,i}(r,k).$$

Note that the complete vector of "counting functions" Ψ can be found in appendix B.1.

Parameters for the random walk of each particle X_σ are a set of statistically node- and time-independent Boolean random variables $\xi_{\sigma,j}(r,k)$, $j = 1,\dots,6$, which govern the shuffling of the node configuration for each species $\sigma \in \{A, I\}$ (cp. sec. 5.4) and with probability distribution

$$P\left(\xi_{\sigma,j}(r,k)=1\right) := p_{\sigma,j}, \qquad P\left(\xi_{\sigma,j}(r,k)=0\right) = 1 - p_{\sigma,j}. \tag{11.11}$$

For isotropy reasons, in the following we investigate only the "unbiased case" $p_{\sigma,j} = 1/6$. The microdynamics for each species can be described by

$$\eta^{\mathrm{M}}_{\sigma,i}(r,k) = \mathcal{R}^{\mathrm{M}}_i\left(\boldsymbol{\eta}_\sigma(r,k)\right) = \sum_{j=1}^{6} \xi_{\sigma,j}(r,k) \sum_{l=1}^{3} \eta_{\sigma,l}(r,k) a^j_{li}, \tag{11.12}$$

where a^j_{li} is a matrix element of the permutation matrix $A_j \in \mathcal{A}_3$ (cp. the example given on p. 117).

Hence, the spatio-temporal evolution of the automaton dynamics can be expressed by the following system of nonlinear microdynamical difference equations. If at time $k+1$ the state of a channel $(r + m_\sigma\, c_i, c_i)_\sigma$ is described by $\eta_{\sigma,i}(r + m_\sigma\, c_i, k+1)$, then

$$\begin{aligned} \eta_{\sigma,i}(r + m_\sigma c_i, k+1) - \eta_{\sigma,i}(r,k) &= \mathcal{R}^{\mathrm{M}}_i\left(\boldsymbol{\eta}^{\mathrm{R}}_\sigma(r,k)\right) - \eta_{\sigma,i}(r,k) \\ &= \mathcal{C}_{\sigma,i}\left(\boldsymbol{\eta}(r,k)\right) \end{aligned} \tag{11.13}$$

for $m_\sigma \in \mathbb{N}$, $\sigma \in \{A, I\}$ and $i \in \{1,2,3\}$.

Simulations of the described automaton, starting from a random distribution of particles X_σ, $\sigma \in \{A, I\}$, exhibit formation of patterns for appropriately chosen parameters p_c, p_d, m_A, and m_I. Examples are shown in fig. 11.4. In the following section we analyze the formation of these patterns. Furthermore, we investigate the lattice-gas cellular automaton (LGCA) model with respect to the concept of diffusion-induced pattern formation in analogy to the Turing conditions (cp. p. 209) for continuous systems.

Lattice-Boltzmann Equation and Its Uniform Steady States. In order to gain more insight into the automaton behavior, we derive the space- and time-discrete lattice-Boltzmann equation from the microdynamical equation (11.13) and perform a linear stability analysis of this mean-field equation (cp. sec. 4.4.3). Recall that the lattice-Boltzmann equation is derived by taking the expected value of both sides of (11.13) (cp.(4.23)) under the mean-field approximation, in which all correlations between occupation numbers $\eta_{\sigma,i}(r,k)$ are neglected. Then with the help of (5.18) we get

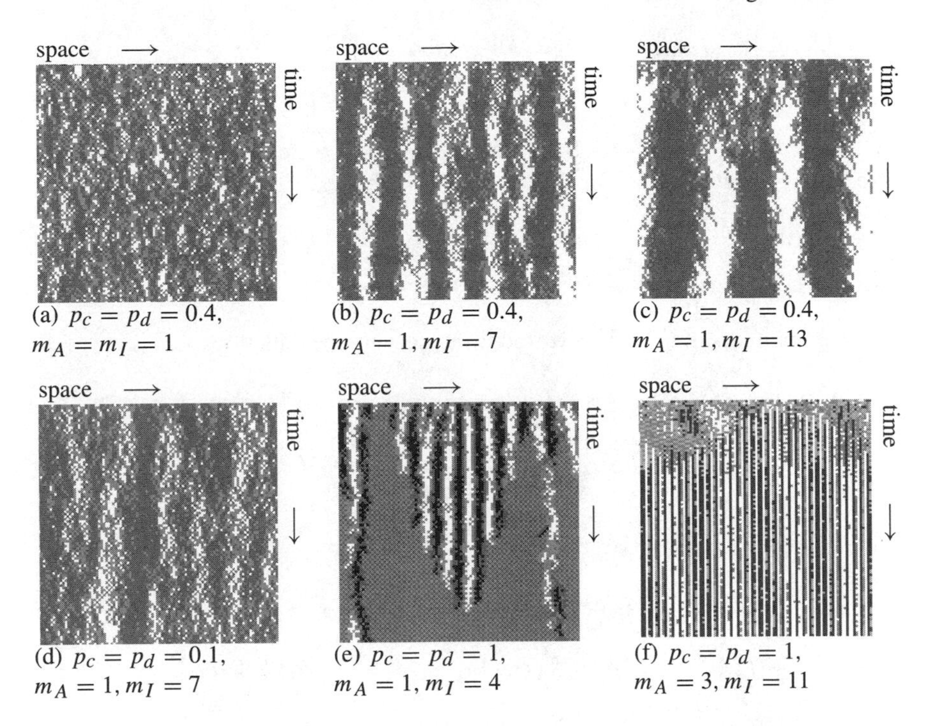

(a) $p_c = p_d = 0.4$, $m_A = m_I = 1$

(b) $p_c = p_d = 0.4$, $m_A = 1, m_I = 7$

(c) $p_c = p_d = 0.4$, $m_A = 1, m_I = 13$

(d) $p_c = p_d = 0.1$, $m_A = 1, m_I = 7$

(e) $p_c = p_d = 1$, $m_A = 1, m_I = 4$

(f) $p_c = p_d = 1$, $m_A = 3, m_I = 11$

Figure 11.4. Space-time plots of activator-inhibitor LGCA simulations; dark gray levels indicate regions of high activator concentration. Parameters: one-dimensional lattice with $L = 100$ nodes and periodic boundaries, random initial conditions, time $k \in \{0, \ldots, 100\}$.

$$
\boxed{
\begin{aligned}
f_{\sigma,i}(r + m_\sigma c_i, k+1) - f_{\sigma,i}(r,k) &= E\left(\mathcal{R}_i^{\mathrm{M}}\left(\boldsymbol{\eta}_\sigma^{\mathrm{R}}(r,k)\right) - \eta_{\sigma,i}(r,k)\right) \\
&= \frac{1}{3}\sum_{l=1}^{3} E\left(\eta_{\sigma,l}^{\mathrm{R}}(r,k)\right) - f_{\sigma,i}(r,k) \\
&= \tilde{\mathcal{C}}_{\sigma,i}\left(\boldsymbol{f}(r,k)\right),
\end{aligned}
}
\tag{11.14}
$$

where

$$
\begin{aligned}
\boldsymbol{f}(r,k) &= \left(f_{A,1}(r,k), f_{A,2}(r,k), f_{A,3}(r,k), f_{I,1}(r,k), f_{I,2}(r,k), f_{I,3}(r,k)\right) \\
&= \left(f_j(r,k)\right)_{j=1}^{6} \in [0,1]^6.
\end{aligned}
$$

Note that $\tilde{\mathcal{C}}_{\sigma,i}\left(\boldsymbol{f}(r,k)\right)$ is a function depending on the reaction parameters p_c and p_d. The expanded form of $E\left(\eta_{\sigma,l}^{\mathrm{R}}(r,k)\right)$ is given in appendix B.1.

The first step in the analysis of pattern formation is the evaluation of the spatially uniform steady states of the lattice-Boltzmann equation. These states are determined by

$$f_{\sigma,i}(r + m_\sigma c_i, k+1) = f_{\sigma,i}(r,k) \qquad \forall\, r \in \mathcal{L},\ \forall\, k \in \mathbb{N},$$
$$\Rightarrow \quad \tilde{\mathcal{C}}_{\sigma,i}\left(\bar{f}\right) = 0 \tag{11.15}$$

for $\sigma \in \{A, I\}$ and $i = 1, \ldots, 3$. Note that the stochastic coupling $\mathcal{R}_i^{\mathrm{M}}$ of the channels ensures homogeneity in (11.15). Hence, assuming that

$$f_{A,1}(r,k) = \bar{f}_{A,1} = f_{A,2}(r,k) = \bar{f}_{A,2} = f_{A,3}(r,k) = \bar{f}_{A,3} =: \bar{f}_A\,,$$
$$f_{I,1}(r,k) = \bar{f}_{I,1} = f_{I,2}(r,k) = \bar{f}_{I,2} = f_{I,3}(r,k) = \bar{f}_{I,3} =: \bar{f}_I\,,$$

finding the solutions of (11.15) corresponds to solving the equations

$$-3(-1+\bar{f}_A)\bar{f}_A\big[(-1+\bar{f}_I)^2\left(2-2\bar{f}_I+\bar{f}_A(-1+4\bar{f}_I)\right)p_c + \bar{f}_I^2\left(-3+\bar{f}_A(3-4\bar{f}_I)+2\bar{f}_I\right)p_d\big] = 0, \tag{11.16}$$

and

$$-3\big[\bar{f}_I p_d + 3\bar{f}_A(-1+\bar{f}_I)^2\left((-1+\bar{f}_I)p_c - \bar{f}_I p_d\right) - 3\bar{f}_A^2\left(1-4\bar{f}_I+3\bar{f}_I^2\right)\left((-1+\bar{f}_I)p_c - \bar{f}_I\,p_d\right) + \bar{f}_A^3\left(1-6\bar{f}_I+6\bar{f}_I^2\right)\left((-1+\bar{f}_I)p_c - \bar{f}_I\,p_d\right)\big] = 0. \tag{11.17}$$

Solutions $\left(\bar{f}_A, \bar{f}_I\right)$ of (11.16) and (11.17) are given by

$$\left(\bar{f}_A, \bar{f}_I\right) \in \{(0,0), (a_1,a_2), (1,1)\}, \tag{11.18}$$

where

$$(a_1, a_2) = (0.5, 0.5) \qquad \text{whenever } p_c = p_d =: p.$$

For $p_c \neq p_d$ the solution (a_1, a_2) is a function of the reactive probabilities p_c and p_d, which can be determined numerically as illustrated in fig. 11.5.

Derivation of the Boltzmann Propagator. In the next step, the stability of these steady states with respect to spatially homogeneous and heterogeneous fluctuations $\delta f_{\sigma,i}(r,k) := f_{\sigma,i} - \bar{f}_\sigma$ is determined. Following the linear stability analysis described in sec. 4.4.3, as a result of the linearization and Fourier transformation we obtain equations (cp. (4.37)) for the growth of each Fourier mode with wave number q, i.e., $F_{\sigma,i}(q,k) = \sum_{r\in\mathcal{L}} \delta f_{\sigma,i}(r,k) e^{-\mathrm{i}(2\pi/L)q\cdot r}$, as

$$\boldsymbol{F}(q,k) = \Gamma(q)^k \boldsymbol{F}(q,0)\,, \tag{11.19}$$

with

$$\boldsymbol{F}^T(q,k) = \left(F_{A,1}(r,k), F_{A,2}(r,k), F_{A,3}(r,k), F_{I,1}(r,k), F_{I,2}(r,k), F_{I,3}(r,k)\right) = \left(F_j(r,k)\right)_{j=1}^{6},$$

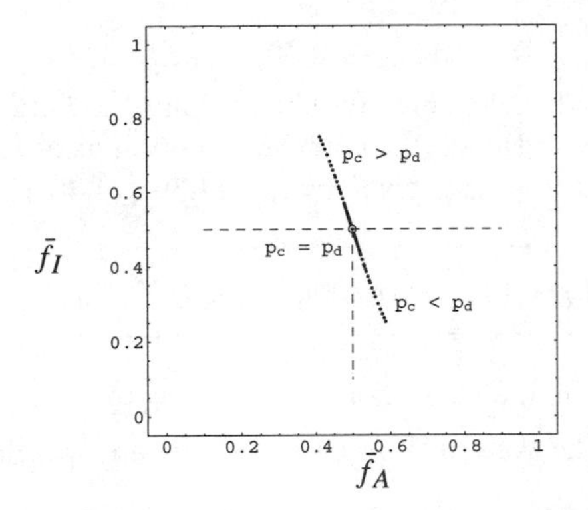

Figure 11.5. Spatially uniform steady states $\left(\bar{f}_A, \bar{f}_I\right)$ for different reaction probabilities $(p_c, p_d) \in \{0.1, 0.2, \ldots, 0.9, 1\}^2$. Dashed lines mark the symmetric case $p_c = p_d$.

and where the Boltzmann propagator (4.36) for this model is given by

$$\Gamma(q) = T\{\boldsymbol{I} + \Omega^0\}, \qquad q = 0, \ldots, L-1, \tag{11.20}$$

with $T = \operatorname{diag}\left(e^{-\mathbf{i}(2\pi/L)q\cdot m_A}, e^{\mathbf{i}(2\pi/L)q\cdot m_A}, 1, e^{-\mathbf{i}(2\pi/L)q\cdot m_I}, e^{\mathbf{i}(2\pi/L)q\cdot m_I}, 1\right)$,

$$\Omega^0 = \left.\begin{pmatrix} \frac{\partial \tilde{C}_{A,1}}{\partial \delta f_1} & \cdots & \frac{\partial \tilde{C}_{A,1}}{\partial \delta f_6} \\ \vdots & & \vdots \\ \frac{\partial \tilde{C}_{A,3}}{\partial \delta f_1} & \cdots & \frac{\partial \tilde{C}_{A,3}}{\partial \delta f_6} \\ \frac{\partial \tilde{C}_{I,1}}{\partial \delta f_1} & \cdots & \frac{\partial \tilde{C}_{I,1}}{\partial \delta f_6} \\ \vdots & & \vdots \\ \frac{\partial \tilde{C}_{I,3}}{\partial \delta f_1} & \cdots & \frac{\partial \tilde{C}_{I,3}}{\partial \delta f_6} \end{pmatrix}\right|_{\bar{\boldsymbol{f}}} \quad \text{and} \quad \boldsymbol{I} + \Omega^0 = \begin{pmatrix} \omega_1 & \ldots & \omega_1 & \omega_2 & \ldots & \omega_2 \\ \vdots & & \vdots & \vdots & & \vdots \\ \omega_1 & \ldots & \omega_1 & \omega_2 & \ldots & \omega_2 \\ \omega_3 & \ldots & \omega_3 & \omega_4 & \ldots & \omega_4 \\ \vdots & & \vdots & \vdots & & \vdots \\ \omega_3 & \ldots & \omega_3 & \omega_4 & \ldots & \omega_4 \end{pmatrix},$$

where $\bar{\boldsymbol{f}} = \left(\bar{f}_A, \bar{f}_A, \bar{f}_A, \bar{f}_I, \bar{f}_I, \bar{f}_I\right)$.

The matrix elements ω_i, $i = 1, \ldots, 4$, are terms depending on p_c, p_d, $\bar{f}_A$, and $\bar{f}_I$. They are given in appendix B, equations (B.2–B.4).

The spectrum of the Boltzmann propagator can be obtained as

$$\Lambda_{\Gamma(q)} = \{\lambda_1(q), \lambda_2(q), 0\},$$

where 0 has a multiplicity of 4 and

$$\boxed{\begin{aligned} \lambda_{1,2}(q) = \frac{1}{2}\Big(& \omega_1 u_A(q) + \omega_4 u_I(q) \\ & \pm\sqrt{4\left(\omega_2\omega_3 - \omega_1\omega_4\right) u_A(q)\, u_I(q) + \left(\omega_1\, u_A(q) + \omega_4\, u_I(q)\right)^2}\Big) \end{aligned}} \tag{11.21}$$

with $u_\sigma(q) := 1 + 2\cos\left(\frac{2\pi}{L} q m_\sigma\right)$, $\sigma \in \{A, I\}$, $q \in \{0, \ldots, L-1\}$.

Furthermore, $\Gamma(q)$ is diagonalizable, since the dimension of the eigenspace of eigenvalue 0 is $6 - \text{rank}\,(\Gamma(q)) = 4$. Hence, the complete solution of system (11.19) for each wave number $q = 0, \ldots, L-1$, is given by (cp. (4.39), (4.40))

$$\boxed{F_{\sigma,i}(q,k) = d_1(q)\, v_{1i}^\sigma(q) \lambda_1(q)^k + d_2(q)\, v_{2i}^\sigma(q)\, \lambda_2(q)^k} \qquad (11.22)$$

for $i = 1, 2, 3$, $\sigma \in \{A, I\}$, and with eigenvectors $v_l(q) = \left(v_{l1}^A(q), \ldots, v_{l3}^I(q)\right) = \left(v_{lj}(q)\right)_{j=1}^6$ for $l = 1, \ldots, 6$. The constants $d_1(q)$ and $d_2(q)$ are specified by the initial condition

$$\sum_{l=1}^{6} d_l(q) v_{lj}(q) = \sum_{r=0}^{L-1} e^{\mathrm{i}\frac{2\pi}{L} q \cdot r} \delta f_j(r, 0)\ .$$

Recall that spatially inhomogeneous structures are determined by undamped modes according to wave numbers $q \in Q^c \supset Q^+ \cup Q^-$ (cp. the definition on p. 96). Since mode $\boldsymbol{F}(q_*, k)$ corresponding to the dominant critical wave number $q_* \in Q^c$ grows fastest, linear stability analysis predicts a spatial pattern with a dominant wavelength of L/q_* for random initial conditions. Later, we will analyze the influence of the initial conditions.

Spatially Homogeneous Perturbations. The stability of the spatially uniform steady states $\left(\bar{f}_A, \bar{f}_I\right)$ with respect to *spatially homogeneous* fluctuations can be analyzed by studying the problem for $q = 0$, since $\sum_{r \in \mathcal{L}} \delta f_{\sigma,i}(r, 0) = F_{\sigma,i}(0, 0)$. In this case the eigenvalues $\lambda_{1,2}$ simplify to

$$\lambda_{1,2}(0) = \frac{3}{2}\left(\omega_1 + \omega_4 \pm \sqrt{4\omega_2\,\omega_3 + (\omega_1 - \omega_4)^2}\right). \qquad (11.23)$$

For the trivial steady states (0, 0) and (1, 1) these eigenvalues become

$$\lambda_1(0) = 1 + 2p_c \quad \text{and} \quad \lambda_2(0) = 1 - p_d \qquad \text{for } \left(\bar{f}_A, \bar{f}_I\right) = (0, 0),$$

and

$$\lambda_1(0) = 1 + 2p_d \quad \text{and} \quad \lambda_2(0) = 1 - p_c \qquad \text{for } \left(\bar{f}_A, \bar{f}_I\right) = (1, 1).$$

Clearly, the spectral radius is always larger than 1, i.e., $\mu(0) \equiv \lambda_1(0) > 1$. Therefore, the homogeneous steady states (0, 0) and (1, 1) are *unstable* with respect to spatially homogeneous perturbations. Hence, spatial diffusion-induced pattern formation close to these homogeneous steady states is not possible since the first Turing condition (see p. 209) is not satisfied.

Now we investigate the stability of the nontrivial steady state (a_1, a_2). If we choose $p_c = p_d := p$, the eigenvalues (11.23) become

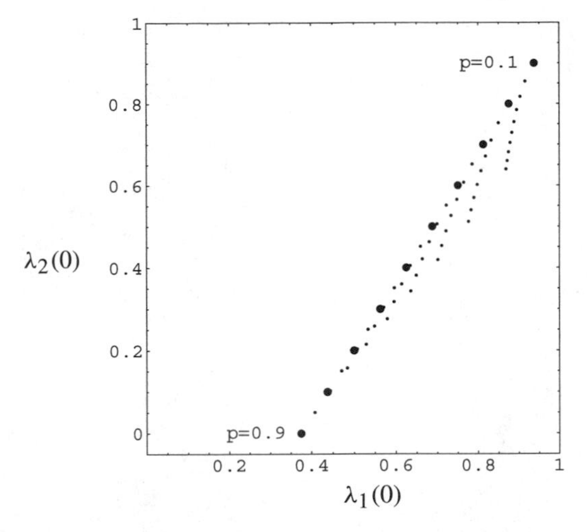

Figure 11.6. Eigenvalues $(\lambda_1(0), \lambda_2(0))$ for different values of the reaction probabilities $(p_c, p_d) \in \{0.1, 0.2, \ldots, 0.9, 1\}^2$; the larger dots represent values for which $p_c = p_d =: p$.

$$\lambda_1(0) = 1 - \frac{5}{8}p \qquad \text{and} \qquad \lambda_2(0) = 1 - p.$$

For other nonequal reaction probabilities p_c and p_d, the eigenvalues can be determined numerically, as is shown in fig. 11.6.

Thus, for any p_c and p_d the spectral radius is given by $\mu(0) \equiv \lambda_1(0) < 1$, and hence the stationary state (a_1, a_2) is *stable* with respect to any spatially homogeneous perturbation. In the next subsection we investigate how a difference in the speed parameters m_A and m_I can act as a destabilizing influence.

Note that as we consider the spatially homogeneous case here, the expected occupation numbers $f_{\sigma,i}(r, k)$ are the same for all nodes r and directions c_i; i.e., $f_{\sigma,i}(r, k) \equiv f_\sigma(k)$. Therefore, the total density $\rho_\sigma(k)$ is given by

$$\rho_\sigma(k) = \frac{1}{3L} \sum_{r=0}^{L-1} \sum_{i=1}^{3} f_{\sigma,i}(r, k) = f_\sigma(k).$$

Fig. 11.7 shows trajectories obtained by iterating the lattice-Boltzmann equation (11.14) at a fixed node r ($m_\sigma = 0$) starting from different initial conditions. The illustrated trajectories confirm the stability results obtained above.

11.1.3 Pattern Formation in One Dimension: Analysis and Simulations

Since we are interested in explaining the formation of diffusion-induced spatial structures, we investigate the case of *spatially heterogeneous* fluctuations, i.e., $q \neq 0$, for the stationary solution (a_1, a_2) only. Furthermore, primary attention will be given to studying the case $p_c = p_d =: p$, and therefore $(a_1, a_2) = (0.5, 0.5)$, because of its "better" analytical tractability. Recall that a significant difference in the diffusion

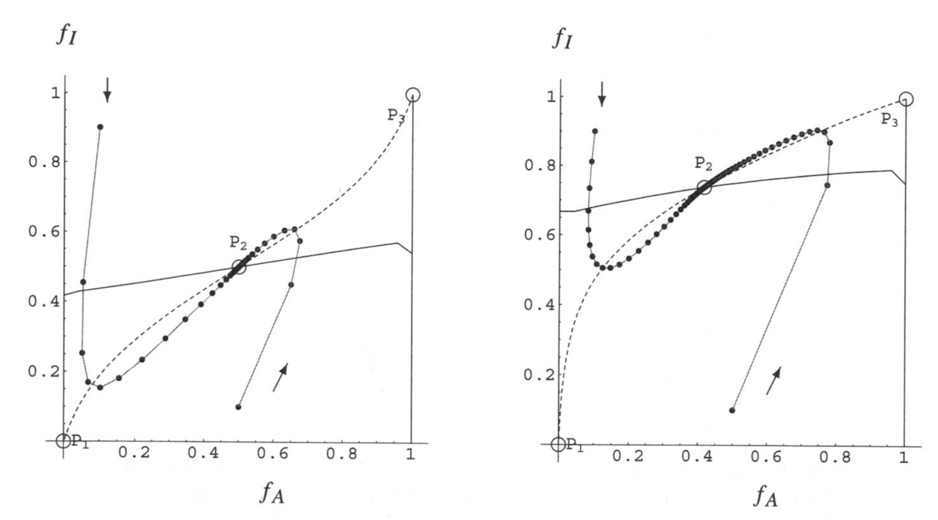

Figure 11.7. Trajectories of the lattice-Boltzmann equation (11.14) (cp. the discussion in sec. 11.1.2) for reaction parameters $p_c = p_d = 0.5$ (left), $p_c = 0.9$, $p_d = 0.1$ (right), and different initial conditions $(f_A(0), f_I(0)) \in \{(0.5, 0.1), (0.1, 0.9)\}$; solid lines refer to (11.16), dashed lines refer to (11.17), stationary points $\left(\bar{f}_A, \bar{f}_I\right)$ are $P_1 = (0, 0)$, $P_2 = (a_1, a_2)$, $P_3 = (1, 1)$.

coefficients of a two-component system is a necessary condition for the evolution of Turing patterns (cp. the Turing condition, p. 209). In our LGCA model the "diffusion coefficients" of species A and I are functions of just the parameters m_A and m_I. When $m_A = m_I$, then both components "diffuse" within the same range. When $m_A < m_I$, then the "diffusion coefficient" of species A is smaller than that of species I, and vice versa.

General Comments on Simulations. All simulations are performed on a periodic lattice with L nodes and with different random seedings of the initial state such that a channel $(r, c_i)_\sigma$ is occupied with probability $P\left(\eta_{\sigma,i}(r, 0) = 1\right) = 0.5$, if not stated otherwise. Spatially averaged occupation numbers, $\boldsymbol{f}^s(k) = \left(f^s_{A,1}, \ldots, f^s_{I,3}\right)$ with $f^s_{\sigma,i}(k) := 1/L \sum_{r=0}^{L-1} \eta_{\sigma,i}(r, k)$, are determined from simulations. They appear if the system has reached on average the steady state $\left(\bar{f}_A, \bar{f}_I\right) = (0.5, 0.5)$. Furthermore, with q^s we denote wave numbers which are found in simulation plots.

We begin with the case when both species move on their lattice with the *same speed*, i.e., $m_A = m_I$. From (11.21) we obtain with $p_c = p_d =: p$ that for each wave number q

$$\lambda_1(q) = \frac{1}{3}\, u_A(q) \left(1 - \frac{5}{8}\, p\right) \quad \text{and} \quad \lambda_2(q) = \frac{1}{3}\, u_A(q)\,(1 - p)\,,$$

where $u_A(q) = u_I(q)$. Hence, the spectral radius is given by

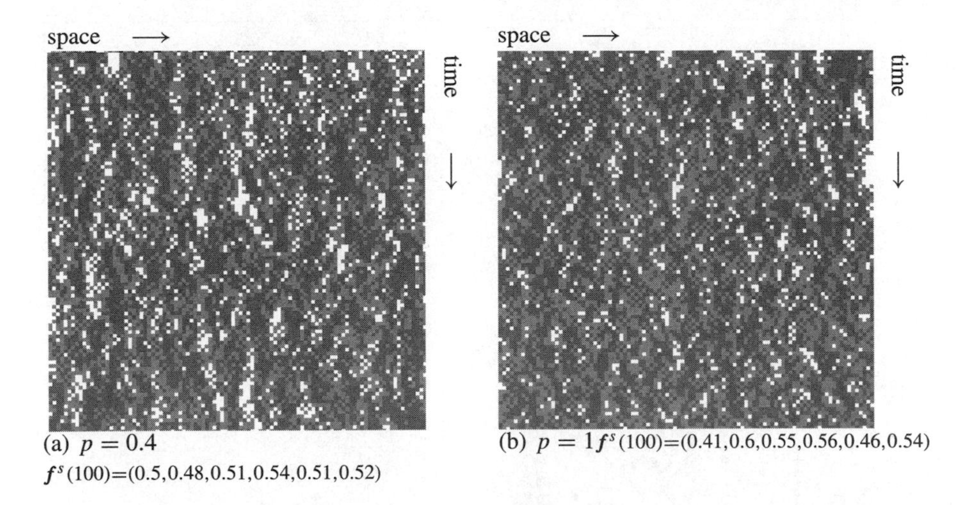

(a) $p = 0.4$
$f^s(100)=(0.5,0.48,0.51,0.54,0.51,0.52)$

(b) $p = 1 f^s(100)=(0.41,0.6,0.55,0.56,0.46,0.54)$

Figure 11.8. Evolution of activator concentration in space and time for equal diffusion coefficients; parameters: $m_A = m_I = 1$, $p_c = p_d =: p$, $L = 100$, $k = 0, \ldots, 100$.

$$\mu(q) \equiv |\lambda_1(q)| = \frac{1}{3}|u_A(q)|\,|\left(1 - \frac{5}{8}p\right)| \leq \frac{1}{3} < 1,$$

which implies that all modes $F(q, k)$ are damped, and local fluctuations $\delta f_{\sigma,i}(r, k)$ decay to zero when time $k \to \infty$. Consequently, the spatially homogeneous steady state $\left(\bar{f}_A, \bar{f}_I\right) = (0.5, 0.5)$ is stable when diffusive transport is present and spatially inhomogeneous structure does not emerge. This coincides with the simulation results shown in fig. 11.8.

Next we study the effects of *unequal diffusion coefficients* on the stability of the steady state $(\bar{f}_A, \bar{f}_I) = (0.5, 0.5)$. Linear stability analysis and simulations show that stationary patterns can emerge in both cases, i.e., when $m_A < m_I$ and $m_A > m_I$. This is in contrast to the traditional Turing-type pattern formation scenario in which patterns can emerge in only one case of the two inequalities but not the other. The second Turing condition (cp. p. 209) requires that the diffusion coefficient ratio $D_b/D_a > 1$, which implies that component b has to diffuse faster than component a. The reason for the different behavior is that contrary to continuous reaction-diffusion systems, large wave numbers are not damped out in LGCA (cp. also sec. 5.4.2).

The Deterministic Case: $p_c = p_d =: p = 1$. First let us look at the automaton dynamics when the reactive interactions are *deterministic*, i.e., $p_c = p_d =: p = 1$. In this case the Boltzmann propagator (11.20) has only one nonzero eigenvalue; i.e., by (11.21) ,

$$\boxed{\lambda_1(q) = \frac{3}{8}\, u_A(q) - \frac{1}{4} u_I(q)} \quad \text{and} \quad \lambda_2(q) = 0, \tag{11.24}$$

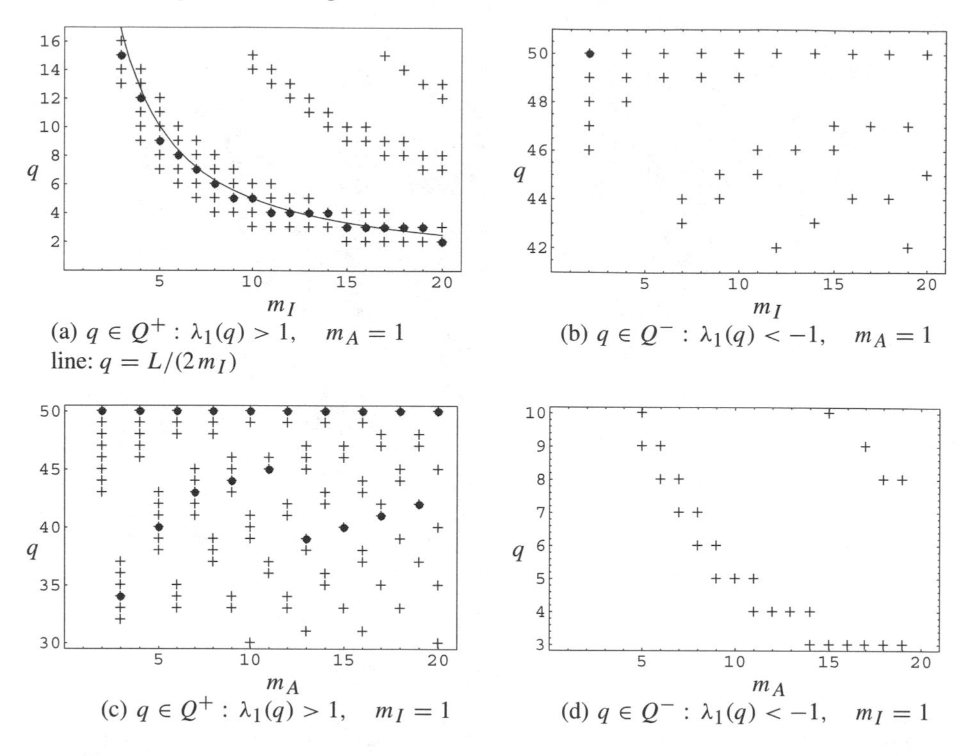

(a) $q \in Q^+ : \lambda_1(q) > 1, \quad m_A = 1$
line: $q = L/(2m_I)$

(b) $q \in Q^- : \lambda_1(q) < -1, \quad m_A = 1$

(c) $q \in Q^+ : \lambda_1(q) > 1, \quad m_I = 1$

(d) $q \in Q^- : \lambda_1(q) < -1, \quad m_I = 1$

Figure 11.9. Critical wave numbers q for which $\mu(q) \equiv |\lambda_1(q)| > 1$; the dots represent the dominant critical wave numbers q_*. Parameters: $L = 100$, $p_c = p_d =: p = 1$.

where $u_\sigma(q) = 1 + 2\cos\left(\frac{2\pi}{L} q m_\sigma\right)$, $q \in \{0, \ldots, L-1\}$. Fig. 11.12 gives an example of the graph of this function.

From this, the sets of critical wave numbers Q^+ and Q^- can easily be determined for fixed speed parameters m_A and m_I. Some examples are shown in fig. 11.9. Note that since $\lambda_1(q)$ is a sinoidal function it is sufficient to consider $q \in \{0, \ldots, \lceil L/2 \rceil\}$.[73]

Hence, if particles of species I *move faster than those of species* A, i.e., if $m_I > 2 > m_A = 1$, then the stationary solution $(\bar{f}_A, \bar{f}_I) = (0.5, 0.5)$ becomes unstable for nonzero "small" wave numbers ($q \leq 16$ if $L = 100$) (cp. fig. 11.9(a)). Moreover, it can be shown that the dominant wave numbers $q_* \in Q^+$ are well approximated by[74]

$$q_* = \left[\frac{L}{2m_I}\right] \qquad \text{for} \quad m_A = 1,\ m_I > 2\,, \tag{11.25}$$

[73] $\lceil y \rceil$ denotes the smallest integer greater than or equal to $y \in \mathbb{R}$.

[74] $[\,y\,]$ denotes the integer closest to $y \in \mathbb{R}$.

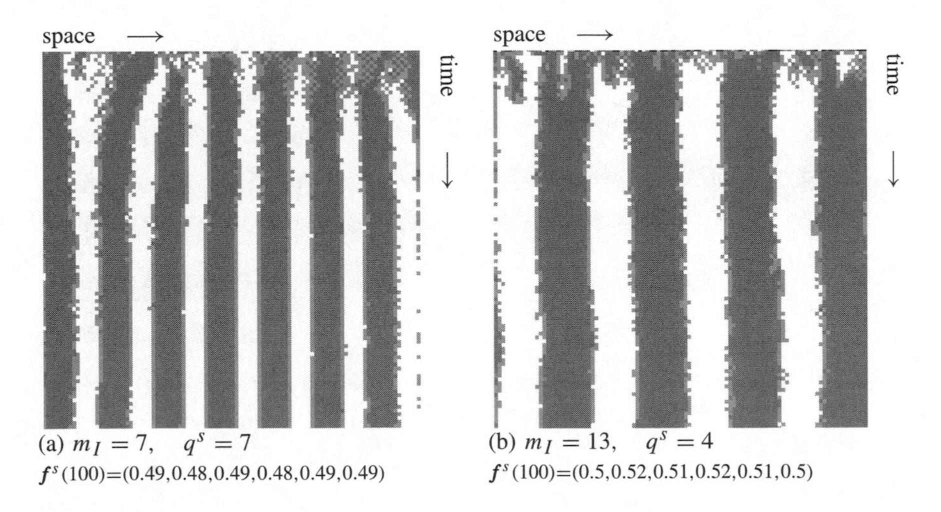

(a) $m_I = 7, \quad q^s = 7$
$f^s(100)$=(0.49,0.48,0.49,0.48,0.49,0.49)

(b) $m_I = 13, \quad q^s = 4$
$f^s(100)$=(0.5,0.52,0.51,0.52,0.51,0.5)

Figure 11.10. Evolution of activator concentration in space and time if $m_I > m_A = 1$. Parameters: $m_A = 1$, $p_c = p_d =: p = 1$, $L = 100$, $k = 0, \ldots, 100$.

because the derivative of $\lambda_1(q)$ (eqn. (11.24)) with respect to q evaluated at this value is almost zero for sufficiently large lattices, i.e.,

$$\partial_q \lambda_1(q)\Big|_{q=\frac{L}{2m_I}} = -\frac{3\pi}{2L}\sin\left(\frac{\pi}{m_I}\right) = \left(-6\pi^2 m_I + 4\pi^2 m_I^3\right)\frac{1}{L^3} + \mathcal{O}\left(\frac{1}{L^7}\right).$$

As an example, space-time plots of two automaton simulations are given in fig. 11.10 for $(m_A, m_I) \in \{(1,4),(1,7)\}$. The wave numbers q^s observed in the plots correspond to the predicted dominant wave numbers q_* shown in fig. 11.9(a).

The *wavelength* of the pattern in the case $(m_A, m_I) = (1,7)$ obtained from linear stability analysis is $L/7 \approx 14.29$, where we assumed $L = 100$. The simulation shown in fig. 11.10(a) confirms this prediction. The observed wavelength is $L/7$, since seven stripes fill the periodic domain. If we take a different lattice size, i.e., $L = 30$, the selected wave number according to the spectral radius and simulation is $q = 2$. Therefore, the wavelength $L/2 = 15$ is conserved up to a small finite size effect imposed by the periodicity of the system. Fig. 11.11 shows pattern evolution if there is no influence of the boundaries in the beginning of the simulation. The simulation was started with one single particle in the middle of the domain, i.e., $\eta_{A,1}(50,0) = 1$, $\eta_{I,i}(50,0) = 0$, and $\eta_{\sigma,i}(r,0) = 0$ for all $\sigma \in \{A, I\}$, $i = 1,2,3$ and $r \neq 50$. Hence, from linear stability analysis "Turing-type" structures are predicted to evolve for parameters $m_I > 2 > m_A = 1$ and $p_c = p_d =: p = 1$.

A pattern in which two spatial wavelengths are simultaneously visible is exhibited in fig. 11.4(f), where we chose $m_A = 3$ and $m_I = 11$. In this case, Q^+ contains wave numbers of very different magnitude, from which $q_1 = 4$ and $q_2 = 32$ are local extrema of the corresponding eigenvalue $\lambda_1(q)$, as can be confirmed in fig. 11.12. Although $q_2 = 32 = q_*$ is the dominant critical wave number, the mode correspond-

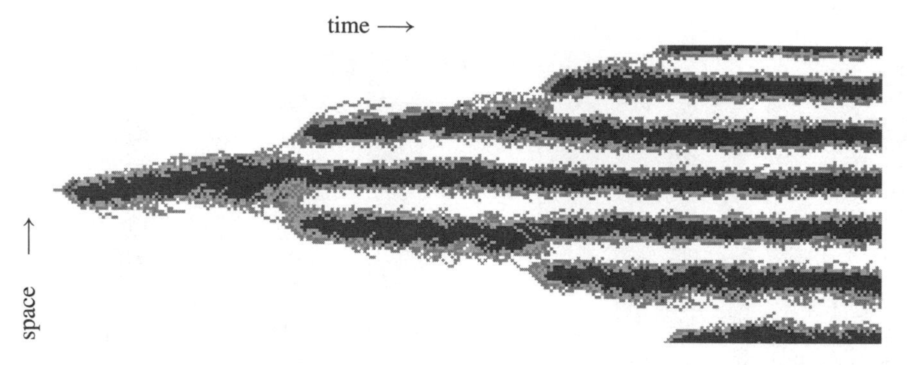

Figure 11.11. Emergence of spatial structure in activator concentration from initial condition $\eta_{A,1}(50, 0) = 1$; all other occupation numbers are zero. Parameters: $m_A = 1$, $m_I = 7$, $p_c = p_d =: p = 1$, $L = 100$, $k = 0, \ldots, 274$.

ing to a wave number of $q_1 = 4$ is also visible in fig. 11.4(f).

If we allow species *A to move faster than species I*, i.e., $m_A > m_I$, then according to linear stability analysis the critical dominant wave numbers are $q_* \in Q^+$ and of "large" magnitude ($q \geq 30$ if $L = 100$), corresponding to wavelengths of less than 4, if $L = 100$ (cp. fig. 11.9(c)). A space-time pattern from a simulation run is shown in fig. 11.13.

Checkerboard Structures. Another type of pattern evolves if the eigenvalue (cp. (11.24)) has a dominant instability at -1, i.e., $q_* \in Q^-$. This situation arises, e.g., in the case $(m_A, m_I) = (1, 2)$, where the mode according to wave number $q = L/2$, i.e., wavelength 2, grows fastest (cp. fig. 11.9(b)). Since this mode develops with an oscillating sign of period 2, a checkerboard-like structure develops, as is shown in fig. 11.14. Note that the appearance of checkerboard patterns can be weakened by the introduction of rest channels, which generally suppress large mode instabilities (cp. sec. 5.4.2).

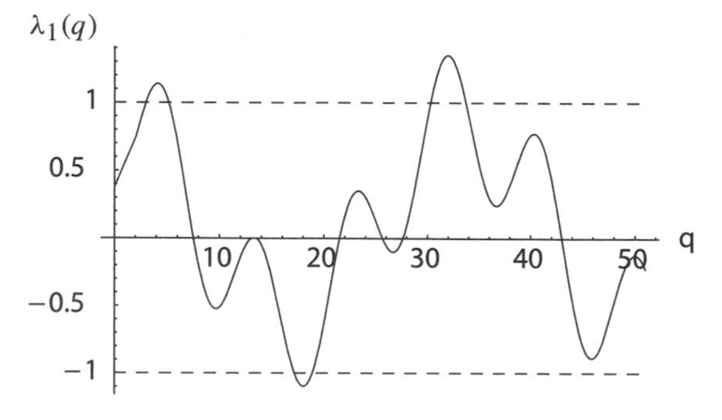

Figure 11.12. Eigenvalue $\lambda_1(q)$ given by (11.24) for $m_A = 3$, $m_I = 11$, and $L = 100$.

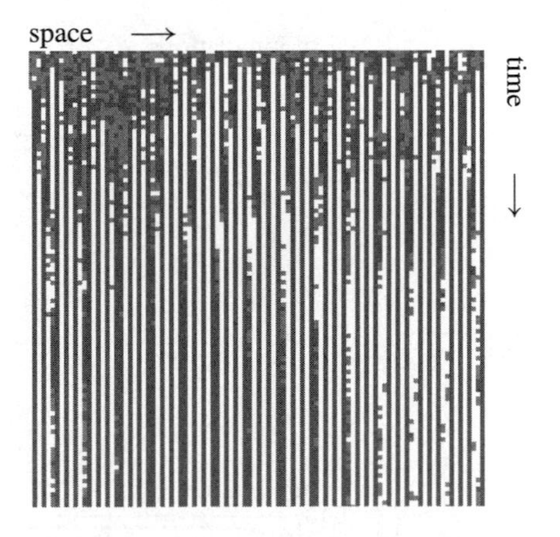

$q^s = 43$ $\quad f^s(100)$=(0.53,0.53,0.54,0.53,0.54,0.53)

Figure 11.13. Evolution of activator concentration in space and time if $m_A > m_I$. Parameters: $m_A = 7, m_I = 1$, $p_c = p_d =: p = 1$, $L = 100$, $k = 0, \ldots, 100$.

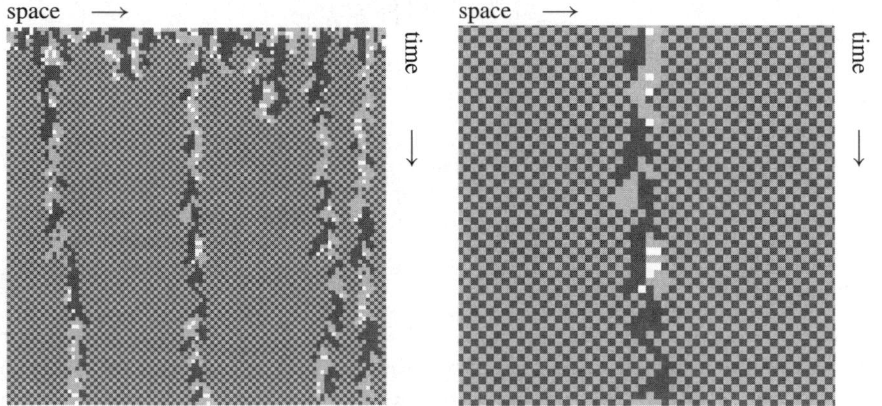

(a) $q^s = 50 f^s(100)$=(0.5,0.46,0.49,0.5,0.48,0.47) (b) Checkerboard pattern: magnified small area of the left figure.

Figure 11.14. Evolution of activator concentration in space and time if the instability $\lambda_1(q) < -1$ is dominant, i.e., $q_* \in Q^-$. Parameters: $m_A = 1, m_I = 2$, $p_c = p_d =: p = 1$, $L = 100$, $k = 0, \ldots, 100$.

Influence of Initial Conditions. We have demonstrated that the evolving pattern in a simulation exhibits the wavelength of the fastest growing mode, i.e., the mode corresponding to the dominant wave number q_*. However, this is not true in general, since the unstable mode selection also depends on the initial distribution of particles.[75] In general, any arbitrary random initial distribution consists of many modes superimposed on each other. The decisive role of initial conditions is illustrated in

[75] As is also the case in continuous reaction-diffusion systems (Murray 2002).

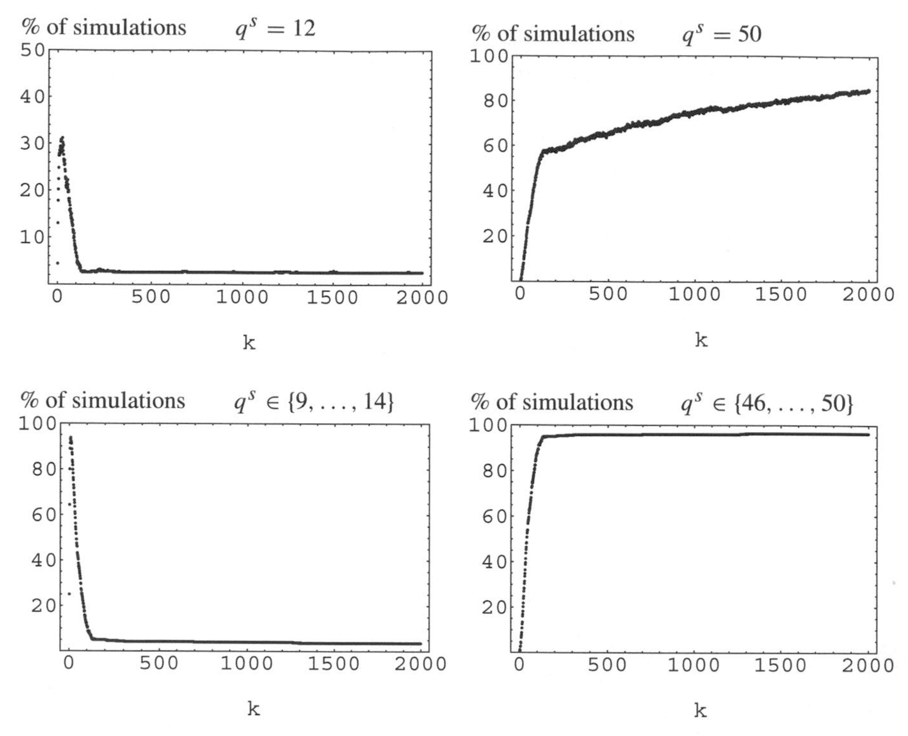

Figure 11.15. Statistics of 500 LGCA simulations: fraction of simulations with dominant critical wave numbers q^s at each time step k starting from different random initial conditions. Parameters: $m_A = 1, m_I = 4, p_c = p_d =: p = 1, L = 100$ and $P\left(\eta_{\sigma,i}(r, 0) = 1\right) = 0.5$ for each $\sigma \in \{A, I\}, i = 1, 2, 3$.

the following, using the parameters $(m_A, m_I) = (1, 4)$, $L = 100$, and $p = 1$. For this parameter set, the dominant critical wave number is $q_* = 12$ (cp. fig. 11.9(a)) with $\lambda_1(12) \approx 1.16778$. But for the (short) wave number $q = 50$, which is also critical, we find $\lambda_1(50) = -1.125$. Hence, mode $F(12, k)$ grows only slightly faster than mode $F(50, k)$. Nevertheless, as is shown in fig. 11.15 in almost 95% of 500 simulations, started with different random initial conditions, modes corresponding to high critical wave numbers determine the pattern, finally.[76] A typical space-time plot for this situation is shown in fig. 11.4(e). It can be seen that the initially growing mode $F(12, k)$, which leads to stripes, becomes "suppressed" by $F(50, k)$, which leads to a checkerboard-like pattern.

A different picture arises if we start with a randomly perturbed initial condition that has a wave number of $q = 12$. Then, as illustrated in fig. 11.16, $F(12, k)$ (stripes) dominates the pattern in almost 96% of 500 simulation runs.

[76] For the statistics shown in fig. 11.15 simulation data have been transformed into Fourier space as $G(q, k) := \left(\sum_{r=0}^{99} e^{\mathbf{i} \frac{\pi}{50} q r} \varrho^s(r, k)\right)^2$, where ϱ^s denotes the mass at each node. At each time step the wave number $q \neq 0$ with maximal $G(q, k)$ is extracted.

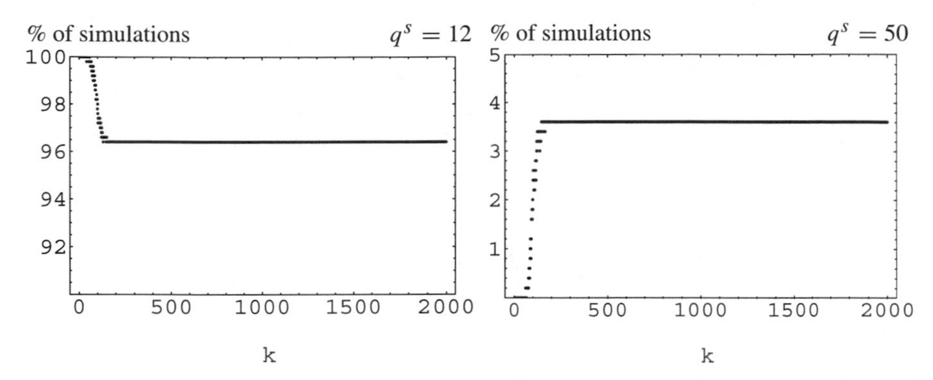

Figure 11.16. Statistics of 500 LGCA simulations: fraction of simulations with dominant critical wave numbers q^s at each time step k starting from different randomly perturbed initial conditions of wave number 12. Parameters: $m_A = 1$, $m_I = 4$, $p_c = p_d =: p = 1$, and $L = 100$.

Effects, as described above, result from an *interaction* of modes. This can*not* be captured by a linear theory where it is assumed that perturbations are "small," implying an independent growth or decay of each of the various modes. Perturbations may soon grow so strongly that the simplifying linearization is no longer appropriate. Hence, in order to take mode interaction processes into account, a *nonlinear analysis* of the Boltzmann equation (11.14) had to be carried out. Segel (1984) gave a good introductory overview of the qualitative features of mode-amplitude equations for nonlinear behavior, devoted to pattern formation in macroscopic reaction-diffusion systems[77].

The Role of Fluctuations: $p_c = p_d =: p < 1$. Now, our objective is to investigate the case of *probabilistic* reactive interactions, i.e., $p_c = p_d =: p < 1$. The nontrivial eigenvalues (11.21) of the Boltzmann propagator for these parameters are given by equation (B.6) in appendix B.2. All space-time patterns introduced so far, emerging from growing modes corresponding to critical wave numbers $q \in Q^+ \cup Q^-$, are conceivable, i.e.,

[77] Cross and Hohenberg 1993; Haken and Olbrich 1978; and Mikhailov 1994 are recommended as well.

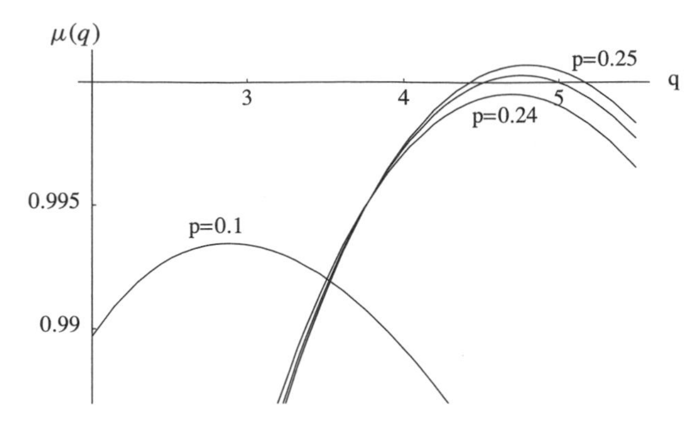

Figure 11.17. Spectral radius $\mu(q)$ for different values of p; the curve between $p = 0.24$ and $p = 0.25$ belongs to the critical reaction parameter $\tilde{p} \approx 0.247$. Parameters: $m_A = 1, m_I = 7$, and $L = 100$.

$$
\begin{aligned}
&q \in Q^+ \text{ and } q \text{ is "small"} && \text{(stripes, e.g., fig. 11.4(b)),}\\
&q \in Q^+ \text{ and } q \text{ is "large"} && \text{(very small narrow stripes,}\\
& && \quad \text{e.g., fig. 11.4(f)),} \quad \text{and}\\
&q \in Q^- \text{ for any magnitude of } q && \text{(checkerboard pattern,}\\
& && \quad \text{e.g., fig. 11.4(e)).}
\end{aligned}
$$

Although linear stability analysis of the deterministic Boltzmann equation can yield very good insight into the automaton dynamics, there are situations in which local fluctuations still play an important role. In particular, if the value of the spectral radius is less than but close to 1 in automaton simulations, a dominant wavelength is present, which contrasts with the results of linear stability analysis. This phenomenon is discussed for parameters $(m_A, m_I) = (1, 7)$ and $L = 100$. For this we calculate from (B.6) the threshold value of the *bifurcation parameter* $\tilde{p}$, for which the spectral radius $\mu(q)$ crosses unity, i.e.,

$$\forall\, p < \tilde{p}: \qquad \mu(q)|_p < 1 \quad \forall\, q \in \left\{0, \ldots, \left\lceil \frac{L}{2} \right\rceil \right\}$$

and

$$\exists\, \tilde{q} \in \left\{0, \ldots, \left\lceil \frac{L}{2} \right\rceil \right\}: \quad \mu(\tilde{q})|_{\tilde{p}} = 1.$$

We find $\tilde{p} \approx 0.247$ and $\tilde{q} = 5$ as indicated in fig. 11.17. Hence, linear stability analysis predicts that no pattern evolves if $p < \tilde{p}$. Simulation results are different! We performed 500 simulation runs with a reactive interaction probability of $p = 0.1 < \tilde{p}$ and identical random initial conditions. From this we find that in about 35% of all simulations the mode corresponding to the wave number $q = 3$ is dominating, and in almost 90% of the simulations modes corresponding to one of the wave numbers $q \in \{2, \ldots, 5\}$ determine the pattern. These results are presented in fig. 11.18.

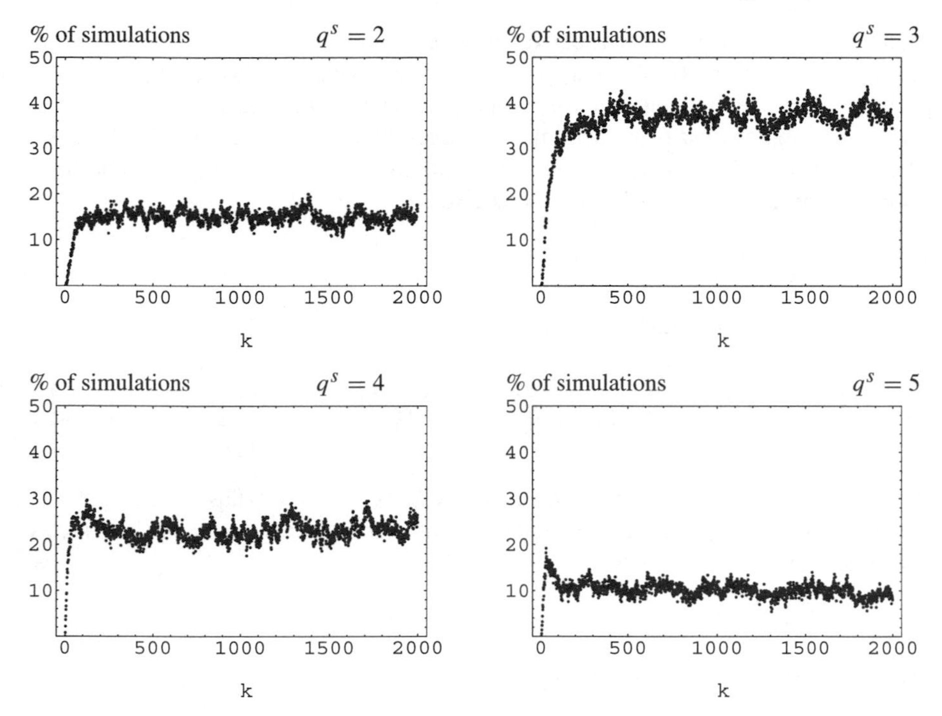

Figure 11.18. Statistics of 500 LGCA simulations: fraction of simulations with dominant critical wave numbers q^s at each time step k starting from the same random initial conditions. Parameters: $m_A = 1, m_I = 7, p_c = p_d =: p = 0.1$, and $L = 100$.

In comparison, a *simulation of the deterministic nonlinear Boltzmann equation* (11.14), started with the same initial condition that we chose for the LGCA simulations, shows that all modes with nonzero wave numbers disappear; i.e., all perturbations vanish (cp. fig. 11.19(a)). On the other hand, a simulation for a reaction parameter $p = 0.25 > \tilde{p}$ shows the theoretically expected wave number $q = 5$ (cp. Figs. 11.17 and 11.19(b)). In this case the dynamics of the nonlinear Boltzmann equation is well captured by linear stability theory. Hence, spatial pattern formation in LGCA simulations is supported by microscopic fluctuations, which lead to continuous perturbations over a wide range of spatial wavelengths. Note that in the "deterministic" LGCA ($p = 1$) no patterns according to linear instability arise if the initial condition is a linear combination of stable modes; i.e., in this case as well, the random initial condition is a necessary condition for pattern formation.

11.1.4 Pattern Formation in Two Dimensions

In the following we investigate Turing pattern formation in two spatial dimensions. What kind of patterns do evolve, and what influence does the lattice geometry have on pattern formation? In this section, we provide answers to these questions for the LGCA activator–inhibitor model with the help of linear stability analysis. Since the mathematical procedure for the derivation of the microscopic LGCA description, the corresponding lattice-Boltzmann equation, and the linear stability analysis are very similar to the one-dimensional case, we do not go into details here.

It is straightforward to extend the activator–inhibitor interaction LGCA model defined in sec. 11.1.2 to two-dimensional lattices: The nearest-neighborhood template $\mathcal{N}_b$ for the *square lattice* ($b = 4$) is given by

$$\mathcal{N}_4 = \{(1,0),\ (0,1),\ (-1,0),\ (0,-1)\},$$

and for the *hexagonal lattice* ($b = 6$)

$$\mathcal{N}_6 = \left\{(1,0),\ \left(\frac{1}{2},\frac{\sqrt{3}}{2}\right),\ \left(-\frac{1}{2},\frac{\sqrt{3}}{2}\right),\ (-1,0),\ \left(-\frac{1}{2},-\frac{\sqrt{3}}{2}\right),\ \left(\frac{1}{2},-\frac{\sqrt{3}}{2}\right)\right\}.$$

For both systems we introduce one rest particle, i.e., $\beta = 1$. Thus, particles X_σ of each species $\sigma \in \{A, I\}$ are distributed in different velocity channels $(r, c_i)_\sigma$, as illustrated in fig. 11.20, where

$$c_i \in \mathcal{N}_b, \quad i = 1, \ldots, b \quad \text{and} \quad c_{\tilde{b}} = (0,0), \quad \tilde{b} = b + 1.$$

Then, the node configuration $\boldsymbol{\eta}(r,k)$ at node $r = (r_1, r_2) \in \mathcal{L}$ will be described as

$$\begin{aligned}\boldsymbol{\eta}(r,k) &= \left(\eta_{A,1}(r,k), \ldots, \eta_{A,\tilde{b}}(r,k), \eta_{I,1}(r,k), \ldots, \eta_{I,\tilde{b}}(r,k)\right)\\ &= (\boldsymbol{\eta}_A(r,k), \boldsymbol{\eta}_I(r,k)) \in \{0,1\}^{2\tilde{b}}.\end{aligned}$$

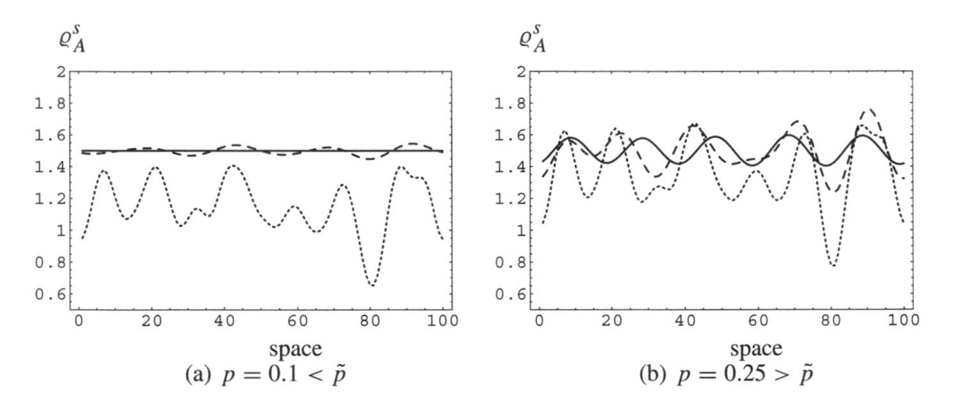

Figure 11.19. Evolution of activator mass $\varrho_A^s(r,k)$ in space and for different times in a simulation of the (deterministic) Boltzmann equation (11.14) (cp. discussion in section 11.1.2) with random initial conditions. Small dashed line: $k = 10$, medium dashed line: $k = 100$, solid line: $k = 1000$. (a) The system has reached the spatially homogeneous equilibrium at time $k = 1000$. (b) The system has reached a spatially inhomogeneous structure with wave number $q^s = 5$ at time $k = 1000$. Parameters: $m_A = 1, m_I = 7, L = 100$, and $\tilde{p} \approx 0.247$.

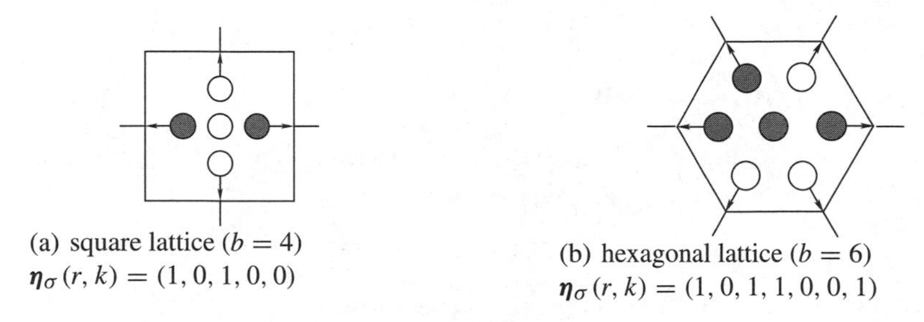

(a) square lattice ($b = 4$)
$\eta_\sigma(r, k) = (1, 0, 1, 0, 0)$

(b) hexagonal lattice ($b = 6$)
$\eta_\sigma(r, k) = (1, 0, 1, 1, 0, 0, 1)$

Figure 11.20. Channels of node r in two-dimensional lattices (one rest channel); gray dots denote the presence of a particle X_σ in the respective channel.

We assume that $|\mathcal{L}| = L_1 \cdot L_2 = L^2$. With an interaction neighborhood of $\mathcal{N}_b^I(r) = \{r\}$ and reaction parameters p_c and p_d defined as in (11.8) we obtain the following reactive interaction rule[78] (R):

$$\mathrm{n}_\sigma^{\mathrm{R}}(r, k) = \begin{cases} \tilde{b} & \text{with prob. } p_c \text{ if } \mathrm{n}_A(r, k) > \mathrm{n}_I(r, k) \geq 0, \\ 0 & \text{with prob. } p_d \text{ if } 0 \leq \mathrm{n}_A(r, k) < \mathrm{n}_I(r, k), \\ \mathrm{n}_\sigma(r, k) & \text{otherwise} \end{cases} \tag{11.26}$$

for each $\sigma \in \{A, I\}$, where the number of particles X_σ at node r at time k is given by

$$\mathrm{n}_\sigma(r, k) = \sum_{i=1}^{\tilde{b}} \eta_{\sigma,i}(r, k).$$

The microdynamical description for the two-dimensional LGCA models and the derivation of the lattice-Boltzmann equation follow along the lines of sec. 11.1.2. Then, the *lattice-Boltzmann equation* is given by an equation of the form

$$\begin{aligned} f_{\sigma,i}(r + m_\sigma c_i, k + 1) - f_{\sigma,i}(r, k) &= \frac{1}{\tilde{b}} \sum_{l=1}^{\tilde{b}} E\left(\eta_{\sigma,l}^{\mathrm{R}}(r, k)\right) - f_{\sigma,i}(r, k) \\ &= \tilde{\mathcal{C}}_{\sigma,i}\left(\boldsymbol{f}(r, k)\right), \end{aligned} \tag{11.27}$$

where

[78] Note that formulation and interpretation of this rule correspond to the reactive interaction rule defined in (11.9).

$$\boldsymbol{f}(r,k) = \left(f_{A,1}(r,k), \ldots, f_{A,\tilde{b}}(r,k), f_{I,1}(r,k), \ldots, f_{I,\tilde{b}}(r,k)\right)$$
$$= \left(f_j(r,k)\right)_{j=1}^{2\tilde{b}} \in [0,1]^{2\tilde{b}}.$$

For both systems, we find that the spatially uniform steady states $\left(\bar{f}_A, \bar{f}_I\right)$, which are solutions of $\tilde{\mathcal{C}}_{\sigma,i}\,(\boldsymbol{f}(r,k)) = 0$, are given by

$$\left(\bar{f}_A, \bar{f}_I\right) = \{(0,0), (a_1,a_2), (1,1)\}, \tag{11.28}$$

where (a_1, a_2) depends on the model and reaction parameters, and will be determined later on. As in the one-dimensional case, the stationary states $(0,0)$ and $(1,1)$ are unstable and (a_1,a_2) is stable with respect to spatially homogeneous perturbations. As a result of the linearization of (11.27) around $\left(\bar{f}_A, \bar{f}_I\right) = (a_1,a_2)$ and Fourier transformation we obtain the *Boltzmann propagator* (4.36):

$$\Gamma(q) = T\{\boldsymbol{I} + \Omega^0\}, \quad q = (q_1,q_2) \quad \text{with} \quad q_1, q_2 = 0, \ldots, L-1, \tag{11.29}$$

where

$$T = \operatorname{diag}\left(e^{-\mathbf{i}(2\pi/L)(c_1\cdot q)m_A}, \ldots, e^{-\mathbf{i}(2\pi/L)(c_{\tilde{b}}\cdot q)m_A}, e^{-\mathbf{i}(2\pi/L)(c_1\cdot q)\,m_I}, \ldots,\right.$$
$$\left. e^{-\mathbf{i}(2\pi/L)(c_{\tilde{b}}\cdot q)\cdot m_I}\right)$$

and[79]

$$\boldsymbol{I} + \Omega^0 = \begin{pmatrix} \omega_1 & \ldots & \omega_1 & \omega_2 & \ldots & \omega_2 \\ \vdots & & \vdots & \vdots & & \vdots \\ \omega_1 & \ldots & \omega_1 & \omega_2 & \ldots & \omega_2 \\ \omega_3 & \ldots & \omega_3 & \omega_4 & \ldots & \omega_4 \\ \vdots & & \vdots & \vdots & & \vdots \\ \omega_3 & \ldots & \omega_3 & \omega_4 & \ldots & \omega_4 \end{pmatrix} \in \mathbb{R}^{2\tilde{b}\times 2\tilde{b}}.$$

The spectrum of the Boltzmann propagator (11.29) is given by

$$\Lambda_{\Gamma(q)} = \{\lambda_1(q), \lambda_2(q), 0\},$$

where 0 has a multiplicity of $2\tilde{b} - 2$ and

$$\boxed{\lambda_{1,2}(q) = \frac{1}{2}\left(\omega_1\, u_A(q) + \omega_4\, u_I(q) \pm\sqrt{4\,(\omega_2\omega_3 - \omega_1\,\omega_4)\, u_A(q)u_I(q) + (\omega_1\, u_A(q) + \omega_4 u_I(q))^2}\,\right)} \tag{11.30}$$

[79] The matrix elements ω_i, $i = 1, \ldots, 4$, are different terms for the square and hexagonal lattice model depending on the reaction parameters.

with $u_\sigma(q) := 1 + \sum_{j=1}^{b} e^{-\mathbf{i}(2\pi/L)(c_j \cdot q)m_\sigma}$, $\sigma \in \{A, I\}$, $q = (q_1, q_2)$, and $q_1, q_2 = 0, \ldots, L-1$.

For both models, $\Gamma(q)$ is diagonalizable and therefore the temporal growth of modes $F_{\sigma,i}(q,k)$ is determined solely by the dominant eigenvalue.

In two-dimensional systems, groups of unstable modes with identical absolute value of the wave number $|q| = \mathrm{q}$ simultaneously start to grow. Therefore, according to linear theory, any superposition of these modes determines the dynamics of the system;[80] i.e., according to (4.42)

$$\delta f_{\sigma,i}(r,k) \sim \sum_{\substack{q \in Q^c \\ |q|=\mathrm{q}}} e^{-\mathbf{i}(2\pi/L)q \cdot r} F_{\sigma,i}(q,k), \tag{11.31}$$

where Q^c is the set of critical wave numbers.

The following analysis will be restricted to two sets of reaction parameters:

$$\text{(i) } p_c = p_d = 1 \quad \text{and} \quad \text{(ii) } p_c = 0.9,\ p_d = 0.1.$$

With these reaction parameters "typical patterns" evolving from the LGCA dynamics are captured. Furthermore, we take $m_A = 1$, $m_I = 11$ and $L_1 = L_2 = L = 100$.

The Square Lattice Model. The stationary states $\left(\bar{f}_A, \bar{f}_I\right) = (a_1, a_2)$ of our LGCA model on a square lattice are given by

$$\text{(i) } p_c = p_d = 1: \quad a_1 = a_2 = 0.5$$

and

$$\text{(ii) } p_c = 0.9,\ p_d = 0.1: \quad a_1 \approx 0.473,\ a_2 \approx 0.727,$$

and the corresponding dominant eigenvalues λ_1 (cp. (11.30)) are shown in fig. 11.21. In order to stress the instability-dependence on the wave number magnitude (cp. (11.31)), the wave numbers are represented in polar coordinate form in figs. 11.21(a) and 11.21(c), i.e.,

$$q = (q_1, q_2) \qquad \text{with } q_1 = |q| \cos(\phi),\ q_2 = |q| \sin(\phi).$$

In case (i), very distinguished collections of critical wave numbers according to various directions ϕ exist (cp. fig. 11.21(a)). The spectral radius $\mu(q) = |\lambda_1(|q|, \phi)|$ is maximal for wave numbers associated with the diagonal directions, i.e., $\phi_d \in \{45°, 135°\}$, and a magnitude of $|q| \approx 6.36$. Fig. 11.21(b) shows that discrete wave numbers (q_1, q_2) of the fastest growing modes are $q_* \in \{(4,5), (5,4)\}$. Consequently, linear stability analysis predicts a spatial pattern with these wave numbers q_* and strong anisotropies in diagonal directions. This result is in good agreement with LGCA simulations. Fig. 11.22(a) shows two snapshots of the spatial distribution of activator concentration after 100 time steps. Both simulations were prepared in a spatially uniform initial state at time $k = 0$. Note that the dominant eigenvalue λ_1 has also instabilities at -1, i.e., $Q^- \neq \varnothing$, which explains the existence of local areas of checkerboard patterns in fig. 11.22(a).

[80] For further reading, see Mikhailov 1994.

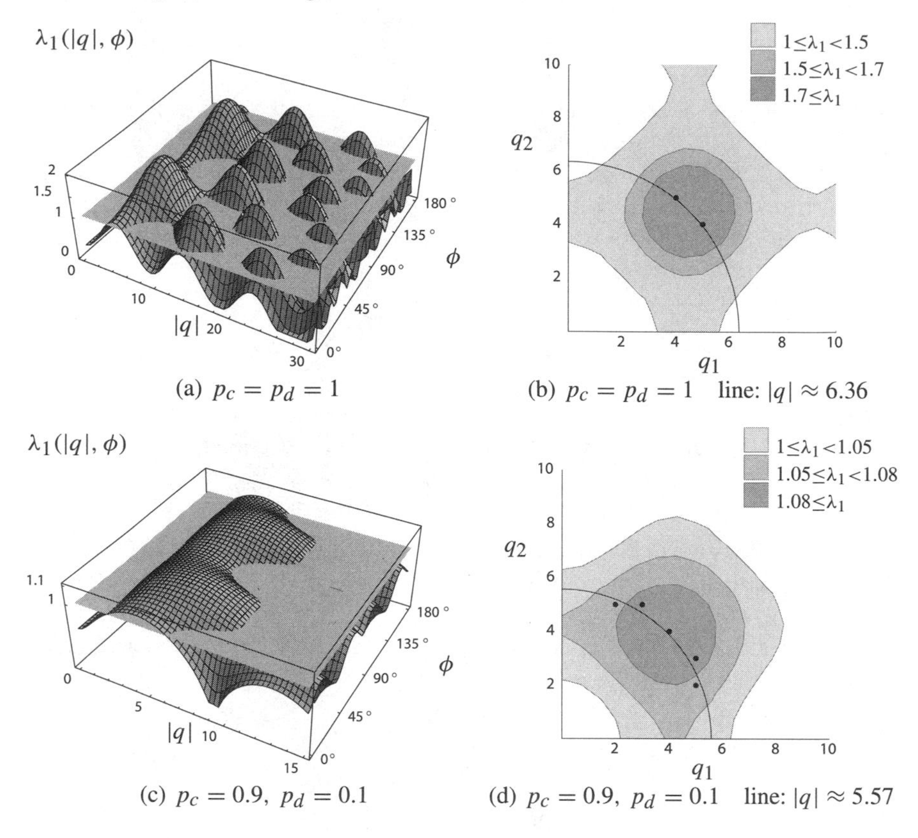

Figure 11.21. Dominant eigenvalue λ_1 (cp. (11.30)) for the *square lattice* dependent on the wave number $q = (q_1, q_2)$, which is represented in polar coordinate form in (a) and (c). The threshold λ_1 is marked by the gray plane. Dots in (b) and (d) represent wave numbers q, whose magnitude is close to the line. Parameters: $m_A = 1$, $m_I = 11$, and $L_1 = L_2 = L = 100$.

In case (ii), as displayed in fig. 11.21(c), instabilities of λ_1 are of much smaller magnitude than in case (i). Furthermore, the patches of critical wave numbers are less distinguished. Dominant critical wave numbers, which refer to the maximal spectral radius $\mu(q) = |\lambda_1(|q|, \phi)|$, have a magnitude of $|q| \approx 5.57$. As can be seen in fig. 11.21(d) this corresponds to a discrete wave number of $q_* = (4, 4)$. Hence, as in case (i), modes associated with diagonal directions grow fastest but, in contrast to case (i), the growth is very slow. Furthermore, since the maximum of the spectral radius is not very distinctive, modes corresponding to other wave numbers, e.g., $\{(2, 5), (3, 5), (5, 2), (5, 3)\}$, might appear (cp. fig. 11.21(d)). But again, the emerging pattern is characterized by a superposition of modes refering to both diagonal directions, as it is shown in fig. 11.22(b).

The Hexagonal Lattice Model. The stationary states $\left(\bar{f}_A, \bar{f}_I\right) = (a_1, a_2)$ for a hexagonal lattice are given by

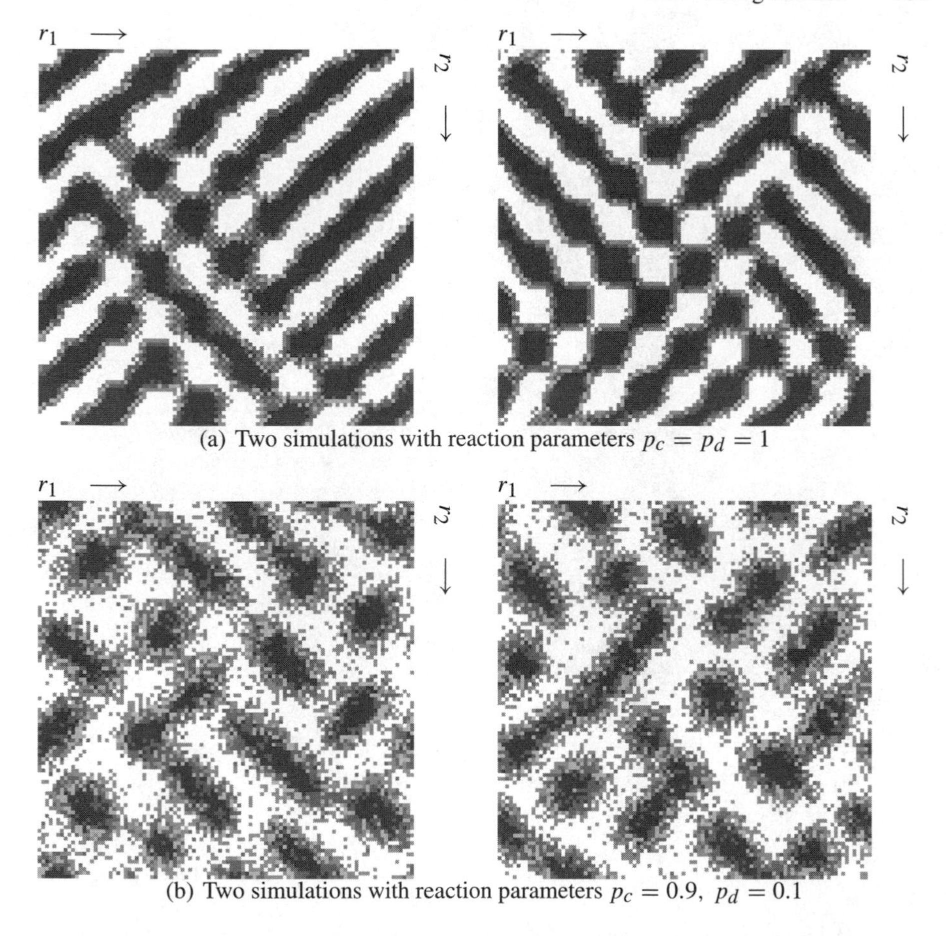

(a) Two simulations with reaction parameters $p_c = p_d = 1$

(b) Two simulations with reaction parameters $p_c = 0.9,\ p_d = 0.1$

Figure 11.22. Simulations: activator concentration in a two-dimensional *square lattice* after $k = 100$ time steps started from different random initial conditions where $P(\eta_{A,i}(r,k) = 1) = \bar{f}_A = a_1$, $P(\eta_{I,i}(r,k) = 1) = \bar{f}_I = a_2$ for each direction c_i. Parameters: $m_A = 1$, $m_I = 11$, $L_1 = L_2 = L = 100$.

(i) $p_c = p_d = 1$: $\quad a_1 = a_2 = 0.5$

and

(ii) $p_c = 0.9$, $p_d = 0.1$: $\quad a_1 \approx 0.505$, $a_2 \approx 0.721$,

and the corresponding dominant eigenvalues λ_1 (eqn. (11.30)) are shown in fig. 11.23, where the wave numbers are represented in polar coordinate form in figs. 11.23(a) and 11.23(c).

Case (i): The dominant eigenvalue λ_1 for the hexagonal lattice (cp. fig. 11.23(a), 11.23(b)), exhibits larger areas of critical wave numbers than the corresponding eigenvalue for the square lattice (cp. fig. 11.21(a), 11.21(b)). However, these are interconnected and less isolated for the hexagonal lattice. Furthermore, the domi-

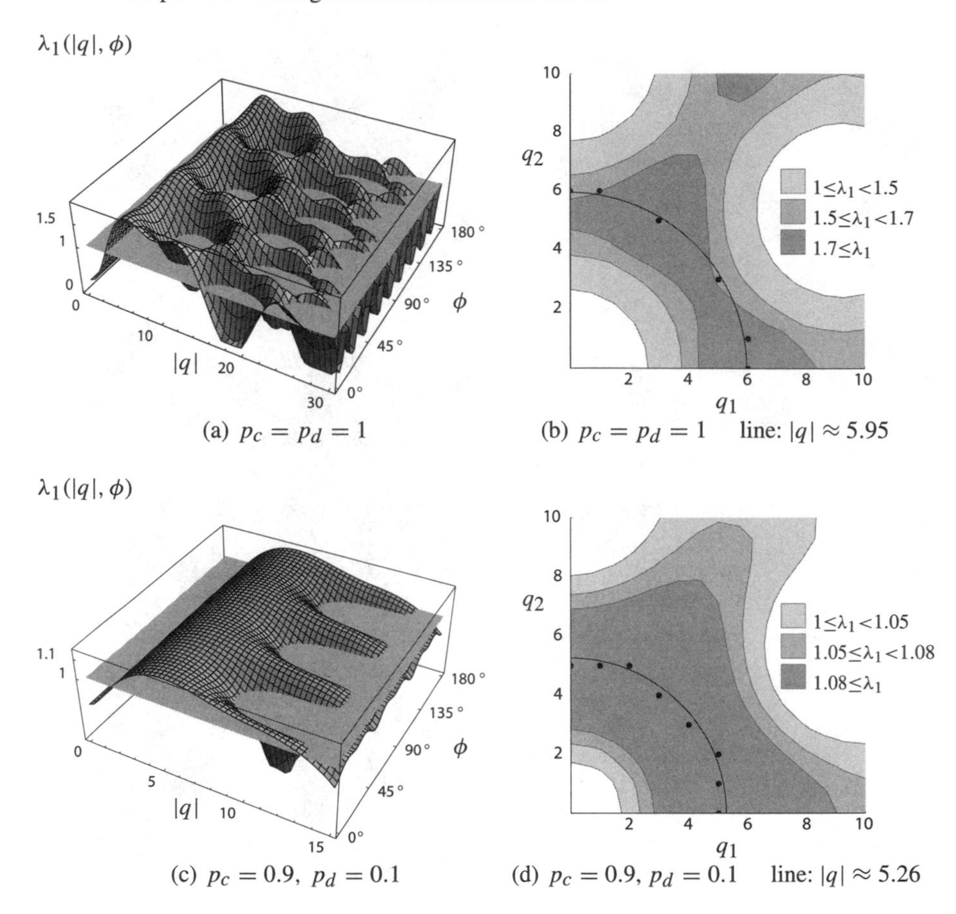

Figure 11.23. Dominant eigenvalue λ_1 (eqn. (11.30)) for the *hexagonal lattice* depends on the wave number $q = (q_1, q_2)$, which is represented in polar coordinate form in (a) and (c); dots in (b) and (d) represent wave numbers q whose magnitude is close to the line. Parameters: $m_A = 1, m_I = 11$, and $L_1 = L_2 = L = 100$.

nant critical wave numbers are associated with directions $\phi_t \in \{0°, 60°, 120°, 180°\}$, rather than with diagonal directions. Hence, the spatial pattern is characterized by a superposition of stripes with directions ϕ_t. An example of this type of pattern is shown in fig. 11.24(a).

Case (ii): From fig. 11.23(c) it can be seen that the dominant eigenvalue λ_1 is nearly *isotropic*; i.e., $\lambda_1(\phi, |q|)$ is equal-valued for all directions ϕ associated with the maximal value $|q| := \mathsf{q} \approx 5.26$. Hence, all modes related to wave numbers with magnitude q start to grow simultaneously (cp. (11.31)). As they grow, nonlinear terms neglected during linearization of the lattice-Boltzmann equations may become important. Then, groups of modes may enhance or suppress the development of themselves and/or other groups. In continuous systems, this process is studied by a nonlinear evolution equation for the amplitudes of unstable modes (Mikhailov 1994).

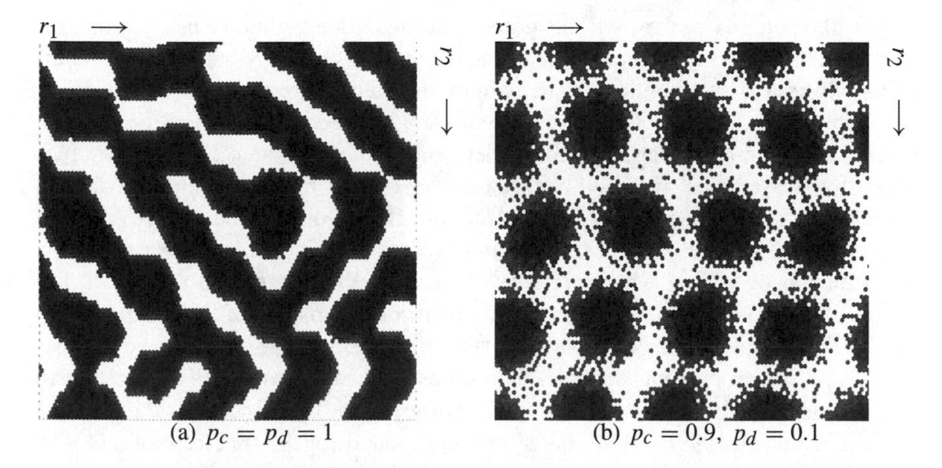

(a) $p_c = p_d = 1$ (b) $p_c = 0.9,\ p_d = 0.1$

Figure 11.24. Simulations: activator concentration in a two-dimensional *hexagonal lattice* after $k = 100$ time steps started from different random initial conditions, in which $P(\eta_{A,i}(r,k) = 1) = \bar{f}_A = a_1$, $P(\eta_{I,i}(r,k) = 1) = \bar{f}_I = a_2$ for each direction c_i. Parameters: $m_A = 1, m_I = 11$, and $L_1 = L_2 = L = 100$.

It turns out that a group of three modes q^0, q^1 and q^2 with equal magnitude exists and that the corresponding vectors of wave numbers form an equilateral triangle (cp. fig. 11.25(a)), i.e.,

$$q^j = (q_1^j, q_2^j) = \mathfrak{q}\left(\cos\left(\phi + j\frac{2\pi}{3}\right), \sin\left(\phi + j\frac{2\pi}{3}\right)\right), \qquad j = 0, 1, 2. \tag{11.32}$$

They mutually enhance each other's growth. The spatial pattern in continuous systems produced by the superposition of these modes is a collection of spots. Since LGCA simulations show similar patterns (cp. fig. 11.24(b)), we suspect the mutual self-enhancement property of mode groups corresponding to wave numbers characterized by (11.32) are also valid for the discrete lattice-Boltzmann model. In the linear regime, the portion of the solution determined by this group of modes is given by

$$\delta f_{\sigma,i}(r,k) \sim A_{\sigma,i}(k) \sum_{j=0}^{2} e^{-\mathbf{i}(2\pi/L)q^j \cdot r}, \tag{11.33}$$

where we defined $A_{\sigma,i}(k) := F_{\sigma,i}(q^j, k)$ for $j = 0, 1, 2$. As illustrated in fig. 11.25(b), the contour line of the real part of the right-hand side of (11.33) is almost circular.

Summary. In this section we showed that in two-dimensional systems the Turing pattern variety is enlarged. Depending on the reaction parameters p_c and p_d, striped or spot-like patterns evolve. We illustrated that the lattice geometry has a strong influence on the evolving pattern and that this behavior is correctly predicted by mean-field analysis. On a square lattice the pattern shows anisotropies in diagonal directions (cp. fig. 11.22), which are not present on a hexagonal lattice (cp. fig. 11.24). A similar pattern dependence on the underlying lattice has been observed and analyzed by Bussemaker 1996, who introduced a LGCA model of random walkers that interact through nearest-neighbor attraction. On the other hand, a LGCA model of "swarming" (Bussemaker, Deutsch, and Geigant 1997; Deutsch 1996) exhibits the formation of streets, which are very similar for the square and hexagonal lattice (cp. ch. 8). This type of pattern results from the formation of "orientational order" by the particles. The corresponding spectrum has a maximum at wave number $q = (0, 0)$, which prevents spatial anisotropies from manifesting as strongly as in LGCA models with nonzero maxima indicating spatial frequencies in the patterns (cp., e.g., Deutsch 1999b). A similar situation is found for the spectrum of a LGCA model for excitable media, which we introduce in the next section (11.2).

11.1.5 Derivation and Analysis of a Macroscopic Description of the Lattice-Gas Cellular Automaton

In section 4.5 we derived macroscopic dynamical equations (4.51), which reduce to

$$\partial_t \varrho_\sigma(x,t) = \hat{a} \sum_{i=1}^{3} \tilde{C}_{\sigma,i}\left(\boldsymbol{f}(x,t)\right) + D_\sigma \partial_{xx} \varrho_\sigma(x,t), \tag{11.34}$$

assuming appropriate space ($x = r\epsilon$) and time ($t = k\delta$) scaling ("diffusive scaling", i.e., $\lim_{\substack{\delta\to 0\\ \epsilon\to 0}} \epsilon^2/\delta = \text{const} := a$). The rate of change of the number of particles scales as $h(\epsilon, \delta) = \hat{a}\delta$, $\hat{a} \in \mathbb{R}^+$. Furthermore,

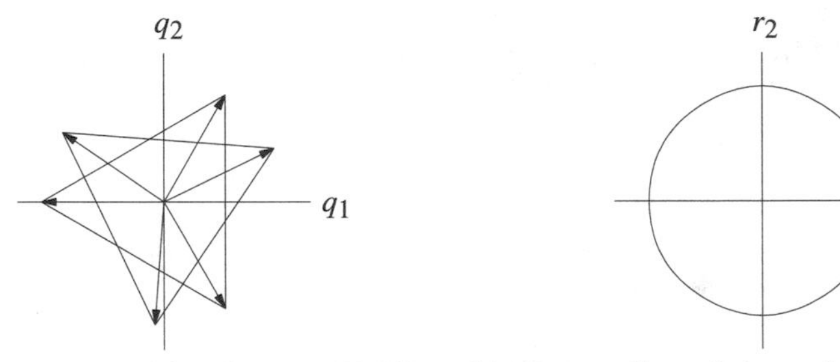

(a) Wave numbers given by eqn. (11.32) (b) Contour line of the real part of the right hand side of eqn. (11.33)

Figure 11.25. Groups of unstable modes associated with wave numbers that form an equilateral triangle (a) and the induced spatial pattern (b).

$$\boxed{D_\sigma := \lim_{\substack{\delta\to 0\\ \epsilon\to 0}} \frac{m_\sigma^2 \epsilon^2}{2\delta} = \frac{1}{2} a m_\sigma^2}$$

is the *diffusion coefficient* of species $\sigma \in \{A, I\}$. Since the LGCA dynamics does not emphasize any particular direction, we assume that the numbers of particles moving to the left or right or at rest are equal, i.e., $f_{\sigma,i}(x,t) = \varrho_\sigma(x,t)/3$, $i = 1, 2, 3$. Then, the reaction part of (11.34) becomes[81]

$$\begin{aligned}
\hat{a}\sum_{i=1}^{3} \tilde{C}_{A,i}(\boldsymbol{f}) = 3\hat{a}\left(1-\frac{\varrho_A}{3}\right)\frac{\varrho_A}{3} \\
\cdot\left(\left(1-\frac{\varrho_I}{3}\right)^2\left(2\left(1-\frac{\varrho_I}{3}\right)-\frac{\varrho_A}{3}\left(1-4\frac{\varrho_I}{3}\right)\right)p_c\right. \\
\left.-\left(\frac{\varrho_I}{3}\right)^2\left(3\left(1-\frac{\varrho_A}{3}\right)-2\frac{\varrho_I}{3}\left(1-2\frac{\varrho_A}{3}\right)\right)p_d\right) \\
=: H_A(\varrho_A, \varrho_I)
\end{aligned} \tag{11.35}$$

and

$$\begin{aligned}
&\hat{a}\sum_{i=1}^{3} \tilde{C}_{I,i}(\boldsymbol{f}) \\
&= \hat{a}\left(1-\frac{\varrho_I}{3}\right)\varrho_A \\
&\cdot\left(3\left(1-\frac{\varrho_I}{3}\right)^2-\varrho_A\left(1-4\frac{\varrho_I}{3}+\frac{\varrho_I^2}{3}\right)+\left(\frac{\varrho_A}{3}\right)^2\left(1-2\varrho_I+2\frac{\varrho_I^2}{3}\right)\right)p_c \\
&-\hat{a}\left(1-\frac{\varrho_A}{3}\right)\varrho_I \\
&\cdot\left(1+\frac{\varrho_A}{3}\left(2\varrho_I-2-\frac{\varrho_I^2}{3}\right)+\left(\frac{\varrho_A}{3}\right)^2\left(1-2\varrho_I+2\frac{\varrho_I^2}{3}\right)\right)p_d \\
&:= H_I(\varrho_A, \varrho_I),
\end{aligned} \tag{11.36}$$

where $\varrho_\sigma = \varrho_\sigma(x,t)$.

Hence, we obtain the following system of LGCA *reaction-diffusion equations*

$$\boxed{\begin{aligned}
\partial_t \varrho_A(x,t) &= H_A(\varrho_A(x,t), \varrho_I(x,t)) + D_A \partial_{xx}\varrho_A(x,t), \\
\partial_t \varrho_I(x,t) &= H_I(\varrho_A(x,t), \varrho_I(x,t)) + D_I \partial_{xx}\varrho_I(x,t),
\end{aligned}} \tag{11.37}$$

where $\varrho_A(x,t), \varrho_I(x,t) \in [0,3]$ for all $x \in [0,l]$ and $t \in \mathbb{R}^+$.

[81] Compare eqns. (B.2) and (B.3) in appendix B.1.

As was mentioned at the beginning of this chapter, continuous reaction-diffusion systems can exhibit diffusion-driven instabilities. These occur when a homogeneous stationary state, which is stable to small, spatially homogeneous perturbations in the absence of diffusion, becomes unstable to small perturbations due to the presence of diffusion (cp. sec. 11.1.1). Next, we investigate if Turing patterns can be observed in the LGCA macroscopic equation (11.37). Again, for the sake of simplicity in the mathematical derivations, primary attention will be given to the special case of equal reaction parameters

$$p_c = p_d =: p \in [0, 1].$$

Following the steps described in sec. 11.1.1, we first determine the stability of spatially uniform stationary states $(\bar{\varrho}_A, \bar{\varrho}_I)$, given by

$$H_A(\varrho_A, \varrho_I) = 0 \quad \text{and} \quad H_I(\varrho_A, \varrho_I) = 0.$$

This is fulfilled, if $(\bar{\varrho}_A, \bar{\varrho}_I) \in \{(0, 0), (3/2, 3/2), (3, 3)\}$.

The Jacobian matrix evaluated at the nontrivial steady state $(\bar{\varrho}_A, \bar{\varrho}_I) = (3/2, 3/2)$ becomes

$$\boldsymbol{J} = \begin{pmatrix} \partial_A H_A & \partial_I H_A \\ \partial_A H_I & \partial_I H_I \end{pmatrix} = \hat{a} p \begin{pmatrix} \frac{1}{8} & -\frac{3}{4} \\ \frac{9}{8} & -\frac{7}{4} \end{pmatrix}, \qquad \hat{a}, p > 0, \tag{11.38}$$

where $\partial_\sigma H_{A/I}$ denotes the partial derivative of $H_{A/I}$ with respect to ϱ_σ. Hence, according to the signs of $\boldsymbol{J}$, the system (11.37) can be identified as a "real activator–inhibitor model" (cp. (11.5)) and therefore $(\bar{\varrho}_A, \bar{\varrho}_I) = (3/2, 3/2)$ is stable with respect to spatially homogeneous perturbations.

Next, we determine the critical diffusion ratio D^c given by (11.3):

$$\begin{aligned} \frac{D_I}{D_A} > D^c &:= \frac{(\det \boldsymbol{J} - \partial_I H_A \partial_A H_I) + 2\sqrt{-\partial_I H_A \partial_A H_I \det \boldsymbol{J}}}{(\partial_A H_A)^2} \\ &\approx 186.952, \end{aligned} \tag{11.39}$$

where $\det \boldsymbol{J} := \partial_A H_A \partial_I H_I - \partial_I H_A \partial_A H_I$. The dominant critical wave number q_*, which is given by equation (11.7), reads

$$q_* = \frac{l}{2\pi} \tilde{q}_* = \frac{l}{2\pi} \sqrt[4]{0.625 \frac{\hat{a}^2 p^2}{D_A D_I}}. \tag{11.40}$$

Thus, only if the diffusion ratio D_I/D_A is larger than D^c, it is possible to destabilize the system (11.37) and a finite range of linearly unstable wave numbers can be found. Depending on the size l of the spatial domain, the allowable discrete wave numbers q_* can be determined. In the example illustrated in fig. 11.26, only one mode becomes unstable, which is the mode corresponding to the wave number $q_* = 3$.

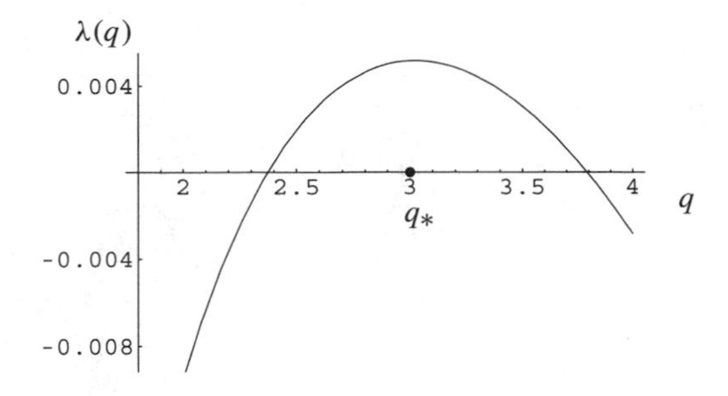

Figure 11.26. Plot of the largest root λ of the dispersion relation (11.6) for the LGCA reaction-diffusion system (11.37). Parameters: $p = 1$, $\hat{a} = 0.5$, $D_A = 0.25$, $D_I = 56.25$, and $l = 58.058$.

Comparison of Lattice-Boltzmann and LGCA Reaction-Diffusion Systems. As demonstrated in sec. 11.1.3, the LGCA lattice-Boltzmann model exhibits Turing-type patterns for $m_I > 2 > m_A = 1$. Hence, with $m_A = 1$, we get

$$D_A = \frac{1}{2}a \quad \text{and} \quad D_I = \frac{1}{2}am_I^2 = D_A m_I^2.$$

Thus, the inequality for the ratio of diffusion coefficients (11.39) becomes

$$m_I^2 > 186.952,$$

which implies that $m_I > 13.673\ldots$ must hold.

An expression for the dominant critical wave number q_* (eqn. (11.40)) in terms of the speed parameter m_I is obtained as

$$\boxed{q_* = \frac{l}{2\pi}\sqrt[4]{\frac{2.5\hat{a}^2 p^2}{a^2 m_I^2}} = \epsilon \frac{L}{2\pi}\sqrt[4]{\frac{2.5\hat{a}^2 p^2}{a^2 m_I^2}}.} \tag{11.41}$$

Therefore, with the parameters p, m_I, and L, we can compute the dominant critical wave number q_* of the LGCA lattice-Boltzmann equation (cp. (11.21)). Then, given the reactive scaling $\hat{a}$, expression (11.41) determines the corresponding space-scaling for the macroscopic reaction-diffusion system (11.37) in terms of ϵ, or in turn fixing the macroscopic system length l, the reactive scaling $\hat{a}$ can be specified.

Example: As an example we choose the following parameters

$$p = 1, \quad m_A = 1, \; m_I = 15, \quad L = 100, \quad a = \lim_{\substack{\delta\to 0\\ \epsilon\to 0}} \frac{\epsilon^2}{\delta} = 0.5, \quad \hat{a} = 0.5$$

and get

$$D_A = 0.25, \; D_I = 56.25, \qquad \frac{D_I}{D_A} = 225 > 186.952 = D^c.$$

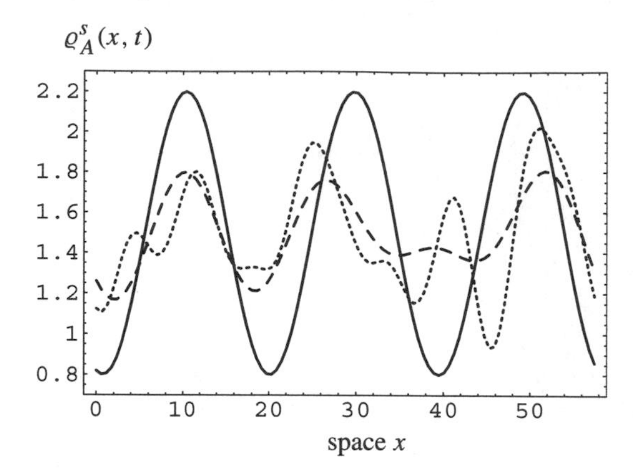

Figure 11.27. Numerical solution of the continuous reaction-diffusion system (11.37) started from random initial conditions. Small dashed line: $t = 0.005$, medium dashed line: $t = 0.05$, solid line: $t = 0.35$. Parameters: $p = 1$, $a = 0.5$, $\hat{a} = 0.5$, $D_A = 0.25$, $D_I = 56.25$, and $l = 58.058$.

From fig. 11.9(a) we obtain the dominant critical wave number of the LGCA lattice-Boltzmann system as

$$q_* = 3.$$

Thus this system corresponds to the continuous reaction-diffusion system (11.37) with a system length of

$$l = q_* 2\pi \sqrt[4]{\frac{a^2 m_I^2}{2.5\hat{a}^2 p^2}} \approx 58.058$$

and a spatial and temporal scaling of

$$\epsilon = \frac{l}{L} \approx 0.581, \quad \delta \approx \frac{\epsilon^2}{a} \approx 0.674.$$

The largest root of the dispersion relation for this example is shown in fig. 11.26. Furthermore, fig. 11.27 illustrates the numerical solution[82] of the continuous reaction-diffusion system (11.37) with a system length of $l = 58.058$ started from random initial conditions. The emerging wavelength of $q^s = 3$ is consistent with our theoretical expectations.

In summary, Turing instabilities in the continuous LGCA reaction-diffusion system (11.37) can occur if the diffusion coefficients fulfill the inequality (11.39). If the system is viewed as an approximation of the LGCA lattice-Boltzmann model (cp. (11.14)), then Turing instabilities are possible for the speed parameters $m_I \geq 14$ if

[82] The C-program for the numerical solution of system (11.37) has been kindly provided by N. Graf, U. Krause, and C. Meyer, University of Osnabrück.

$m_A = 1$, and hence for the corresponding diffusion coefficients $D_A = 1/2\,a$ and $D_I = D_A\, m_I^2$. In contrast, the LGCA lattice-Boltzmann model already shows Turing patterns for speed parameters $m_I \geq 3$, if $m_A = 1$. Furthermore, we derived an appropriate spatial and temporal scaling relation between both models in the "Turing regime," i.e., for $m_I \geq 14$.

It turned out that the variety of pattern formation processes in the lattice-Boltzmann model is larger than in the continuous reaction-diffusion model. In particular, the spectrum of the reaction-diffusion model always comprises at most one group of unstable modes, because it is determined by the parabolic dispersion relation (cp. (11.6)). In contrast, the spectrum of the lattice-Boltzmann model may involve many different groups of unstable modes (cp. fig. 11.9), depending mainly on the speed parameters m_A and m_I. This is caused by the imposed particle-motion process, which leads to a spectral radius dependence on sums and products of waves with different frequencies (cp. (11.21)).

11.2 Excitable Media

11.2.1 Introduction

The topic of this section is the spatio-temporal dynamics of *spiral waves*. Spiral waves are dominant patterns in a variety of physical, chemical, and biological systems. For example, Hassell, Comins, and May (1991) (see also Comins, Hassell, and May 1992) introduced a spatial model for host–parasitoid interactions which exhibits formation of spiral waves. Furthermore, plankton populations exhibiting spiral waves have been observed in the ocean on a kilometer scale (Wyatt 1973), and modeled, e.g., by Medvinskii et al. (2000) who used Pascual-Scheffer's model of phytoplankton-zooplankton interactions with randomly moving fish. The best known example of a spiral-generating chemical system is the Belousov-Zhabotinskii reaction (cp. fig. 2.7; Winfree 1972; Zhabotinskii and Zaikin 1970), while the chemical signaling system of the cellular slime mold *Dictyostelium discoideum* provides a further biological example (Newell 1983).

Some of these examples can be represented as "excitable" systems. The concept of *excitable media* was introduced by Wiener and Rosenbluth (1946) in order to explain heart arrhythmias caused by spiral waves. They invented the notations of *refractory*, *excitable*, and *excited states*. The defining characteristics of excitable systems are: (i) starting at a stable equilibrium (resting state) a stimulus above a certain threshold generates (ii) a burst of activity (excited state) followed by (iii) a refractory period (recovery state). The behavior of a typical trajectory in an excitable system is summarized in fig. 11.28. Due to the activity initiated by a supercritical perturbation, traveling excitation waves of various geometries occur, including ring and spiral waves (Winfree 1987). Excitable media can be constructed by, e.g., a two-component system of nonlinear partial differential equations of the type

$$\partial_t a = \frac{1}{\tau}\, F(a, b) + D_a \nabla^2 a,$$

$$\partial_t b = G(a, b) + D_b \nabla^2 b, \tag{11.42}$$

where ∇^2 is the Laplacian operator and D_a and D_b are diffusion coefficients, for which $D_a \neq 0$ and D_b may be 0 (Murray 2002). τ is a small positive constant ($0 < \tau \ll 1$), determining the different time scales for the faster component a ("excitation" variable) and the slower component b ("recovery" variable) which typically exist in excitable media. While G can be a monotonic or even linear function, F has to be a nonlinear function with sigmoidal shape.

The relevant information about the nullclines $F(a, b) = 0$ and $G(a, b) = 0$ is summarized in fig. 11.29. The intersection point S of the two nullclines defines the steady state, which is a unique, asymptotically stable resting state of the medium. Small perturbations from this resting state to a value a to the left of point D are immediately damped out, but large perturbations to a value a to the right of point D trigger a long excursion before returning to the resting state S. The time for a to change from its value at S to that of A (B to C) is relatively short compared to the time for a change from A to B (C to S). The diffusive transport in (11.42) reinforces and excites (neighboring) states that are close to the equilibrium.

CA models of excitable media attempt to reduce an excitable medium to its simplest possible form, as is shown in fig. 11.28 (Ermentrout and Edelstein-Keshet 1993). A very "simple" deterministic CA model with "excited," "refractory," and "resting" states was suggested by Greenberg, Hassard, and Hastings (1978). In its simplest form, the evolution of the cells in a two-dimensional square lattice is characterized by the following rule: A "resting" cell ($s(r) = 0$) becomes "excited" ($s(r) = 2$) if at least one of its neighbors is "excited," otherwise it remains "resting"; an "excited" cell becomes "recovering" ($s(r) = 1$) and a "recovering" cell becomes "resting" in the next time step; i.e.,

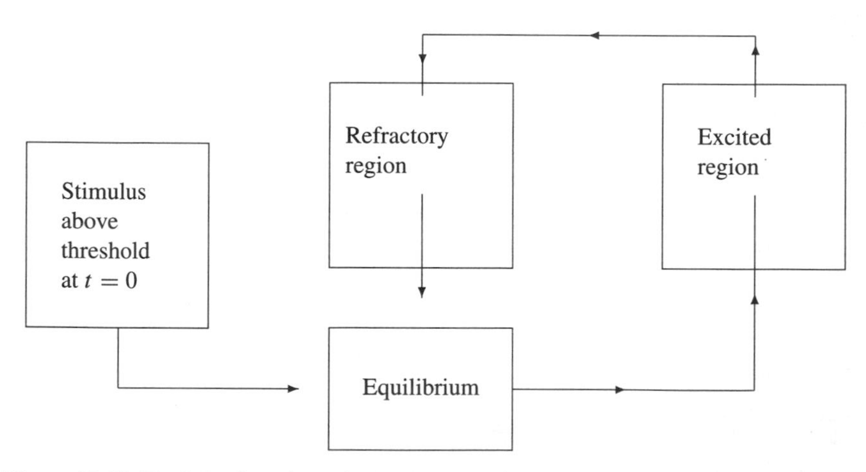

Figure 11.28. The behavior of a typical trajectory in an excitable system (Greenberg, Hassard, and Hastings 1978).

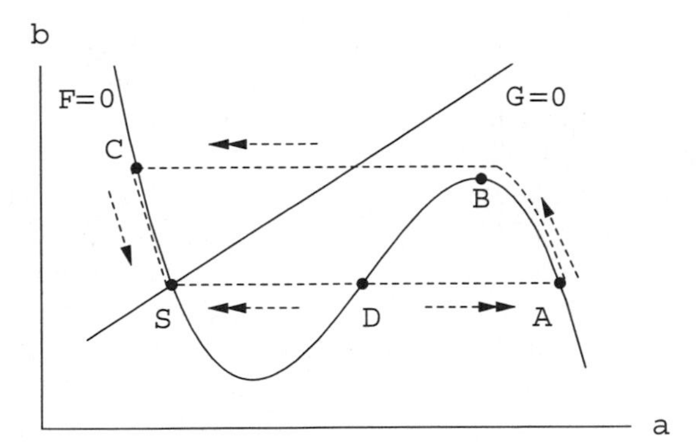

Figure 11.29. Typical schematic nullclines of excitable kinetics; the rest state S is excitable for perturbations larger than threshold D, and a trajectory occurs during which a increases rapidly (DA), causing a slower and temporary increase in b (AB), followed by a rapid extinction of a (BC) and a slow decrease of b back to the rest state (CS).

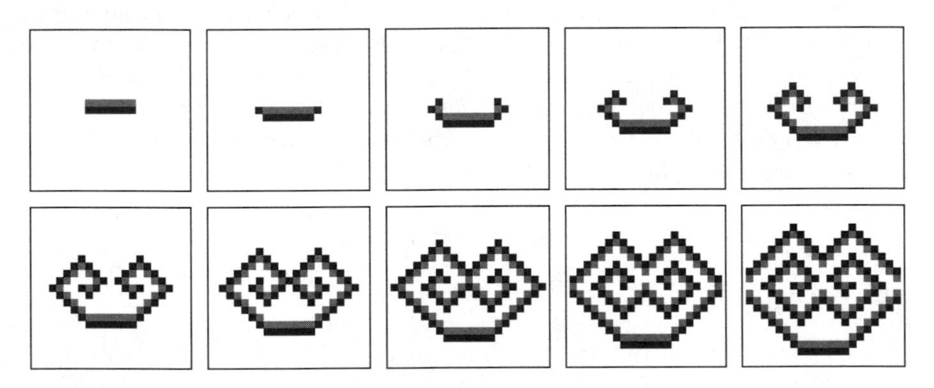

Figure 11.30. Evolution of the Greenberg-Hastings CA for times $k = 0, \ldots, 10$ in a lattice with 21×21 cells. White: resting cell ($s(r) = 0$), gray: recovering cell ($s(r) = 1$), black: excited cell ($s(r) = 2$).

$$s(r, k+1) = \mathcal{R}\left(\boldsymbol{s}_{\mathcal{N}(r)}(k)\right) = \begin{cases} 0 & \text{if } s(r,k) = 1 \text{ or} \\ & \text{if } s(r,k) = 0 \wedge \forall \tilde{r} \in \mathcal{N}_4^I(r) : s(\tilde{r}, k) \neq 2, \\ 1 & \text{if } s(r,k) = 2, \\ 2 & \text{if } s(r,k) = 0 \wedge \exists\, \tilde{r} \in \mathcal{N}_4^I(r) : s(\tilde{r}, k) = 2, \end{cases}$$

where the interaction neighborhood template $\mathcal{N}_4^I$ is taken to be the von Neumann neighborhood (cp. the definition on p. 70). Fig. 11.30 shows the evolution of a spiral wave generated with this rule.

This approach was extended later, i.e., by Markus and Hess (1990), Gerhardt et al. 1990a, and Fast and Efimov 1991.[83] Problems of these discrete models are related

[83] For further reading, Kapral et al. 1991 and Schönfisch 1993 are recommended.

to curvature effects,[84] lack of dispersion,[85] spatial anisotropy and computational complexity. Another type of CA model can be constructed through a finite-difference approximation of the partial differential equation model for excitable media given by (11.42) (Weimar, Tyson, and Watson 1992).

In the following we introduce a LGCA interaction-diffusion model, based on the concept summarized in fig. 11.28, which exhibits ring and spiral waves in a two-dimensional space. Snapshots of simulations (cp. fig. 11.33) show very "natural" spiral waves. Furthermore, we show that a crucial difference between our LGCA model and the corresponding lattice-Boltzmann model exists: although the LGCA model mimics an excitable medium the lattice-Boltzmann model does not!

11.2.2 Definition of the Automaton Rules

The LGCA consists of three components, $\sigma \in \{A, B, C\}$, of which A represents an excitation variable, which can be excited ($\eta_{A,i} = 1$) or unexcited ($\eta_{A,i} = 0$), whereas B and C are viewed as "information carriers" which control the duration of excitation of A. The system is defined on a two-dimensional square ($b = 4$) lattice $\mathcal{L}_\sigma = \mathcal{L} = \mathcal{L}_1 \times \mathcal{L}_2$ with $L_1 = L_2 = L$ nodes in each space direction. Particles X_σ are distributed in four velocity channels $(r, c_i)_\sigma$, $i = 1, \ldots, 4$, and one rest channel $(r, c_5)_\sigma$, and the interaction neighborhood for each species σ is taken as $\mathcal{N}^I_{4,\sigma}(r) = \{r\}$. Hence, the node configuration becomes

$$\boldsymbol{\eta}(r,k) = (\boldsymbol{\eta}_A(r,k), \boldsymbol{\eta}_B(r,k), \boldsymbol{\eta}_C(r,k)) \in \{0,1\}^{15}$$

with

$$\boldsymbol{\eta}_\sigma(r,k) = \left(\eta_{\sigma,1}(r,k), \ldots, \eta_{\sigma,5}(r,k)\right) \quad \sigma \in \{A, B, C\}.$$

The **automaton interaction rule** consists of two subrules: a reactive interaction rule (R) and the local shuffling rule for an isotropic random walk (M), which we introduced before (cp. sec. 5.4). The reactive interaction rule (R) specifies the creation or destruction of particles X_σ in each channel $(r, c_i)_\sigma$ and solely depends on the states in that channel (r, c_i), i.e.,

$$\eta^{\mathrm{R}}_{\sigma,i}(r,k) = \mathcal{R}^{\mathrm{R}}_{\sigma,i}\left(\eta_{A,i}(r,k), \eta_{B,i}(r,k), \eta_{C,i}(r,k)\right).$$

Let $h(\theta, \boldsymbol{\eta}_A)$ be a threshold function with the threshold value $\theta \in \{0, \ldots, 5\}$ defined as

$$h(\theta, \boldsymbol{\eta}_A) := \sum_{l=\theta}^{5} \Psi_l(\boldsymbol{\eta}_A) = \begin{cases} 1 & \text{if } \sum_i^5 \eta_{A,i} \geq \theta, \\ 0 & \text{else,} \end{cases}$$

[84] The propagation speed depends on the curvature of the wave front.

[85] A wave front lacks dispersion if it can propagate only into resting medium but not into partially recovered medium.

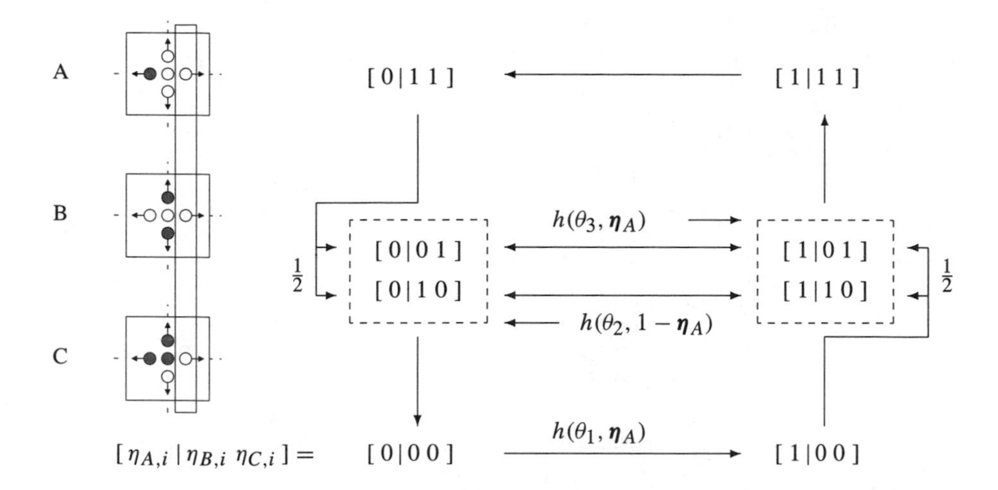

Figure 11.31. Schematic representation of the *reactive interaction rule* R with threshold variables θ_1, θ_2, and θ_3. The sequences represent all possible configurations of $[\eta_{A,i}(r,k) \mid \eta_{B,i}(r,k)\eta_{C,i}(r,k)]$. For instance, a transition from $[1 \mid 01]$ to $[0 \mid 01]$ is only possible if the total number of free channels of species A exceeds θ_2, i.e., if $h(\theta_2, 1-\boldsymbol{\eta}_A) = 1$; a transition from $[1 \mid 00]$ to $[1 \mid 01]$ occurs with probability 1/2.

with the previously defined counting functions Ψ_l (cp. p. 139). Furthermore, let $\{\zeta(r,k) : r \in \mathcal{L}, k \in \mathbb{N}\}$ be a sequence of node- and time-independent identically distributed Bernoulli type random variables, such that $P(\zeta(r,k) = 1) = 1/2$. Then, the interaction rule is described by the following equations:

$$\begin{aligned}
\mathcal{R}^{\mathrm{R}}_{A,i} &= \eta_{A,i} + \check{\eta}_{A,i}\check{\eta}_{B,i}\check{\eta}_{C,i}\, h(\theta_1, \boldsymbol{\eta}_A) + \check{\eta}_{A,i}\left(\eta_{B,i}\check{\eta}_{C,i} + \check{\eta}_{B,i}\eta_{C,i}\right) h(\theta_3, \boldsymbol{\eta}_A) \\
&\quad - \eta_{A,i}\eta_{B,i}\eta_{C,i} - \eta_{A,i}\left(\eta_{B,i}\check{\eta}_{C,i} + \check{\eta}_{B,i}\eta_{C,i}\right) h(\theta_2, 1-\boldsymbol{\eta}_A),
\end{aligned}$$

$$\begin{aligned}
\mathcal{R}^{\mathrm{R}}_{B,i} &= \eta_{B,i} + \zeta\, \check{\eta}_{B,i}\eta_{A,i}\check{\eta}_{C,i} + \check{\eta}_{B,i}\eta_{A,i}\eta_{C,i}\,(1 - h(\theta_2, 1-\boldsymbol{\eta}_A)) \\
&\quad - \zeta\, \eta_{B,i}\check{\eta}_{A,i}\eta_{C,i} - \eta_{B,i}\check{\eta}_{A,i}\check{\eta}_{C,i}\,(1 - h(\theta_3, \boldsymbol{\eta}_A)) ,
\end{aligned}$$

$$\begin{aligned}
\mathcal{R}^{\mathrm{R}}_{C,i} &= \eta_{C,i} + \zeta\, \check{\eta}_{C,i}\eta_{A,i}\check{\eta}_{B,i} + \check{\eta}_{C,i}\eta_{A,i}\eta_{B,i}\,(1 - h(\theta_2, 1-\boldsymbol{\eta}_A)) \\
&\quad - \zeta\, \eta_{C,i}\check{\eta}_{A,i}\eta_{B,i} - \eta_{C,i}\check{\eta}_{A,i}\check{\eta}_{B,i}\,(1 - h(\theta_3, \boldsymbol{\eta}_A)) ,
\end{aligned}$$

where we dropped the node and time dependence and used the shortcuts $\check{\eta}_{\sigma,i} := (1 - \eta_{\sigma,i})$ and $1 - \boldsymbol{\eta}_A = \left(1 - \eta_{A,1}, \ldots, 1 - \eta_{A,5}\right)$. The threshold values θ_1 and θ_3 define the minimum number of particles of species A at node r, which are necessary to activate ($\eta_{A,i} = 1$) new X_A particles; θ_2 gives the minimum number of deactivated ($\eta_{A,i} = 0$) X_A particles on the node which is needed to deactivate an activated X_A particle. These rules are summarized in fig. 11.31. Note that the reactive interaction rule R is symmetrical with respect to the components B and C.

The stable equilibrium state at node r according to this reactive interaction rule is given by $\boldsymbol{\eta}(r) = (0, \ldots, 0)$. Any stimulus of this state with respect to the compo-

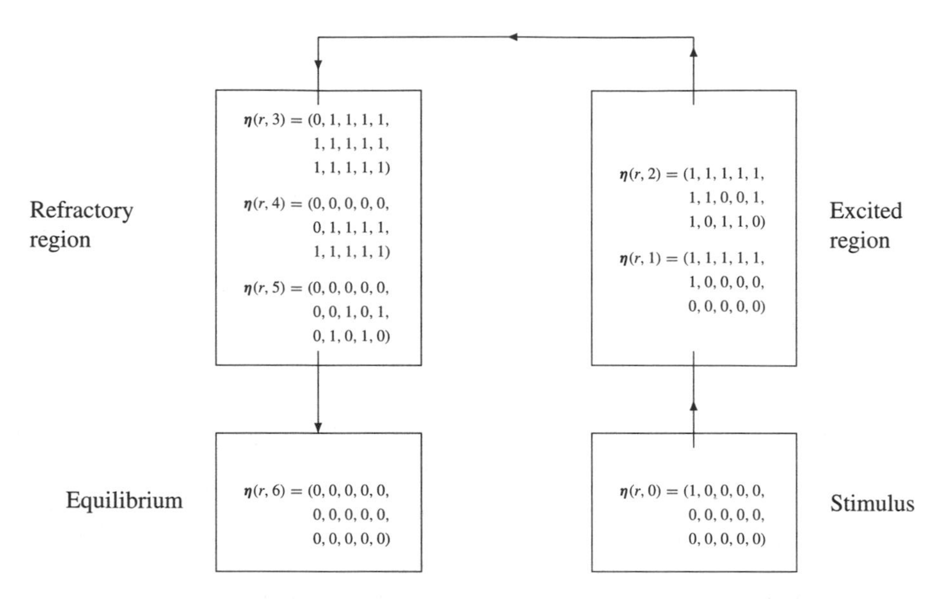

Figure 11.32. Example of a possible trajectory resulting from the application of the reactive interaction rule R using threshold parameters $\theta_1 = 1$, $\theta_2 = \theta_3 = 3$; $\boldsymbol{\eta}(r,k) = (\boldsymbol{\eta}_A(r,k), \boldsymbol{\eta}_B(r,k), \boldsymbol{\eta}_C(r,k))$.

nents B or C leads back to the equilibrium state in one time step, if $\theta_1, \theta_2, \theta_3 > 0$. An example of a trajectory initiated by a stimulus with respect to component A and threshold values $\theta_1 = 1, \theta_2 = \theta_3 = 3$ is illustrated in fig. 11.32. With these threshold parameters the local interaction rule defines an *excitable system*. In the following, examples are based on these parameters, i.e., $\theta_1 = 1, \theta_2 = \theta_3 = 3$.

The complete spatio-temporal evolution of the automaton dynamics for reactive interaction, shuffling and propagation (P ∘ M ∘ R) is described by the following microdynamical difference equation:

$$\eta_{\sigma,i}(r + m_\sigma c_i, k+1) - \eta_{\sigma,i}(r,k) = \mathcal{R}_i^{\mathrm{M}}\left(\boldsymbol{\eta}_\sigma^{\mathrm{R}}(r,k)\right) - \eta_{\sigma,i}(r,k) \tag{11.43}$$

for $m_\sigma \in \mathbb{N}, \sigma \in \{A, B, C\}$ and $i \in \{1, \ldots, 5\}$. This LGCA model leads to a "spiral-shaped" concentration profile of component A in the lattice, as it is shown in fig. 11.33. The simulation was started with a single seed in the center of the lattice, with threshold parameters $\theta_1 = 1$, $\theta_2 = \theta_3 = 3$ and speed parameters $m_A = m_B = m_C = 1$. During the initial transient, several individual spirals evolve from the seed and divide the system into corresponding domains.

11.2.3 Lattice-Boltzmann Equation and Its Uniform Steady States

Again, we follow the lines of sec. 4.4.3 and derive the *lattice-Boltzmann equation* from (11.43) as

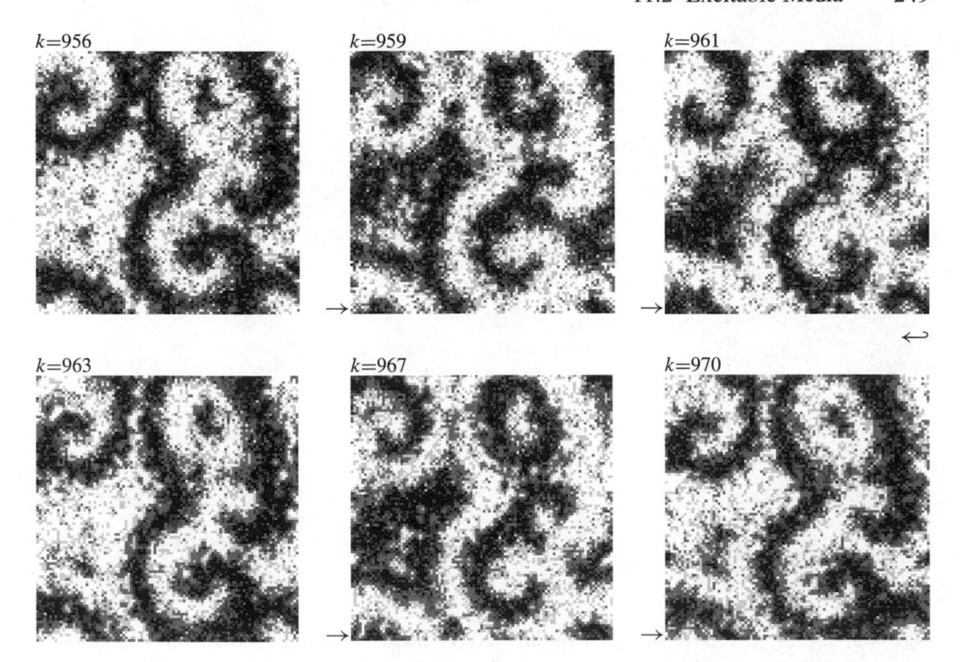

Figure 11.33. Snapshot of concentration of component A in a two-dimensional square lattice with periodic boundary conditions, started from a seed in the center, i.e., $\eta_{A,1}((50,50),0) = 1$ and $\eta_{\sigma,i}(r,0) = 0$ else. Parameters: $L = 100, \theta_1 = 1, \theta_2 = \theta_3 = 3, m_A = m_B = m_C = 1$.

$$
\begin{aligned}
f_{\sigma,i}(r + m_\sigma c_i, k+1) - f_{\sigma,i}(r,k) &= \frac{1}{5}\sum_{l=1}^{5} E\left(\eta^{\mathrm{R}}_{\sigma,l}(r,k)\right) - f_{\sigma,i}(r,k) \\
&= \tilde{\mathcal{C}}_{\sigma,i}\left(\boldsymbol{f}(r,k)\right),
\end{aligned}
\tag{11.44}
$$

where

$$
\boldsymbol{f}(r,k) = (\boldsymbol{f}_A(r,k), \boldsymbol{f}_B(r,k), \boldsymbol{f}_C(r,k)) = \left(f_j(r,k)\right)_{j=1}^{15} \in [0,1]^{15}.
$$

Spatially uniform stationary states $\left(\bar{f}_A, \bar{f}_B, \bar{f}_C\right)$, where $\bar{f}_\sigma = f_{\sigma,i}$, are solutions of $\tilde{\mathcal{C}}_{\sigma,i}\left(\boldsymbol{f}(r,k)\right) = 0$. Hence, for threshold parameters $\theta_1 = 1$ and $\theta_2 = \theta_3 = 3$, these are evaluated as

$$
\left(\bar{f}_A, \bar{f}_B, \bar{f}_C\right) \in \{(0,0,0), (a_1, a_2, a_3)\},
\tag{11.45}
$$

where

$$
a_1 \approx 0.496, \qquad a_2 = a_3 \approx 0.492.
$$

In order to compare the dynamics of the LGCA and lattice-Boltzmann model we follow the dynamics of the particle number $\mathrm{n}_\sigma(r,k) = \sum_{i=1}^{5} \eta_{\sigma,i}(r,k)$ and mass $\varrho_\sigma(r,k) = \sum_{i=1}^{5} f_{\sigma,i}(r,k)$ at a node r, when *particle motion is excluded*, i.e.,

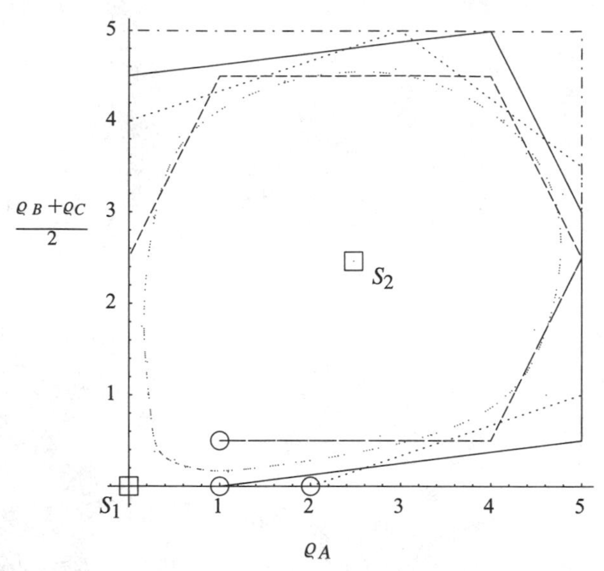

Figure 11.34. Trajectories at a node r of the LGCA (lines) and lattice-Boltzmann (dots) model for different initial conditions, which are marked with circles; the solid line refers to the example given in fig. 11.32. Although $S_1 = (0,0)$ is the stable equilibrium state of the LGCA model, it is an unstable equilibrium state in the lattice-Boltzmann model. $S_2 = (2.48, 2.46)$ is the second stationary state of the lattice-Boltzmann model. Parameters: $\theta_1 = 1, \theta_2 = \theta_3 = 3$, $m_A = m_B = m_C = 0$.

$m_\sigma = 0$. The resulting equation for the lattice-Boltzmann model in terms of ϱ_σ is given by

$$\varrho_\sigma(r, k+1) = \varrho_\sigma(r, k) + \sum_{i=1}^{5} \tilde{\mathcal{C}}_{\sigma,i}(\boldsymbol{f}) , \tag{11.46}$$

where the expanded form is given in appendix C (cp. eqn. (C.1–C.3), p. 285). Fig. 11.34 shows trajectories for both models starting from the initial conditions

$$\begin{aligned}\boldsymbol{\eta}(r,0) \in \big\{&(1,0,0,0,0,0,0,0,0,0,0,0,0,0,0),\\ &(1,0,0,0,0,1,0,0,0,0,0,0,0,0,0),\\ &(0,1,0,0,0,1,0,0,0,0,0,0,0,0,0),\\ &(1,1,0,0,0,0,0,0,0,0,0,0,0,0,0)\big\} ,\end{aligned}$$

and hence

$$\begin{aligned}(\mathrm{n}_A(r,0), (\mathrm{n}_B(r,0) + \mathrm{n}_C(r,0))/2) &= (\varrho_A(r,0), (\varrho_B(r,0) + \varrho_C(r,0))/2)\\ &\in \{(1,0), (1,0.5), (2,0)\},\end{aligned}$$

where we set $f_{\sigma,i}(r,0) = \eta_{\sigma,i}(r,0)$.

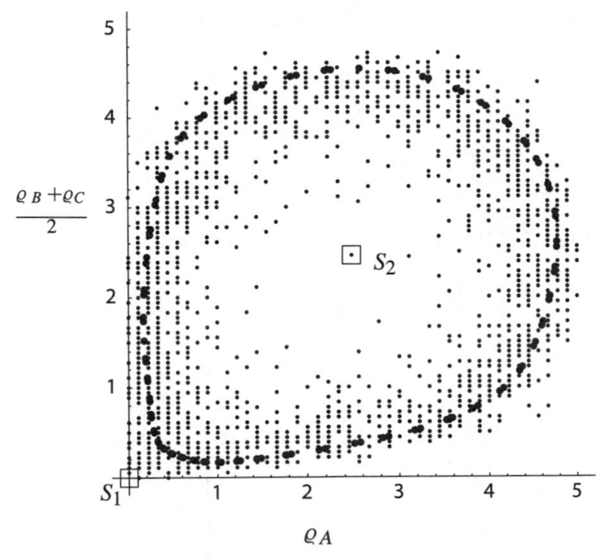

Figure 11.35. Trajectories of a simulation run averaged over all nodes $r \in \{(r_x, r_y) : r_x = 50 \pm 1, r_y = 50 \pm 1\}$ of the LGCA (small dots) and lattice-Boltzmann (bold dots) model for an initial condition located at $S_2 = (2.48, 2.46)$. Parameters: $L = 100$, $k = 100, \ldots, 2000$, $\theta_1 = 1, \theta_2 = \theta_3 = 3, m_A = m_B = m_C = 1$.

The stationary state $S_1 = (0, 0)$ which is stable in the LGCA model is unstable in the lattice-Boltzmann model.[86]

Furthermore, from linear stability analysis with respect to the second stationary state of the lattice-Boltzmann model (eqn. (11.46)), we obtain that $S_2 = (2.48, 2.46)$ is also unstable. Hence, the circle shown in fig. 11.34 is a stable attractor ("invariant circle"). Therefore, in contrast to the LGCA model, the system described by the lattice-Boltzmann equation can be viewed as an *oscillatory medium*, which consists of self-oscillating elements, coupled with the neighbors (Mikhailov 1994).

However, the dynamics of the LGCA model *with particle motion* can be captured locally by the lattice-Boltzmann approximation, as is indicated in fig. 11.35. The data are taken from a LGCA simulation with a random initial condition corresponding to the stationary state S_2 at each node. The (small) dots represent averaged values over all nodes $r \in \{(r_x, r_y) : r_x = 50 \pm 1, r_y = 50 \pm 1\}$ in a lattice with 100×100 nodes. In order to derive the period of oscillation, a linear stability analysis of the lattice-Boltzmann equation (11.44) has to be performed.

11.2.4 Stability Analysis of the Lattice-Boltzmann Equation

As before, the stability of the spatially homogeneous stationary solutions (11.45) $(\bar{f}_A, \bar{f}_B, \bar{f}_C)$ with respect to fluctuations $\delta f_{\sigma,i}(r, k) = f_{\sigma,i}(r, k) - \bar{f}_\sigma$ is determined

[86] This can be confirmed by a linear stability analysis of (11.46).

by the spectrum of the Boltzmann propagator (4.36). For the parameters $\theta_1 = 1$, $\theta_2 = \theta_3 = 3$, and $m_A = m_B = m_C = 1$ the Boltzmann propagator is of the form

$$\Gamma(q) = T\{\boldsymbol{I} + \Omega^0\}, \quad q = (q_1, q_2) \quad \text{with} \quad q_1, q_2 = 0, \ldots, L-1\,, \tag{11.47}$$

where

$$\begin{aligned} T = \operatorname{diag}\Big(& e^{-\mathbf{i}(2\pi/L)(c_1\cdot q)\, m_A}, \ldots, e^{-\mathbf{i}(2\pi/L)(c_5\cdot q)\, m_A}, \\ & e^{-\mathbf{i}(2\pi/L)(c_1\cdot q)\, m_B}, \ldots, e^{-\mathbf{i}(2\pi/L)(c_5\cdot q)\cdot m_B}, \\ & e^{-\mathbf{i}(2\pi/L)(c_1\cdot q)\, m_C}, \ldots, e^{-\mathbf{i}(2\pi/L)(c_5\cdot q)\cdot m_C}\Big) \end{aligned}$$

and[87]

$$\boldsymbol{I} + \Omega^0 = \begin{pmatrix} \omega_1 \ldots \omega_1 & \omega_2 \ldots \omega_2 & \omega_2 \ldots \omega_2 \\ \vdots \quad\; \vdots & \vdots \quad\; \vdots & \vdots \quad\; \vdots \\ \omega_1 \ldots \omega_1 & \omega_2 \ldots \omega_2 & \omega_2 \ldots \omega_2 \\ \omega_3 \ldots \omega_3 & \omega_4 \ldots \omega_4 & \omega_5 \ldots \omega_5 \\ \vdots \quad\; \vdots & \vdots \quad\; \vdots & \vdots \quad\; \vdots \\ \omega_3 \ldots \omega_3 & \omega_4 \ldots \omega_4 & \omega_5 \ldots \omega_5 \\ \omega_3 \ldots \omega_3 & \omega_5 \ldots \omega_5 & \omega_4 \ldots \omega_4 \\ \vdots \quad\; \vdots & \vdots \quad\; \vdots & \vdots \quad\; \vdots \\ \omega_3 \ldots \omega_3 & \omega_5 \ldots \omega_5 & \omega_4 \ldots \omega_4 \end{pmatrix} \in \mathbb{R}^{3\cdot 5\times 3\cdot 5}.$$

The spectrum of the Boltzmann propagator (11.47) is given by

$$\Lambda_{\Gamma(q)} = \{\lambda_1(q), \lambda_2(q), \lambda_3(q), 0\},$$

where 0 has a multiplicity of 12 and

$$\begin{aligned} \lambda_{1,2}(q) &= \frac{1}{2}\, u(q) \left(\omega_1 + \omega_4 + \omega_5 \pm \sqrt{8\omega_2\,\omega_3 + (\omega_4 + \omega_5 - \omega_1)^2}\right), \\ \lambda_3(q) &= (\omega_4 - \omega_5) u(q) \end{aligned}$$

with

$$u(q) := 1 + \sum_{j=1}^{b} e^{-\mathbf{i}(2\pi/L)c_j\cdot q} = 1 + 2\left(\cos(\frac{2\pi}{L} q_1) + \cos(\frac{2\pi}{L} q_2)\right) \in [-3, 5].$$

[87] The matrix elements ω_i, $i = 1, \ldots, 5$, are different terms for both stationary states $(\bar{f}_A, \bar{f}_B, \bar{f}_C)$ given by (11.45).

To be more specific, the eigenvalues with respect to each stationary state are given by:

$$\left(\bar{f}_A, \bar{f}_B, \bar{f}_C\right) = (0, 0, 0): \tag{11.48}$$

$$\boxed{\lambda_1(q) = u(q), \qquad \lambda_2(q) = \lambda_3(q) = 0,}$$

$$\left(\bar{f}_A, \bar{f}_B, \bar{f}_C\right) \approx (0.496, 0.492, 0.492): \tag{11.49}$$

$$\boxed{\lambda_{1,2}(q) \approx (0.152 \pm 0.186\mathbf{i})\, u(q), \qquad \lambda_3(q) \approx 0.062\, u(q).}$$

Thus, in both cases the spectral radius $\mu(q) = |\lambda_1(q)|$ depends on the vector of wave numbers $q = (q_1, q_2)$ via $|u(q)| \in [0, 5]$. In contrast to the two-dimensional "activator–inhibitor model" (cp. sec. 11.1.4), no special direction is preferred, as illustrated in fig. 11.36.

In case (11.48) the spectral radius is always a real number and its maximum instability refers to wave numbers with magnitude $|q| = 0$. In the second case (11.49) the dominant eigenvalue has a nonzero imaginary part. The *period of oscillation* Δk can be determined from

$$\lambda_1(q) = e^{\alpha + \mathbf{i}\omega} = |\lambda_1(q)| \left(\cos(\omega) + \mathbf{i}\, \sin(\omega)\right) \approx 0.152 u(q) + \mathbf{i} 0.186 u(q),$$

where $\alpha \approx \ln\left(0.241 |u(q)|\right)$ and $\omega \approx \arccos\left(0.632[u(q)/|u(q)|]\right)$, as

$$\Delta k = \frac{2\pi}{\omega} \approx \begin{cases} 7.086 & \text{if } u(q) > 0, \\ 2.786 & \text{if } u(q) < 0. \end{cases}$$

Hence, oscillations in the density of all components resulting from spatially homogeneous fluctuations, i.e., $|q| = 0$, have a period of $\Delta k \approx 7$. This result is confirmed in fig. 11.37, where it can be seen that global and local densities perform regular oscillations with the predicted period.

Fig. 11.36(b) shows that $Q^c = \{q = (q_1, q_2) : |q| \leq 15\}$. Although modes with wave numbers of magnitude $|q_*| = 0$ grow fastest, propagating soundlike modes with "speed" $(L/|q|)\Delta k^{-1}$ grow almost as fast for small $|q|$, since (for $L = 100$) it holds that

$\lvert q\rvert$:	0	1	2	3	4
$\mu(\lvert q\rvert, 0)$:	1.2029	1.2019	1.1991	1.1943	1.1878
$\mu(0,0) - \mu(\lvert q\rvert, 0)$:		0.0010	0.0038	0.0086	0.0151

$\lvert q\rvert$:	5	6	7	8	9
$\mu(\lvert q\rvert, 0)$:	0.0236	1.1793	1.1691	1.1571	1.1434
$\mu(0,0) - \mu(\lvert q\rvert, 0)$:	0.0338	0.0458	0.0595	0.0749	0.0919.

According to the analysis regarding the role of fluctuations in a LGCA model performed in sec. 11.1.3, we claim that traveling modes, which are observable in simulations, can be explained by our mean-field stability analysis. Fig. 11.33 shows that

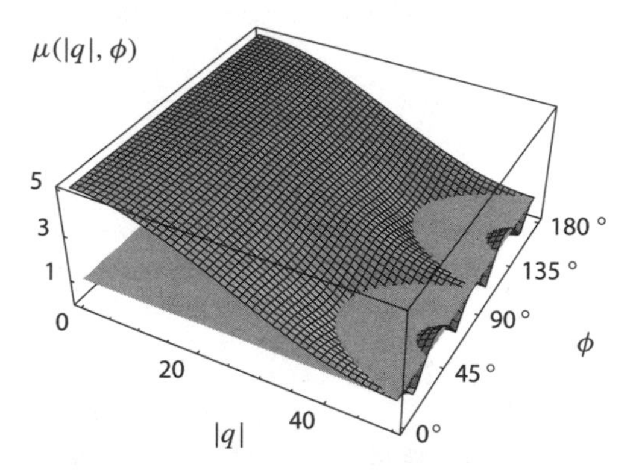

(a) $(\bar{f}_A, \bar{f}_B, \bar{f}_C) = (0, 0, 0)$ (cp. case (11.48))

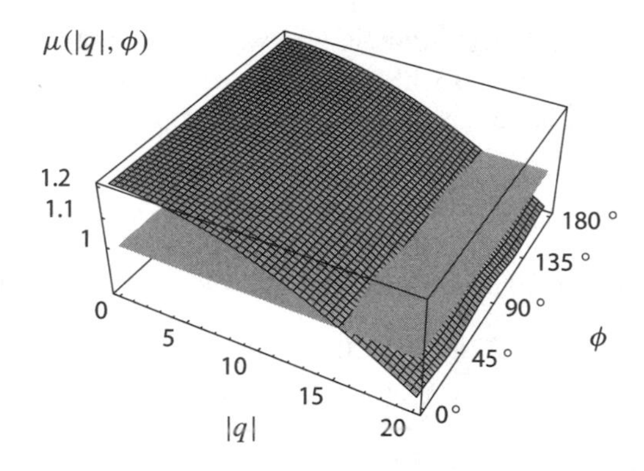

(b) $(\bar{f}_A, \bar{f}_B, \bar{f}_C) \approx (0.496, 0.492, 0.492)$ (cp. case (11.49))

Figure 11.36. Spectral radius $\mu(q) = |\lambda_1(q)|$ dependent on the wave number $q = (q_1, q_2) = (|q|\cos(\phi), |q|\sin(\phi))$ represented in polar coordinate form. Parameters: $L = 100$, $\theta_1 = 1$, $\theta_2 = \theta_3 = 3$, $m_A = m_B = m_C = 1$.

the rotation time for all spirals coincides with the predicted period of oscillation Δk. The snapshot taken at time $k = 956$ closely resembles the ones taken at times $k+7 = 963$ and $k+14 = 970$. In order to confirm these results, a detailed statistical analysis of the LGCA model, i.e., of the wavelength of spirals in simulations, as well as a nonlinear analysis of the lattice-Boltzmann[88] model, has to be performed. As a first step in this direction, a Karhunen-Loéve decomposition (Killich et al. 1994) of the spatial patterns of a simulation has been performed.[89] Preliminary results support

[88] See, e.g., Sepulchre and Babloyantz 1995 (discontinuous media) and Mikhailov 1994 (continuous media).

[89] By Uwe Börner, Max Planck Institute for the Physics of Complex Systems, Dresden.

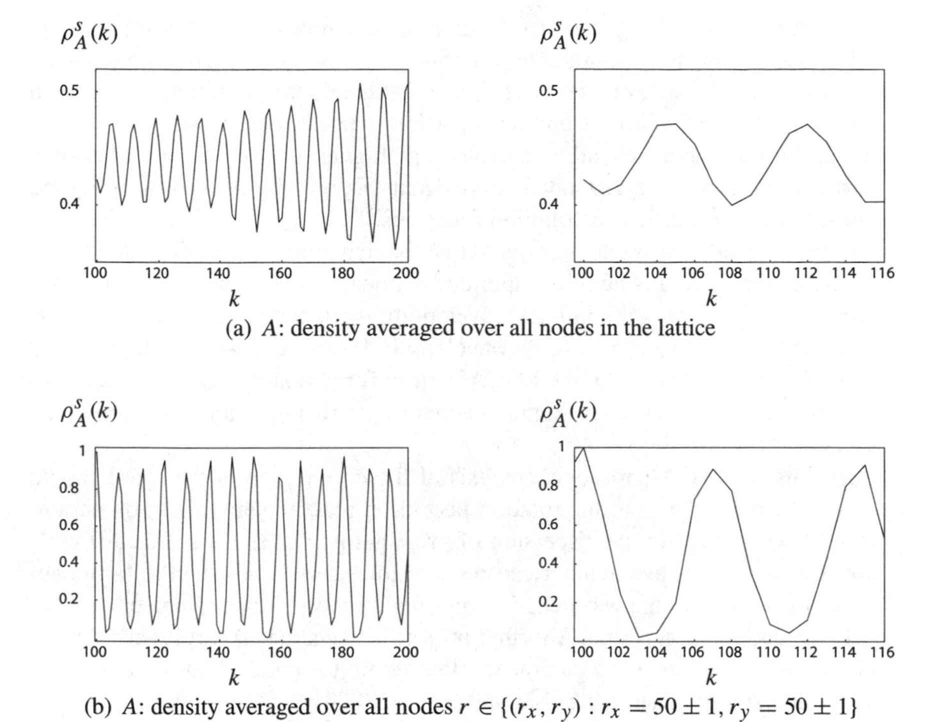

Figure 11.37. Averaged densities of component A in a simulation with an initial condition located at $\left(\bar{f}_A, \bar{f}_B, \bar{f}_C\right) \approx (0.496, 0.492, 0.492)$; oscillations exhibit a temporal period of $\Delta k \approx 7$. Each figure on the right is a magnified part of the one on the left. Parameters: $L = 100, \theta_1 = 1, \theta_2 = \theta_3 = 3, m_A = m_B = m_C = 1$.

the predicted rotation time of $\Delta k \approx 7$ and indicate a wave number of magnitude 4 or 5.

Summary. In this chapter it has been shown by means of a Turing-type and an excitable interaction that CA are well suited to model and analyze pattern formation dynamics in reaction-diffusion systems. The presented CA models are basic modules that can be extended to include more particle types, further interactions, and other boundary conditions. Trying to understand the precise relations between particle-based models (e.g., LGCA) and continuous approaches (e.g., partial differential equations) leaves many challenging problems for the future.

11.3 Further Research Projects

1. **Extended approximations:** Especially in the case of the excitable LGCA model, study extended approximations that include higher order (on-node) correlations. These approximations can be derived from the Chapman-Kolmogorov

description of the CA. The lattice-Boltzmann equations, on which we focus in this book, represent the simplest approximation of the system. They are obtained by describing the states in terms of single particle distribution functions, which represent the probability of finding a particle with a given velocity at a given node. With this approach, all pair, triplet, and higher order correlations between particles are neglected. Extended approximations may be derived by including two-, three-, etc. particle distribution functions.[90]

2. **Nonlinear analysis:** We demonstrated that the dynamics of the LGCA model for excitable media cannot be fully captured by linear stability analysis, which predicts unstable modes with infinite wavelength, while the LGCA model exhibits spiral patterns with a certain finite wavelength. Hence, it is a research project to extract better predictions of the LGCA pattern formation dynamics by a nonlinear analysis of the lattice-Boltzmann equations[91] (using renormalization tools, see, e.g., Haken 1978b).
3. **Spiral dynamics:** Microscopic analysis of the core region of the spiral should reveal the dependence of the rotation period on reaction and migration parameters. Combined with the dispersion of propagating waves the selected (by the core) wavelength/wave number can be determined self-consistently. In the absense of dispersion the wavelength simplifies to the product of universal wave velocity and rotation period. A further project is to characterize rules and threshold values of the involved parameters that exhibit secondary instabilities, i.e., spiral breakup or meandering (Bär and Brusch 2004).
4. **Multicomponent systems:** A challenging research project is to extend the analysis described in this chapter to multicomponent systems, e.g., by considering a second inhibitor species. The question is whether, the behavior in corresponding lattice-gas models is similar to partial differential equation systems. In particular, patterns of isolated excitation (traveling spots or "quasi particles") have been observed in two-dimensional partial differential equation systems with an activator and two inhibitor species (Schenk et al. 1997). Such behavior is not possible for diffusively coupled systems with only a single inhibitor. Furthermore, a fast second inhibitor acts similarly to a nonlocal spatial coupling of the activator (with only a single inhibitor) which enables the coexistence and competition of spatial domains with Turing patterns or traveling waves, respectively. The domain boundaries can be mobile themselves, giving rise to stable and coherent "drifting pattern domains" (Nicola et al. 2002).
5. **Higher order interactions:** A further project is to expand the analysis to LGCA interaction rules of higher order as in corresponding partial differential equation systems. For example, the Barkley model (Barkley 1991) for activator A and inhibitor I species with $\partial_t A = A(1-A)[A-((b+I)/a)]/\epsilon + \delta A$, $\partial_t I = A - I$ shows spiral waves but no space-time chaos, whereas both behaviors are observed if the inhibitor dynamics depends cubically on the activator (Bär and Eiswirth 1993).

[90] See also sec. 4.4.2 and Bussemaker 1995.

[91] In general, the lattice-Boltzmann equations are nonlinear.

12

Discussion and Outlook

"If the facts don't fit the theory, change the facts."
(Albert Einstein)

Contrary to "continuous systems" and their canonical description with partial differential equations, there is no standard model for describing discrete systems, particularly interacting discrete biological cells. In this book, cellular automata (CA) are proposed as models for spatially extended systems of interacting cells. CA are neither a replacement for traditional (continuous) mathematical models nor preliminary mathematical models but constitute a proper class of discrete mathematical models: discrete in space, time, and state space, for which analytical tools already exist or can be developed in the future.[92]

While reaction-diffusion models are the appropriate tool for describing the spatio-temporal dynamics of signaling molecules or large cell populations, microscopic models at the cellular or subcellular level have to be chosen, especially if one is interested in the dynamics of small populations. Interest in such "individual-based" approaches has recently grown significantly, also because more and more (genetic and proteomic) cell data are available. Important questions arise with respect to the mathematical analysis of microscopic individual-based models and the particular links to macroscopic approaches. While in physico-chemical processes typically the macroscopic equation is known *a priori*, the master equations in biological pattern formation are often far from clear. Many important questions are open for future research. It is a challenge for the future to systematically link pattern formation models as presented in this book to intracellular genetic and signaling net-

[92] So we clearly reject the following statement quoted from a review paper on CA modeling of biological systems: "*We do not believe that CA should be viewed as a replacement for rigorous mathematical models. Instead, they should be considered as a first step in the modeling process. Once it has been established that the CA implementation of one's hypothesis produces the desired results, then one must proceed toward deriving a traditional mathematical model. For then and only then is it possible to bring to bear tools from analysis such as stability theory, bifurcation theory, and perturbation methods*" (from Ermentrout and Edelstein-Keshet 1993).

works. This implies covering a whole range of cellular and molecular scales and will hopefully be possible in the future since the experimental data needed for the mathematical modeling already exists or can now be collected. A first step is the *hybrid cellular automaton* presented in this book as a model for avascular tumor growth. The model incorporates discrete cells and continuous nutrient and signal molecule concentrations (ch. 10).[93]

In the following discussion we focus on a critical evaluation of the modeling potential of CA. The discrete modeling idea is repeated. Artefacts can arise due to the discrete nature of the automaton; strategies for their avoidance will be summarized (sec. 12.1.2). Then, morphogenetic processes are characterized that can be investigated by CA models (sec. 12.2). In particular, CA are valuable instruments for the analysis of local cell interactions. In this book we have investigated certain examples, particularly automaton models of cell dispersal, cohort movement (swarming), aggregation, and pigment, tissue, and tumor pattern formation based on local interactions. We critically discuss the biological relevance of the model results and their limitations. Furthermore, our results are compared with other microscopic and macroscopic models (secs. 12.3–12.8).

Extensions of the CA concept are straightforward. Prospective applications and the elaboration of more advanced analytical tools are described in the "Outlook" (sec. 12.9).

12.1 Cellular Automaton Characterization

We have distinguished deterministic, probabilistic, and lattice-gas cellular automata (LGCA) and have presented numerous examples for each type. However, most of the examples are LGCA since they simultaneously allow for the modeling of morphogenetic motion and the analysis of spatio-temporal pattern formation. *Lattice-gas models* were originally introduced as discrete microscopic caricatures of hydrodynamical systems. It was shown that simple collision rules for discrete particles can give rise to the intricate structures of hydrodynamic flow as long as the rules conserve mass and momentum (Frisch, Hasslacher, and Pomeau 1986; Kadanoff 1986). Lattice-gas automata as models of local biological cell–cell interaction maintain the discrete particle nature of lattice-gases, but incorporate additional degrees of modeling freedom; in particular, less conservative constraints are imposed (e.g., momentum or energy conservation are abandoned). This is justified since biological cells typically perform an active creeping motion with negligible small inertia and with dissipative friction.

12.1.1 Cell-Based Instabilities and Cellular Self-Organization

Viewing cellular automata as interacting cell systems allows us to apply analytic tools from statistical mechanics. In particular, corresponding Boltzmann equations

[93] More recently a hybrid CA model has been suggested that additionally takes into account the effects of vasculature (Alarcon, Byrne, and Maini 2003).

may be derived for LGCA. We have focused on the "linear stability analysis" of CA and have demonstrated how elementary cell–cell interactions can be investigated in the framework of appropriately constructed automaton models. One can trace back pattern initiation to *cell-based instabilities* in LGCA: these automata are prototypes of cellular self-organization. Contrary to the classical Turing instability, e.g., which is associated with (macroscopic) morphogen transport properties (diffusion coefficients),[94] cell-based instabilities are connected with (microscopic) *cell properties* (e.g., cell affinity and cell sensitivity; cp. particularly, the discussion in secs. 12.4 and 12.5).

At the *microscopic level* CA represent only a caricature of the forces between the cells in the real biological system. A molecular dynamics approach taking into account the detailed form of molecular and intracellular interactions is much more appropriate to obtain information at the smallest scales and to model exact cellular shapes.[95] However, in the intermediate *mesoscopic regime* focusing on the dynamics of cell–cell interactions, CA are a promising model tool. As in the hydrodynamic lattice-gas automata the examples presented in this book show that for many important cell–cell interactions the coarse-grained perspective of CA suffices to capture the self-organized system behavior at the macroscopic level.

12.1.2 Discreteness and Finite Size Effects

A CA is a discrete model: discrete in space, time, and state space. The discreteness is partly wanted, but may also be responsible for undesired model artefacts. In the following, we discuss some of the discreteness consequences that may be present in CA models, and strategies how to avoid or reduce them.

Discrete State Space and Neighborhood. The discreteness of the state space in CA (cp. the definition in ch. 4) reflects the discrete nature of biological cells and is a wanted property. In lattice-gas automata, the maximum number of migrating cells per node, b, depends on the lattice (coordination number) and the number of different cell speeds: it is $b = 4$ for the square lattice and $b = 6$ for the hexagonal lattice, if only cells with identical speed are considered (there exists no limit for the number of resting cells, however). In addition, interactions are defined with respect to a discrete local neighborhood in the CA. Since a neighborhood in the automaton (e.g., the von Neumann interaction neighborhood) can host only a discrete number of cells, the CA is a natural tool to model cell interactions of a limited number of spatially adjacent cells, which characterizes the biological situation of local cell interaction. This property of *discrete and local interaction* is more difficult to achieve in other microscopic approaches and impossible to describe in the framework of continuous models.

[94] In ch. 11 we presented an automaton model based on a microscopic Turing-like interaction, in which "cell velocity" is the important microscopic parameter.

[95] A modeling strategy for a dynamic phenomenological description of cell shapes is provided by the extended Potts model (see, e.g., Glazier and Graner 1993; Marée and Hogeweg 2001).

Discrete Time. A synchronous update is inherent in the dynamics of LGCA and an exclusion principle is imposed on the state space. The exclusion principle with respect to the individual velocity channels implies that at any time at most one cell is allowed in each channel at every lattice node. An asynchronous update would break the exclusion principle because a cell could migrate to a node that already hosts a cell with the same orientation leading to a state space conflict (cp. the definition in ch. 4). It is, however, possible to allow for asynchronous update in modified lattice-gas automaton models (see, e.g., Börner et al. 2002). Synchronous and asynchronous updates in CA models have been systematically investigated in Schönfisch and de Roos 1999. There is a close relation between asynchronous probabilistic CA and interacting particle systems (Durrett and Levin 1994a, see also Voss-Böhme and Deutsch 2004).

Discrete Lattice Space. The discrete structure of the underlying lattice is also a wanted property of the automaton model since this allows a straightforward modeling of hard-core cell repulsion. In the automaton, cells are distributed at the nodes of a regular discrete lattice, which introduces a natural cellular distance defined by the lattice spacing. However, the discrete lattice may induce artefacts, and care has to be taken to choose an appropriate lattice. In particular, the lattice may induce *spurious modes* and *spatial anisotropies*. It is an important advantage of the CA concept that many artefacts can be identified already by a linear approximation, i.e., the Fourier spectrum of the underlying Boltzmann equation. In the applications described in this book, these artefacts may be avoided, in particular by the introduction of resting cells (cp. sec. 5.4.2) and the use of hexagonal lattices (see examples in secs. 12.3, 12.4, and 12.5). Nevertheless, in other systems a hexagonal lattice might be the wrong choice and cause, e.g., "frustration," a phenomenon known from antiferromagnets and spin glasses (Fischer and Hertz 1993): Suppose you want to put spins on a lattice such that any nearest-neighbor pair has opposite spin. On the square lattice this is easy: just put "spin up" on the even sublattice and "spin down" on the odd sublattice. However, on the hexagonal lattice there is no homogeneous way to arrange the spins: there is frustration.

We considered the checkerboard parity mode in the diffusive CA (cp. discussion in sec. 12.3), which indicates an unwanted conserved quantity. Here, conditions for the occurence of the checkerboard parity, which is a geometric artefact induced by the underlying lattice, can be derived. In particular, checkerboard parity modes are observed on square lattices with even lattice size and periodic boundary conditions since then the CA consists of two totally independent subsystems corresponding to the odd and even sublattices, respectively. A possible strategy to couple the subsystems (and to avoid the checkerboard parity) is the introduction of a "leaking probability" that couples the previously uncoupled subsystems or, alternatively, the introduction of "resting cells," which are cells that do not migrate (cp. fig. 5.10). It is commonly assumed that spurious modes have no counterpart in biological systems. However, there might exist biological effects corresponding to spurious modes.

Spatial artefacts can be expected if Fourier analysis indicates dominance of particular lattice directions. This is, e.g., the case in the adhesive and activator–inhibitor

LGCA (cp. ch. 7 and discussion in sec. 12.4). Here, diagonal directions are prevailing in square lattice simulations, while bias of any directions is much less pronounced in hexagonal simulations (cp. fig. 7.4). The Fourier spectrum, however, also allows us to explain such "spatial artefacts." For example, in the adhesive and activator–inhibitor LGCA, the "anisotropic" distribution of maxima in the spectrum corresponding to the square lattice is responsible for "diagonal anisotropies" in the square simulations, which are nearly absent on the hexagonal lattice (cp. figs. 7.6, 11.22, and 11.24).

Finite Size Effects and Boundary Conditions. All the examples introduced in this book operate on finite lattices. In physical models, periodic boundary conditions are typically imposed to mimic an infinite system. However, a biological embryo or a tissue culture are definitely finite systems and special care has to be taken to choose appropriate boundary conditions in a corresponding model. We have typically applied periodic boundary conditions in the models introduced in this book. This is justified whenever a model focuses on a local developmental aspect, e.g., the transition from an unordered cell distribution to the formation of relatively small (oriented) clusters. However, if the pattern of interest is highly dynamic, as, e.g., a collection of migrating swarms, simulations have to be performed on very large lattices in order to avoid artefacts and to mimic biologically relevant behaviors. If one wants to simulate a whole developmental sequence, realistic lattice sizes and more natural boundary conditions have to be implemented. We have provided an example, namely a model that simulates pigment pattern formation of salamander larvae, which assumes hybrid (open and reflecting) boundary conditions (cp. ch. 9 and discussion below).

12.2 Cellular Automata as a Modeling Tool

Cellular Automata: Local Fields and Coarse-Grained Interaction. Morphogenetic motion describes translocation of individual cells during development. Thereby, cell motion is influenced by the interaction of cells with elements of their immediate surrounding (haptotaxis and differential adhesion, interaction with the basal lamina and the interstitial matrix, contact guidance, contact inhibition, mechanical guidance within preestablished channels) and processes that involve cellular response to signals that are propagated over larger distances (e.g., chemotaxis, galvanotaxis; Armstrong 1985). It turns out that the LGCA methodology offers a very flexible modeling tool. LGCA are CA supplied with a particular set of "morphogenetic local rules" defining *interaction* and *convection* of polarized cells. In a LGCA a discrete-time evolution step consists of a (stochastic) *interaction* and a deterministic *migration step*, respectively. During (active) migration all cells move simultaneously to nodes in the direction of their polarization. Due to interaction, cell configurations at individual lattice nodes r are assumed to instantaneously change from $\boldsymbol{\eta}(r)$ to $\boldsymbol{\eta}^{\mathrm{I}}(r)$, which are pre- and postinteraction state, respectively. The essential modeling idea is the appropriate formulation of a local field characterizing a given cell interaction. It is assumed that the specificity of a particular cell interaction can be transformed into a corresponding local field $\boldsymbol{G} = \boldsymbol{G}\left(\boldsymbol{\eta}_{\mathcal{N}(r)}\right)$, which is a functional determined from the neighborhood configuration $\boldsymbol{\eta}_{\mathcal{N}(r)}$ in the automaton. In this book we have, for simplicity, focused on linear interaction functionals. The scalar product of a prospective local flux $\boldsymbol{J} = \boldsymbol{J}\left(\boldsymbol{\eta}^{\mathrm{I}}\right)$ and the field, $-\Delta E = \boldsymbol{G} \cdot \boldsymbol{J}$, can then be interpreted as a corresponding change in *interaction energy*. The interaction rules are not defined by a rigorous minimization of ΔE; instead the postinteraction state $\boldsymbol{\eta}^{\mathrm{I}}$ is chosen probabilistically according to "Boltzmann weights" $e^{-\Delta E}$. The biological interpretation is that a cell attempts to minimize the work against the surrounding field by a biased random change of polarization and not by a deterministic approach to a local energy minimum.

The applications in this book provide examples how elementary cell interactions, particularly differential adhesion, alignment, contact inhibition, haptotaxis, chemotaxis, or contact guidance, can be translated into corresponding local fields. As an example we consider *adhesive cell interaction*: a motile cell is a highly sensitive detector of small adhesive differences and moves steadily up an adhesive gradient. The influence of the cell surrounding can be formulated as a *density gradient field* in the CA

$$\left[\boldsymbol{G}\left(\boldsymbol{\eta}_{\mathcal{N}(r)}\right)\right] := \sum_{p=1}^{b} c_p \mathrm{n}\left(r + c_p\right)$$

(ch. 7, eqn. (7.1); cp. also the discussion in sec. 12.4). Another illustration is *contact guidance* directing cells towards the orientations of the surrounding substratum (extracellular matrix). The corresponding field in the automaton is the *director field* determining the mean orientation of surrounding matrix material (cp. the discussion in sec. 12.5). We have also investigated an *orientation-induced type of interaction* between cells that can be viewed as an extension of contact guidance to cell neighbors (cp. sec. 8.1). A further example in the book is a local field modeling the effect of chemotaxis (cp. 10.1). Our modeling approach is "modul-oriented". The book focuses on the definition and analysis of moduls for different elementary cell interactions, in particular on the definition of corresponding local fields. Typically, for modeling a specific morphogenetic problem, e.g., pigment pattern formation, various moduls have to be combined (see, e.g., ch. 9).[96]

12.3 Diffusive Behavior and Growth Patterns

For the evaluation of cell interaction effects on pattern formation, it is crucial to understand a (hypothetical) cellular system with no interactions. Corresponding "interaction-free" CA can be interpreted as *diffusive models* of cell dispersal

---

[96] Note that a similar modul-oriented modeling strategy can be chosen in applications of the extended Potts model that was originally introduced as a model of differential adhesion (Glazier and Graner 1993). Meanwhile, the Potts model formalism has also been applied to chemotactic interaction (Marée and Hogeweg 2001; Savill and Hogeweg 1997) and to consideration of the pattern-forming effects of different cell shapes, particularly of rod-shaped bacterial cells (Starruss and Deutsch 2004).

(sec. 5.4). We provided deterministic, probabilistic, and lattice-gas models of random dispersal. The local rule in the LGCA describes mixing of cells at single lattice nodes. Corresponding microdynamical (Boltzmann) equations of the automaton are linear. Stability analysis shows that all Fourier modes are stable; accordingly, no spatial patterns can be expected. In addition, Fourier analysis yields the existence of spurious "checkerboard parity modes" (cp. the discussion in sec. 12.1.2).

One can break the stability of the diffusive CA by coupling the diffusive rule to an (antagonistic) "sticking reaction" in which moving cells irreversibly transform into resting cells when encountering resting cells in their vicinity. This can be interpreted as a "degenerate interaction" of moving and resting cells (sec. 5.5). The "sticking rule" uses an idea originally proposed by Witten and Sander (1981) and is described in Chopard and Droz (1998). Automaton simulations exhibit diffusion-limited aggregation (DLA)-like growth patterns (cp. fig. 5.11).

Diffusion-limited growth processes coupled with (long-range) chemotactic cell communication are claimed to be responsible for the formation of certain bacterial aggregations (cp. fig. 4.3; Ben-Jacob et al. 1994). Snow crystal formation out of cold steam or metal solidification in an undercooled melt are physical examples of diffusion-limited growth. In these cases, pattern formation is due to an instability arising from the interplay of diffusion with microscopic forces (in particular, surface tension) and crystal anisotropy; as a result a stable phase moves into a metastable phase. Corresponding moving boundary problems have been primarily analyzed by continuum methods (e.g., phase-field methods; Caginalp and Fife 1986; Langer 1980). The DLA-LGCA, however, contains no analogue to surface tension or crystal anisotropy. Accordingly, the DLA instability is different from snow crystals, for example. The DLA instability in the automaton arises from a competition of diffusive motion and sticking.

In addition, we have studied growth rules in probabilistic, deterministic, and LGCA formulations (ch. 6). It is straightforward to define probabilistic CA rules that induce growing aggregates. The temporal growth dynamics sensitively depends on the choice of the rule. A mean-field approximation is only reasonable when growth is combined with random particle motion, since correlations arising from growth are partly destroyed by particle movement. It is possible to derive probabilistic CA and LGCA rules for random motion and growth leading to identical equations for the evolution of total particle densities. However, on the mean-field level there is an important difference between the formulations: the mean-field approximation of LGCA models leads to time- and space-dependent difference equations, while the spatial dependence is lost in the mean-field description of probabilistic CA.

12.4 Adhesive Interactions and Cell Sorting

The irreversible sticking rule in the DLA model (see above) is a crude implementation of *adhesion-driven cell interaction*. Adhesive interactions of cells and between cells and the extracellular matrix (ECM) are essential for stabilization, namely the maintenance of mechanical tissue contiguity. Stabilizing adhesive interaction may be mediated by cell junctions (desmosomes, gap, and tight junctions) or by components of the extracellular matrix (e.g., fibronectin, laminin, collagen) and their attachment to matrix receptors (e.g., integrins). Since the experiments performed by Holtfreter it has become clear that adhesive cell interaction is also responsible for the *dynamics of tissue formation*. Holtfreter (1939) demonstrated that characteristic rearrangements of tissues are formed, in particular sorting out and engulfment patterns, when randomly mixed tissue fragments are allowed to reaggregate.

Migrating cells do not install fixed intercellular junctions but prefer flexible adhesive interactions. In particular, it is known that cell-surface glycoproteins (cadherins) interact with each other even over small gaps (in the order of 10–20 nm) characterizing adjacently moving cells. Such contacts do not immobilize the cells but determine dynamic adhesive cell properties. Meanwhile it is well established that a migrating cell is able to detect even very small adhesive differences in its surrounding. For example in *Hydra* cell aggregates, there is experimental evidence that differential adhesion is the main source of cell sorting (Technau and Holstein 1992). A moving cell extends microspikes and lamellipodia in all directions and appears to engage in a "tug-of-war" with its immediate environment consisting of cells (of the same or other types) and the ECM. The result of the competition causes the polarization of the cell and, furthermore, orientation and movement into the "most adhesive direction" of its surrounding.

A refined automaton model of adhesive cell–cell interactions presented in the book is motivated by this microscopic picture of cellular interaction (ch. 7). In a single-cell-type model based on the tug-of-war behavior, it is assumed that a cell tends to orient into the direction of a local field, the *density gradient field*. An *adhesivity* parameter that can be viewed as a *bulk adhesivity* determines the sensitivity of the cell to the field and is the only free parameter in the local stochastic rules defining the state transition in the CA. We have furthermore introduced a CA that allows us to model differential adhesion of an arbitrary number of cell types (ch. 7). Thereby, interaction is based on the *density gradient fields* of various cell types and particularly allows us to model *cross adhesivities* between different cell types.

Adhesive Interaction of a Single Cell Type. Biologically, the single cell-type-model corresponds to an experiment in which cells are randomly distributed on a two-dimensional substratum and their adhesive dynamics of local monolayer formation and movement is observed. It turned out that a similar model was previously introduced as a model for phase segregation in a quenched binary alloy (Alexander et al. 1992). The microdynamical equation of the automaton is nonlinear. Stability analysis of the linearized Boltzmann equation shows that in the stable regime the dominant mode is diffusive. Conditions for a *cell-based, adhesion-driven instability* can be derived: this instability occurs if cell adhesion or cell density are sufficiently large (fig. 7.3). Then, the dominant mode destabilizes, implying initial periodic spatial pattern formation with a characteristic wavelength that can be deduced from the mean-field analysis.

The model suggests a mechanism for periodic pattern formation based on local interactions of a single cell type without chemical communication mediated by reaction-diffusion dynamics. However, the periodicity of the automaton model is only maintained within the short time scale of the linear regime, which may comprise days or hours in corresponding tissue experiments (fig. 7.4; $k = 100$). The initial local aggregation is due to the antidiffusive dynamics of the automaton. Accordingly, aggregation neighborhoods become void of cells, implying spatially heterogeneous patch formation. On a larger time scale, coalescence of neighboring aggregation patches with corresponding growth of single-phase domains and increasing wavelength of the patch distribution are observed (cp. fig. 7.4; $k \geq 1000$), and a scaling law for the growth of average domain sizes ($R(k) \sim k^{1/3}$) has been derived in Alexander et al. 1992.

In contrast to the adhesive automaton model, the Turing mechanism (of diffusion-driven instability) in reaction-diffusion systems—yielding a similar dispersion relation as the adhesive automaton (cp. fig. 7.5)—is able to stabilize an initially formed periodic pattern that may, eventually, lead to a bounded, stationary, spatially nonuniform, steady state (cp. Murray 2002). The stabilization is due to an appropriately chosen nonlinear reaction dynamics. Note that in Turing's original model (Turing 1952), the reaction dynamics was assumed to be linear. This particularly means that if a uniform steady state becomes unstable, then the chemical concentrations would grow exponentially, which is, of course, biologically (and chemically) unrealistic. Since then a number of nonlinear reactions have been proposed that can stabilize initial spatial heterogeneities (see Maini 1999 for an overview).

Nevertheless, the adhesive automaton model (cp. fig. 7.4) is relevant for the explanation of animal patterns, e.g., stripe patterns of some fish, which have previously been attributed to long-range communication in reaction-diffusion systems (Maini, Painter, and Chau 1997). The necessary assumption is that the pattern formation process is "frozen" in a particular state, which is justified by biological observations. In particular, it is known that nonjunctional cell–surface adhesion proteins (e.g., cadherins) that dominate the migratory phase may also induce tissue-specific cell–cell adhesion, which is furthermore stabilized by the assembly of cell junctions. Since many transmembrane cadherins (in particular, glycoproteins) are used for junctional as well as nonjunctional adhesive contacts—they are able to accumulate at sites of cell–cell contact—these molecules are candidates for inducing a transition from the migratory to the stationary state. Further processes characterizing a turnover from dynamic to static interactions and contributing to the stabilization of tissue organization are contact inhibition of pseudopodal activity and reversible loss of the cell's contractile machinery (Armstrong 1985). However, for a more realistic modeling of animal pigment pattern formation, it is important to consider the effects of interactions with other cell types and with the extracellular matrix (cp. the discussion below in sec. 12.6).

Differential Adhesion. As has been stated before, rearrangement of dissociated cells of different types was investigated in experiments performed by Townes and Holtfreter 1955 (cp. fig. 7.1). In order to explain these experiments, a hypothesis based on *differential cell adhesion* was proposed (Steinberg 1964). Steinberg suggested that cells interact through a physically motivated "interaction potential." Steinberg examined various scenarios leading to sorting out and engulfment, in particular. Thereby, he could reproduce the final cell configurations observed in Townes' and Holtfreter's experiments. Steinberg's differential adhesion hypothesis is based on an equilibrium principle—active cell migration plays no role in his argument, and the final cell arrangement is supposed to correspond to a minimum value for the tissue interfacial free energies.

Cells have, however, an autonomous motility that is not considered in Steinberg's model. The nonequilibrium adhesive CA model introduced here has been used to test the implications of active migration modulated by adhesive interactions (ch. 7). In particular, we could identify scenarios leading to sorting out and engulfment pattern formation. There are six parameters in the model: besides the adhesivities $\alpha_{11}, \alpha_{22}, \alpha_{12}, \alpha_{21}$, these are the average cell densities $\bar{\rho}_1, \bar{\rho}_2$ of cell types 1 and 2, respectively. A condition for sorting out arrangements is that the cross-adhesivities $(\alpha_{12}, \alpha_{21})$ are smaller than the adhesion between equal cell types $(\alpha_{11}, \alpha_{22})$, which can even be zero (fig. 7.8). Engulfment patterns arise if $\alpha_{11} < \alpha_{12} = \alpha_{21} < \alpha_{22}$, which corresponds to Steinberg's engulfment scenario. The CA also allows us to investigate asymmetric situations in which $\alpha_{12} \neq \alpha_{21}$. For example, if $\alpha_{12} = 0$ but $\alpha_{21} > 0$, then cell type 1 follows 2, while 2 is not influenced by 1. An example for the formation of encapsulation patterns with asymmetric cross-adhesivities is $\alpha_{12} < \alpha_{22} < \alpha_{21} < \alpha_{11}$ (cp. fig. 7.9). Asymmetric interactions can also be found in frustrated spin systems, in particular in spinglass and antiferromagnetic models (Fischer and Hertz 1993).

It is possible to perform a stability analysis of the two-cell-type model similar to the single-cell-type model that would explain the initial wavelength in the linear regime (cp. figs. 7.4 and fig. 7.8, $k = 100$; cp. also suggestions for "Further Research Projects" in sec. 7.5). Furthermore, one can derive a scaling law for the dynamics of patch growth in order to characterize differences observed in single- and two-species automaton simulations (cp. figs. 7.4 and fig. 7.8, $k \geq 100$). While there is basically only one growth scenario possible in the single-cell-type model, which is due to the self-aggregation of cells, the two-cell-type model offers more flexibility. For example, in the simulation shown in fig. 7.8, the aggregation tendency is not caused by self-aggregation of cells but, contrary, by repulsion of different cell types (self-adhesivities are zero, cross-adhesivities negative). As a consequence of repulsion cells try to reduce contacts with the other cell type. Accordingly, meanders are formed instead of the islands that characterize the single-cell-type self-aggregation patterns. A meandering structure is maintained for long times since this reduces contacts between different cell types. Islands and meanders cannot be distinguished at the level of linear stability analysis. However, one can discriminate their growth dynamics by comparing the length of contact lines (between cells and medium in the single-cell-type model and between cells of different types in the two-cell-type model) which are much smaller in the repulsion model than in the self-aggregation model (cp. figs. 7.4, and fig. 7.8, $k \geq 100$). While one can observe sorting out (meandering patterns) even at small time scales, engulfment patterns arise only at large time scales (cp. fig. 7.9, $k \geq 1000$).

There are physical models that exhibit, phenomenologically, similar patterns as observed in cell rearrangement experiments. Sorting out (demixing) is, for example, found in a lattice-gas model of immiscible fluids such as one finds in a mixture of oil and water (Adler, d'Humiéres, and Rothman 1994). Particle types are viewed here as different "colors." The rules depend on the particle configurations at neighboring sites and define the uphill diffusion of colors. Besides particle densities there are no further free parameters and only a single phase separation scenario is possible within the framework of this model, namely segregation due to repulsion. The two particle types separate spontaneously: complete phase separation visible as two separated "particle bands" is always observed for long time scales.

Encapsulation-like patterns are found in a lattice-gas model of microemulsions that form, e.g., in an oil-water mixture in the presence of a surfactant (Boghosian, Coveney, and Emerton 1996). The model adds a third "amphiphilic species" with a vector degree of freedom to the immiscible two-phase lattice-gas rules. In simulations, the mixture initially starts to separate phases. The amphiphilic particles cluster at the color interfaces and make it harder for the bubbles to coalesce. Rather than achieving a full phase separation for long time scales as in the immiscible fluid model, parameters can be chosen such that the "microemulsions" eventually reach a critical size and stop growing. The resulting pattern is a homogeneous distribution of "micelles," if the ratio of colored and amphiphilic particles is appropriately chosen. A phenomenological analogy can be drawn between microemulsions and biological cells if one views the amphiphilic species as a membrane constituent. Note that a differential-adhesion-based model has recently been suggested for the formation of pigment patterns in zebra fish (Moreira and Deutsch 2004). In addition, a systematic study of differential adhesion has also been carried out in an interacting-particle-based approach (Voss-Böhme and Deutsch 2004).

Further Models of Cellular Rearrangement. A different hypothesis to explain the cell sorting experiments of Townes and Holtfreter has been proposed in a taxis-diffusion model (Edelstein 1971). Here it was assumed that each cell type is able to secrete a type–specific chemical. The chemical stimuli are supposed to cause a directed cellular response. However, there is no experimental evidence for cell type-specific chemotactical agents in the cell sorting experiments of Townes and Holtfreter. Chemotaxis plays a role in, e.g., the formation of the nervous system of vertebrates and invertebrates. Various neurotrophic factors that may serve as chemotactic attractants for growth cones have been identified (e.g., NGF). In sec. 10.3 we introduced a hybrid LGCA model that exhibits taxis-based cell aggregation. As the adhesive LGCA, this model can be extended to multiple cell types (cp. the research project on p. 206).

Various other models of cellular rearrangement have been suggested. A many-particle model of potential-driven interactions between single motile cells has been introduced in Drasdo 1993. This model permits, in addition, the simulation of cell replication. A model based on Delaunay tesselation has been suggested for tissue simulations (Schaller and Meyer-Hermann 2004). Further many-particle systems are the so-called "active Brownian walker systems" which have been predominantly used for the analysis of "indirect cell interactions," in particular by means of chemical messengers produced by the cells (Schimansky-Geier, Schweitzer, and Mieth

1997; cp. also sec. 3.2.2). The latter models are based on a (physically-oriented) microscopic picture of cell movement that can be modified by an interaction potential or a self-generated field (e.g., chemical concentration field), respectively.

An alternative view on the generation of cell motion has been proposed in a computer model that treats cells as mobile two-dimensional polygons (Weliky and Oster 1990; Weliky et al. 1991). In this model cell motion and rearrangement are based on two different mechanisms: mechanical stress resulting from stretching or deformation of the cell sheet by external forces and internally generated protrusive or contractile forces caused by a specifically activated cell subpopulation. The model allows, for example, a realistic modeling of gastrulative motion. An extended Potts model (Glazier and Graner 1993) was introduced earlier in which biological cells are modeled as collections of lattice sites, what allows to consider cellular shape changes. An extended Potts model has for example been proposed as a model for essential phases in the life cycle of the slime mold *Dictyostelium discoideum* (Marée and Hogeweg 2001; Savill and Hogeweg 1997) and for "cell-shape based pattern formation" in myxobacteria (Starruss and Deutsch 2004).

12.5 Collective Motion and Aggregation

Orientation-Induced Interaction. The local gradient field defining adhesive interactions is determined by the density distribution of neighboring cells. There are, however, examples of cell interactions that are primarily driven by orientations. If cells align to the shape of the substratum, this behavior is known as contact guidance. In many cases of interest (when the lack of ECM and the contact with a rigid substrate does not allow the building up of mechanical tension fields) cell alignment can be a result of the interaction with the cell's nearest neighbors. There are experiments in which cells align towards the mean polarization of surrounding cells, e.g., in tissue experiments with homogeneously distributed fibroblast cells. Cells are approaching confluence after a couple of days, which can be viewed as a cellular example of swarm formation (fig. 8.2; Elsdale 1973). Also, in *in vitro* trajectories of cells obtained from the surgical specimen of a highly malignant *Glioblastoma multiforme* brain tumor, at high cell densities a partially ordered migration is apparent (Hegedus 2000). A further example is provided by street formation and aggregation of myxobacteria (Dworkin 1996).

A LGCA based on orientation-dependent interaction yields a model of swarm initiation, i.e., orientation-induced pattern formation (ch. 8). In the LGCA it is assumed that cell orientations can change according to the local "director field" defining the mean cell orientation in a cell's vicinity. Stability analysis of the linearized automaton Boltzmann equation helps to identify a dominant "orientational mode." This mode can become unstable, which allows us to characterize the onset of swarming as a phase transition depending on average cell density ($\bar{\rho}$) and the sensitivity (α) to the director field. An appropriate order parameter is the "spatially averaged velocity" ($\bar{\phi}$). For $\alpha < \alpha_c$ one has $\bar{\phi} = 0$ corresponding to a spatially homogeneous distribution. When the sensitivity parameter α reaches a critical value, the spatially homogeneous state becomes unstable, leading to a breaking of rotational symmetry, and a state where $\bar{\phi} \neq 0$.

We have determined critical "swarming sensitivities" through mean-field analysis of the linearized automaton Boltzmann equation (cp. fig. 8.5). These values are in good agreement with the values obtained from simulations (cp. fig. 8.6). Linear stability analysis can only explain short time scale effects, in particular the formation of oriented patches as observed in fig. 8.4 ($k = 100$). The number of possible macroscopic directions of patch migration is restricted by the underlying lattice: 4 and 6 in the square and hexagonal lattice, respectively (cp. figs. 8.4 and 8.8). These anisotropies can be distinguished by plotting $z(q_1, q_2) := \ln(\lambda(q_1, q_2))$ as a surface plot (as has been done for the adhesive automaton; cp. fig. 7.6 and the discussion in sec. 12.1.2). Simulations on finite lattices produce artefacts for long time scales since an artificial clustering is induced by repetitive waves of coincidence caused by the imposed periodic boundary conditions.

A macroscopic description of orientation-induced cell interaction in actively migrating cell populations can be given in terms of transport equations of Boltzmann type, in which the density distribution represents the population of cells as a function of space coordinate, orientation angle, and time (Alt and Pfistner 1990). Interactions are typically formulated by appropriately constructed integral kernels that can model the influence of neighboring cell densities. A one-dimensional version was introduced as a model of myxobacterial behavior within swarms, in which cells are gliding almost in one direction (Pfistner 1990). If the "interaction integrals" express an alignment tendency, stability analysis allows us to determine parameter regimes that characterize the onset of swarming. A similar model framework has been used to account for alignment and the rippling phenomenon (Lutscher 2002; Lutscher and Stevens 2002). These (deterministic) continuum models assume that cells can perceive and instantaneously act according to the density distribution of surrounding cells. In contrast, in our CA, cells behave stochastically according to the distribution of a finite number of neighboring cells, which is, biologically, a more appropriate assumption.

In Deutsch and Lawniczak (1999) "scaled CA" have been introduced to discriminate convective and interactive scaling. This allows us to distinguish the time scales of cell shape changes (e.g., lamellipodium formation), which is typically of the order of less than a minute and cell translocation in the order of some minutes depending on the particular situation. Furthermore, the macroscopic behavior of a one-dimensional version of the scaled automaton has been analyzed. In particular, a hyperbolic partial differential equation, which is of the same type as that introduced in Pfistner (1990), has been derived from the microdynamical automaton equations. Our analysis shows how the interactions, which can be identified at the macroscopic level, depend on the imposed scaling of space coordinates, time, and interaction probabilities. For example, if biological interactions happen on a slower scale than migration (for which convective or Euler scaling is imposed, i.e., time and space scale at the same rates), they can be distinguished at the automaton, but not at the PDE level (Deutsch and Lawniczak 1999).

Orientation-induced interactions have, furthermore, been investigated in diffusively migrating cell populations. The appropriate macroscopic description is in this case formulated as integro-partial differential equations (Mogilner and Edelstein-Keshet 1996b; Mogilner et al. 1997). The continuous description allows a distinction of spatial and angular diffusion; the ratio of these diffusivities determines which pattern formation scenario occurs. In particular, linear stability analysis permits us to decide among three different bifurcation scenarios: (A) formation of angular order in a spatially homogeneous distribution (low angular/high spatial diffusivity), (B) aggregation and formation of regular clusters without common orientation (low spatial/high angular diffusivity), and (C) development of "orientation waves" in patches of aligned objects (low angular and low spatial diffusivity; Mogilner et al. 1997). In contrast, there is no distinction between spatial and angular diffusion in our CA, which shows either diffusive (for $\alpha \approx 0$) or swarming behavior (for $\alpha > \alpha_c$); a supercritical alignment sensitivity ($\alpha > \alpha_c$) corresponding to low angular diffusion induces, simultaneously, a low spatial diffusion. Accordingly, only scenario (C) can occur in the automaton. It is, however, desirable to extend the automaton rule and to combine it with a diffusive rule, as discussed, e.g., in sec. 12.3. This increases the number of parameters but yields a more realistic model of cell migration since it permits us to distinguish angular and spatial diffusion as a precondition to simulate also scenarios (A) and (B).

Further extensions of our single-cell-type automaton based on orientation-induced cell interactions are necessary in order to develop specific models of cell motion and swarming with respect to particular biological systems. In particular, cellular adhesivity has to be incorporated (cp. suggestions for research projects in sec. 8.3).

Myxobacteria: Street Formation, Aggregation and Rippling. Here, we focus on the discussion of a biological example in which direct cell communication is well established: myxobacterial street (swarm) formation. Myxobacteria sense the relative velocity of neighboring cells with respect to their own velocity vector, eventually inducing subsequent alignment of one cell to the (majority of) orientations of neighboring cells as well as adjustment of corresponding migration speeds (Bender 1999). The interaction mechanisms are not known certainly: candidates for mechanical transductions of cell–cell interaction are external organelles known as pili (fimbriae) (Dworkin 1993; Dworkin and Kaiser 1993; Kaiser 1979), or the extracellular slime which is constantly produced by the cells and used as a local force mediator between adjacent cells. In the myxobacterium *Myxococcus xanthus*, the motility and interaction behavior is controlled by two distinct genetic systems, the adventurous (A) and social (S) systems (Hodgkin and Kaiser 1979). Both A and S motilities are modulated by interactions between cells. A-motile cells (A^+S^-) are capable of migrating autonomously, although their swarm expansion rate does increase up to sixfold with increasing cell density (Kaiser and Crosby 1983). Contrarily, S-motile cells (A^-S^+) do not migrate without neighbors and their motility increases gradually with cell density.

The CA introduced in ch. 8 can be interpreted as a model of myxobacterial adventurous and social cell motion, which assumes that cells autonomously migrate and can stochastically change their orientation as a result of encounters with neighboring cells. The *swarming instability* of the automaton which, particularly, appears if the density increases beyond a critical level, provides an explanation for the formation of oriented cell patches (cohorts) based on merely local orientation-induced cell interactions without the necessity of changing the behavioral strategy (from uncooperative to social behavior). In Deutsch (2000) a model was proposed that includes resting cells, which permit a more realistic simulation of patch growth and cohort motion also for longer time scales. Simulations and analysis show that a mere increase of cell density ($\bar{\rho}$) can induce the formation of aligned cell cohorts. The parameter diagram demonstrates that a transition from the stable region (corresponding to individual (adventurous) motion) into the unstable region (aligned patches) can be achieved by two strategies—either by an increase of the sensitivity parameter ($\alpha > \alpha_c$, which would describe a strategy change from uncooperativity to cooperativity), but also by just achieving a critical cell density ($\bar{\rho} > \bar{\rho}_c$; cp. the parameter diagram in fig. 8.5). The latter scenario can explain the natural transition from reproductive feeding to street formation in adventurous A^+S^- strains of *Myxococcus xanthus* (Kaiser and Crosby 1983). In particular, this developmental situation is accompanied by an increase of cell density, which induces a higher probability of random cell encounters and subsequent alignment.

Bacterial streets culminate in aggregation centers serving as initiation spots for fruiting body formation. The automaton also yields an admittedly rather simplified model of aggregation, if one includes an aggregation sensitivity, which determines the tendency of moving cells to transform into resting cells. We have demonstrated that aggregation centers can form as a result of subtle changes in aggregation sensitivity parameters, i.e., direct communication (Deutsch 2000). These aggregation centers mainly consist of stationary cells. In myxobacteria, however, aggregation and subsequent fruiting body formation are accompanied by spiral cell migration (O'Connor and Zusman 1989). More sophisticated models are necessary in order to incorporate such dynamic behaviors. In particular, models with a better resolution of cell orientations have to be developed which can be achieved by considering hexagonal lattices or by extending the neighborhood in square lattice versions.

There are other models of myxobacterial aggregation (many-particle systems and CA), which can explain the formation of streets and aggregation centers but which assume that cell communication is achieved indirectly through sensing concentrations of diffusible chemoattractants or nondiffusible slime, respectively, whose concentrations are explicitly modeled (Stevens 1992; Stevens and Schweitzer 1997). The importance of chemotactic communication for myxobacterial development is experimentally doubtful (Dworkin and Eide 1983; Kearns and Shimkets 1998), justifying to analyze the implications of direct cell–cell communication in myxobacteria. A CA based on direct cell–cell communication has also been suggested as a model of rippling pattern formation preceding the development of fruiting bodies in myxobacteria (Börner et al. 2002).[97]

---

[97] Recently, the rippling phenomenon has attracted further modelers and various hypotheses have been proposed (Igoshin et al. 2001; Lutscher and Stevens 2002).

12.6 Pigment Pattern Formation

As an example we have addressed the formation of stripes in larvae of the axolotl *Ambystoma mexicanum* and developed a corresponding automaton model. Periodic pigment patterns of *Ambystoma mexicanum* larvae consist of alternating "vertical stripes" of melanophores and xanthophores (Olsson and Löfberg 1993; fig. 1.6). Pattern formation results from adhesive interactions between melano- and xanthophores and from contact guidance induced by the ECM. We have designed a CA that incorporates two pigment cell types and their interactions with the ECM, which is modeled as a separate "stationary cell type" introducing an orientational anisotropy. Again, the microscopic cell-based picture is a moving cell engaged in a tug-of-war interaction with neighboring cells and the surrounding matrix, which induces stochastic orientation changes. One lattice side is viewed as the "neural crest" and characterized by permanent cell creation. We have demonstrated that the interplay of contact inhibition (of different pigment cell types) and contact guidance (along the direction imposed by the ECM) is essential for pattern formation. Neither of these processes can produce patterns in itself. Stripe width (3–5 cell lengths) and stripe number (approximately 5–6) are remarkably similar in both the real larvae and the appropriately scaled simulation pattern (cp. sec. 9.3).

The simulations show that a CA model based on cell–cell interactions and migration allows us to explain periodic stripe pattern formation as observed in larvae of the axolotl *Ambystoma mexicanum*. It would only be a technical difficulty to characterize corresponding cell-based instabilities along the lines of the analysis that we have performed for other cellular automata (cp. chs. 5–8). In particular, such analysis could substantiate the influences of the stationary ECM and the migrating pigment cells on the pattern formation process. Our automaton offers a model for experimental situations, in which the interplay of migrating cells and ECM is of major interest—wound healing is an important example. Recently, a CA based on differential adhesion has been introduced as a model for pigment pattern formation in zebra fish (Moreira and Deutsch 2004).

12.7 Tumor Growth

It has become clear that mathematical modeling can contribute to a better understanding of the still largely unknown cancer dynamics (Gatenby and Maini 2003). In particular, a number of CA models for different phases of tumor growth have been suggested (see review in Moreira and Deutsch 2002). In this book we have presented a CA for the avascular growth phase in multicellular spheroids. A better understanding of *in vitro* avascular tumor dynamics might allow one to design treatments that transfer a growing tumor into a saturated (nongrowing and therefore nondangerous) regime by means of experimentally tractable parameter shifts. The hybrid CA approach implements a lattice-gas automaton picture on the level of the cells and explicitly takes into account mitosis, apoptosis, and necrosis, as well as nutrient consumption and a diffusible signal that is emitted by cells becoming necrotic.

All cells follow identical interaction rules. The necrotic signal induces a chemotactic migration of tumor cells towards maximal signal concentrations. Starting from a small number of tumor cells automaton simulations exhibit the self-organized formation of a layered structure consisting of a necrotic core, a ring of quiescent tumor cells and a thin outer ring of proliferating tumor cells.

Future extensions include application of the proposed model to clinical tumor data, including angiogenesis and interactions between the tumor cells and the immune system, which particularly implies the introduction of realistic initial and boundary conditions into the model. Simulations of the hybrid automaton model can be extended to follow the development of macroscopic tumors. Simulated tumor sizes are not limited given enough computing power and a sufficiently large lattice size with a corresponding large lattice/tumor size ratio which guarantees that boundary effects do not influence the modeling of the diffusion process. The presented two-dimensional hybrid CA approach is a first step towards development of three-dimensional models.

The main objective of the two-dimensional model presented here is to reproduce experimental data from multicellular spheroids grown under *in vitro* conditions. In these experiments, a monoclonal cell population is assumed. Accordingly, in the simulations we have followed the fate of a single cell population. However, it is straightforward to include cellular heterogeneity in the model as, e.g., in the model described in (Kansal, Torquato, Chocas, and Deisboeck 2000). Model extensions will incorporate genetic and epigenetic cell heterogeneity.

The general advantage of using individual-based approaches over locally averaged continuum models is that local properties of cells on small length scales, such as the detachment of a single cell from the primary tumor that may precede metastasis formation, or (de-)differentiation and apoptosis (if only a small number of cells at special positions are concerned), cannot be described appropriately by a continuum approach. The idea of the presented hybrid LGCA based on realistic cell and signal kinetic parameters can be easily adapted to model spatio-temporal pattern formation in any biological system of discrete cells interacting by means of diffusing signals (and nutrients; see, e.g., Walther et al. 2004).

12.8 Turing Patterns and Excitable Media

In ch.11 we presented microscopic models for interactions that were originally analyzed in (macroscopic) reaction-diffusion systems. We show that our Boltzmann analysis can also be used in these systems to extract the underlying instabilities and to predict the spatio-temporal patterns seen in the simulations. The LGCA models for activator–inhibitor interactions and excitable media are strategic models, in the sense that the set of rules is chosen "as simple as possible" in order to capture *essential interactions* in many-particle systems. Even the simple LGCA rule of the activator–inhibitor model (ch. 11) leads to a variety of patterns in one (cp. fig. 11.4) and two (cp. figs. 11.22 and 11.24) space dimensions, which are controlled by the reaction probabilities p_c, p_d and the speed parameters m_A and m_I, respectively. We analyzed the corresponding lattice-Boltzmann equations with regard to the capability of spatial pattern formation. For this, we performed a stability analysis of space- and time-dependent difference equations and analyzed the spectrum of the linearized and Fourier-transformed discrete lattice-Boltzmann equations. It turned out that this method leads to very good predictions of the spatio-temporal LGCA dynamics. The calculations are straightforward and can be followed, in particular to get familiar with the linear stability analysis of LGCA. Furthermore, we showed the limits of this approach. These arise on the one hand from the nonlinearity of the lattice-Boltzmann equations and on the other hand from the local fluctuations inherent in the LGCA model, which cannot be captured by the deterministic lattice-Boltzmann equations. We demonstrated in the activator–inhibitor LGCA that fluctuations can support spatial pattern formation processes.

The discrete lattice-Boltzmann equations form a basis for the derivation of corresponding continuous partial differential equations for the mass of each component (cp. (11.37)). A linear stability analysis of this continuous system leads to a critical diffusion ratio for which it is possible to obtain diffusion-induced instabilities ("Turing regime"). Moreover, we derived an appropriate reactive, spatial, and temporal scaling relation between the system of partial differential equations and the lattice-Boltzmann model in the "Turing regime," i.e., for $m_I \geq 14$, $m_A = 1$ (cp. (11.41)). One apparent difference between the continuous reaction-diffusion model and the discrete lattice-Boltzmann model is the capability of pattern formation. The lattice-Boltzmann model exhibits a variety of patterns for many different values and relations of the speed parameters m_A and m_I, resulting from the imposed particle motion process in the LGCA model, and which can be explained by the spectrum shape. In contrast, the parabolic spectrum of the reaction-diffusion model predicts only one particular type of pattern in the "Turing regime" (cp. (11.6)).

In section 11.2 we introduced a LGCA model with a three-component interaction which mimics the behavior of excitable media. Typical patterns found in simulations are traveling rings and rotating spiral waves (cp. fig. 11.33). The excitability of the medium is characterized by means of the phase space dynamics of the system in the absence of diffusion. In the motion-free case, a mean-field (Boltzmann) approximation of the LGCA model is expected to be inadequate, as we have demonstrated for growth processes (cp. ch. 6). This might explain why the lattice-Boltzmann approximation of the automaton characterizes (theoretically) an oscillatory medium, while the LGCA rules lead to excitable dynamics without particle motion. On the other hand, including particle motion, the complete LGCA dynamics of the (local and global) density oscillations (cp. fig. 11.37) can be explained by the spectrum of the linearized Boltzmann equations, which indicates the occurence of traveling waves isotropically rotating in the medium with a fixed spatial period.

12.9 Outlook

The research strategy presented in this book is "modular"; i.e., starting from "basic interaction modules" (modul examples are, e.g., adhesion, alignment, contact repulsion, chemotaxis-based interactions) coupling of the modules is required to design models for specific morpohogenetic problems. The focus of future activities is the analysis of extended and combined interactions, which are not necessarily restricted to cells but could also comprise interactions at the organismic or the subcellular level. The resulting models contain a multitude of spatial and temporal scales and impose enormous difficulties for analytic treatment. We sketch a couple of possible applications and give perspectives for further analysis.

12.9.1 Further Applications

So far we have predominantly considered one- and two-dimensional LGCA models. Three-dimensional versions are useful as models, for example, of fruiting body formation characterizing *myxobacteria* (cp. fig. 1.2) or the analysis of tumor growth (cp. fig. 10.7). In myxobacteria a close correlation of cell positioning and specific gene expression profiles has been experimentally observed (Sager and Kaiser 1993). It is therefore worthwhile to develop and analyze CA that incorporate the effects of "pattern-driven" *cell differentiation* and reorientation.

Cell alignment can induce street formation, an example of swarming behavior. We have presented a primitive swarm model based on mere orientation-induced interactions (ch. 8). From fish it is known that schooling persists even under extensive maneuvres (Bleckmann 1993). Thus, in order to school, fish not only have to respond to changes in the directions of neighbors but also to be able to recognize and to react to velocity changes. Lateral organs of fish allow for velocity and acceleration detection of the surrounding flow field, which seems to be an important requirement for schooling (Bleckmann 1993). With the help of appropriately constructed CA, in which the state space has to be extended to account for multispeed "particles" and the dynamics now has to incorporate at least two preceding time steps in order to be able to describe temporal changes ("accelerations"), one should generate more realistic *fish schooling* models. Physical multispeed models can be found in (Wolfram 1986a; Qian, d'Humières, and Lallemand 1992). It is also feasible to include the effects of the surrounding flow field more directly and to analyze "hydrodynamic swarming cellular automata." Design and analysis of multispeed models will also have indications for a better understanding of morphogenetic cell motion.

Coupling the dynamics of intra- and intercellular chemical concentration fields with cellular distributions is a necessary precondition for multiscale models. Such models could, e.g., be used as models of bacterial and yeast pattern formation or chemotactic aggregation of the cellular slime mold *Dictyostelium discoideum*. The particular challenge is to incorporate different scales of cellular and molecular dynamics which is possible by connecting differently "scaled CA." In the book we have presented a chemotactic LGCA and a LGCA model for avascular tumor growth that can be viewed as first steps into this direction (ch. 10).

A promising application field is immunobiology, since organ-selective cell migration and spatial pattern formation are basic properties of the cellular immune system. For example, germinal centers serve the function of improving the affinity of B-cells to antigen attacks and are characterized by a spatial zonation (Meyer-Hermann,

Deutsch, and Or-Guil 2001; Meyer-Hermann 2002). Certain autoimmune diseases as polymyositis can be detected since they go ahead with particularly dense cellular patterns (immune cell infiltrates) that have an organ-destructive impact (Schubert, Masters, and Beyreuther 1993). The specificity of this dynamics seems to be due to combinations of receptor and adhesion molecules on the cell surface (Schubert 1998). It is an open problem to analyze connections of well-defined (specific) cellular adhesive properties with corresponding macroscopic aggregation patterns. A first step in this direction was presented in Grygierzec et al. 2004, in which a "multi-cell-type adhesive" CA model was suggested for a systematic investigation of the interplay of differential adhesion, aggregation, and cell migration. So far, in CA only the case of one and two cell types has been investigated with respect to the influence of differential adhesion (ch. 7). For the immune system, however, the interplay of a higher number of cell types differing in their adhesive properties is essential and has to be investigated in the future. Also, dynamic changes of adhesion and migration properties are feasible and have to be analyzed, e.g., in the pattern formation of germinal centers to improve our understanding of the cellular immune system (Meyer-Hermann 2002).

Three-dimensional versions of the three-species adhesive CA could be studied as models of *gastrulative motion*. It is challenging also here to analyze how "morphogenetic movements" are determined by changes of adhesive properties. In some applications it might prove useful to relax the exclusion principle or to increase the size of the interaction neighborhoods.

A particularly challenging application field is the modeling of *in vitro* and *in vivo* tumor growth. Tumorous tissues are prototypes of evolutionary systems: the interacting components (particularly tumor, healthy, and immune cells) can partially change their properties at rather short time scales. Cancer arises from the accumulation of usually somatic mutations in individual cell lines. Mutated cells may gain some competitive advantage over nonmalignant neighbors, being able to reproduce faster and invade territories usually reserved for "normal" cells. In this perspective, cancer is the antithesis of embryological development, characterized by a nondeterministic sequence of events leading to the disruption of an orderly multicellular organismic architecture. Thus, ideas from the modeling of developmental systems can be used but, in addition, evolutionary models have to be incorporated for a better understanding of tumor dynamics, which can be viewed as an evolving spatially and temporally interacting cell system.

12.9.2 Further Analysis

The master equation that fully specifies a probabilistic CA is a Chapman-Kolmogorov equation. We have shown how a Boltzmann equation can be derived if all correlations are neglected and the system dynamics is expressed solely by means of single particle distribution functions. The Boltzmann equation is typically nonlinear due to the interactions imposed. In this book we have restricted attention to the linear analysis of the Boltzmann equation; i.e., we have analyzed the Fourier spectrum of the linearized and Fourier-transformed Boltzmann equation. However, it is possible to refine the approximation by including higher order *correlations*.

The Boltzmann equation arises under the assumption that the probability of finding two cells at specific positions is given by the product of corresponding single particle distribution functions; i.e., any correlations are neglected and distributions fully factorize. By including two-, three-, etc. particle distribution functions the effect of correlations can be systematically studied. In particular, if pair correlations are taken into account, but third and higher order correlations are neglected, a *generalized Boltzmann equation* for the single particle distribution function is obtained, coupled to the so-called *ring equation* describing the evolution of the pair correlation function. For the adhesive (density-dependent) CA (Alexander model) the ring equation has been successfully evaluated (Bussemaker 1996). It is a challenge to determine corresponding equations for the other CA presented in this book. This analysis could particularly improve the understanding of long time behavior.

Furthermore, it is straightforward to directly simulate the Boltzmann equation arising from the mean-field assumption as described for example in Czirók et al. 2003. Contrary to the stochastic CA, the lattice Boltzmann equation is deterministic. A comparison of numeric solutions of the Boltzmann equation with automaton simulations has also been shown in ch. 11, where we have presented stochastic automaton simulations and simulations of the Boltzmann equation. Traditionally, *lattice Boltzmann methods* have been used as numerical methods to solve the Navier-Stokes equation. There is, however, an essential difference between Navier-Stokes lattice-Boltzmann methods and lattice-Boltzmann versions of CA models described in this book, since in the latter we are not primarily interested in a numerical scheme for solving a given macroscopic equation but in a better understanding of collective behavior arising in microscopic interaction models. The most simple Navier-Stokes lattice-Boltzmann models use a single relaxation time or lattice-BGK approximation (BGK stands for Boghosian, Coveney, and Emerton). This approach is equivalent to the exclusive consideration of the dominant mode in the Fourier spectrum of the linearized Boltzmann equation. Accordingly, a systematic *bifurcation theory* for CA models of cellular interaction might be developed.

Contrary to the early lattice-gas models, which were primarily constructed as a microscopic caricature of macroscopically known Navier-Stokes-like fluid dynamics, the LGCA automata introduced in this book start with a microscopic system description without *a priori* knowledge of macroscopic dynamics. We have demonstrated how to construct corresponding macroscopic equations from a microdynamical picture (cp. sec. 4.5, sec. 11.1.5, and Arlotti, Deutsch, and Lachowicz 2003; Deutsch and Lawniczak 1999). In some applications, however, macroscopic equations have been studied independently from the underlying stochastic process. In particular, significant analysis has been performed on integro-partial differential equations as models of orientation-induced pattern formation (Geigant 1999). The further comparison of microscopic and macroscopic perspectives is a challenging future project.

The *Boltzmann approach* focuses on the bulk behavior of cells. In this book, we have applied the Boltzmann strategy to discrete CA. This strategy could be equally well used in macroscopic or stochastic lattice-free models. For example, corresponding models of chemotactic pattern formation could be compared to existing models based on the (fully stochastic) Langevin or Brownian approach (Stevens and Schweitzer 1997).

In the orientation-dependent automaton (cp. ch. 8) a phase transition was detected on the basis of mean-field theory. Scaling properties have been analyzed (Bussemaker, Deutsch, and Geigant 1997) but could be further investigated with the help of *renormalization* tools. However, the precise scaling exhibited by theoretical (CA) models is very difficult to justify in a concrete biological situation.

Note that the *importance of space and stochasticity* has also been realized in ecological and evolutionary dynamics (see Boerlijst 1994; de Roos, McCauley, and Wilson 1998; Durrett and Levin 1994a for examples). It was, for instance, shown that multiple stable states and stochastic colonization events may lead to coexistence of competitors that are incompatible in deterministic "(mean-field) models" (Levin 1974, 1992). Furthermore, the interaction of spatial pattern formation with Darwinian selection has been examined. It could be demonstrated that only in a "spatial model" may hypercycles be resistant to parasites (Boerlijst 1994). These ideas might have implications for the better understanding of tumor evolution in individual patients. Appropriately constructed CA might allow for direct analysis of morphogenetic and evolutionary tumor dynamics in the future.

There is an interesting link from CA to interacting particle systems (Liggett 1985, 1999), which can be viewed as asynchronous CA. Important questions concern appropriately constructed stochastic models of cell interaction that can be analyzed exploiting the methods and techniques that have been developed for interacting particle systems (see, e.g., Voss-Böhme and Deutsch 2004).

The main intention of a CA model is the study of global behavior arising from given local rules. The inverse problem of deducing the local rules from a given global behavior is extremely difficult. There have been some efforts using evolutionary algorithms, but there are no general methods that allow one to find a CA rule which reproduces a set of observations (Capcarrere, Tettamanzi, and Sipper 1998, 2002). Based on the variability in the local dynamics, we demonstrated that CA modeling provides an intuitive and powerful approach to capture essential aspects of complex phenomena on various scales. Furthermore, due to the simple structure and unconditionally numerical stability of CA, a tool is available that is open for multifaceted experiments in various applications. Because of the discrete nature of CA, care has to be taken to detect and avoid artificial model behavior. Although there are almost no limitations in designing local rules, strategic LGCA models should be designed in order to allow an analytical treatment of the CA.

In conclusion, there are both challenging future perspectives with regards to interesting applications of the CA idea and possible refinements of analytical tools for the investigation of CA. It is our hope that the potential of CA for modeling essential aspects of biological systems will be further exploited in the future.

A

Growth Processes: A Mean-Field Equation

In this appendix we derive the mean-field equations of the following lattice-gas growth model (see (6.26)): *All* empty channels (r, c_i) simultaneously gain a particle with probability γ, if at least $B < \tilde{b} = 5$ particles are present at the node. The transition probability to reach a node configuration $\mathbf{z} \in \{0, 1\}^5$ given the node configuration $\boldsymbol{\eta}(r, k)$ is expressed by

$$W(\boldsymbol{\eta}(r,k) \to \mathbf{z}) = \begin{cases} 1 & \text{if } \boldsymbol{\eta}(r,k) = \mathbf{z} \in \{(0,0,0,0,0),(1,1,1,1,1)\} \\ & \text{or if } \boldsymbol{\eta}(r,k) = \mathbf{z} \text{ and } \mathrm{n}(r,k) < B, \\ \gamma & \text{if } \mathrm{n}(r,k) \in [B, \tilde{b}-1] \\ & \text{and } \boldsymbol{\eta}(r,k) \neq \mathbf{z} = (1,1,1,1,1), \\ 1-\gamma & \text{if } \mathrm{n}(r,k) \in [B, \tilde{b}-1] \\ & \text{and } \boldsymbol{\eta}(r,k) = \mathbf{z} \notin \{(0,0,0,0,0),(1,1,1,1,1)\}. \\ 0 & \text{else.} \end{cases} \tag{A.1}$$

Therefore, based on (4.21), we obtain the mean-field equation

$$\begin{aligned} f_i(r + mc_i, k+1) &= E\left(\mathcal{R}^{\mathrm{M}}\left(\boldsymbol{\eta}^{\mathrm{G}}(r,k)\right)\right) \\ &= \frac{1}{5}\sum_{l=1}^{5}\left[\sum_{\mathbf{z}\in\mathcal{E}}\sum_{\boldsymbol{\eta}\in\mathcal{E}} z_l W(\boldsymbol{\eta}(r,k) \to \mathbf{z})\, P(\boldsymbol{\eta}(r,k))\right] \\ &= \frac{1}{5}\sum_{l=1}^{5}\left[P((1,1,1,1)) + \sum_{\substack{\boldsymbol{\eta}|\eta_l=1\in\mathcal{E} \\ \sum_i \eta_i < B}} P(\boldsymbol{\eta}(r,k)) \right. \end{aligned}$$

$$
\left. + (1-\gamma+\gamma) \sum_{\substack{\boldsymbol{\eta}|_{\eta_l=1}\in\mathcal{E} \\ \sum_i \eta_i \in [B,\tilde{b}-1]}} P(\boldsymbol{\eta}(r,k)) + \gamma \sum_{\substack{\boldsymbol{\eta}|_{\eta_l=0}\in\mathcal{E} \\ \sum_i \eta_i \geq B}} P(\boldsymbol{\eta}(r,k)) \right]
$$

$$
= \frac{1}{5}\sum_{l=1}^{5} \left[\sum_{\boldsymbol{\eta}|_{\eta_l=1}\in\mathcal{E}} P(\boldsymbol{\eta}(r,k)) + \gamma \sum_{\substack{\boldsymbol{\eta}|_{\eta_l=0}\in\mathcal{E} \\ \sum_i \eta_i \geq B}} P(\boldsymbol{\eta}(r,k)) \right]
$$

$$
= \frac{1}{5}\sum_{l=1}^{5} \big[f_l(r,k) + \gamma E\big((1-\eta_l(r,k))(\Psi_B(\boldsymbol{\eta}(r,k)) + \cdots + \Psi_4(\boldsymbol{\eta}(r,k)))\big)\big], \tag{A.2}
$$

where the indicator functions Ψ_a are defined in (6.22).

B

Turing Patterns

Equations that are relevant for an analytic investigation of the LGCA model with activator–inhibitor interactions introduced in ch. 11 are provided in this appendix.

B.1 Complete Interaction Rule

Recall that the microdynamical equation (11.10) for the one-dimensional two-component lattice-gas model introduced in sec. 11.1.1 is given by

$$\begin{aligned} \eta^{\mathrm{R}}_{\sigma,i}(r,k) &= \mathcal{R}^{\mathrm{R}}\left(\boldsymbol{\eta}(r,k)\right) \\ &= \Psi\left(\boldsymbol{\eta}_A(r,k)\right) M_\sigma \left(\eta_{\sigma,i}(r,k)\right) \Psi^T\left(\boldsymbol{\eta}_I(r,k)\right). \end{aligned} \tag{B.1}$$

The vector of counting functions Ψ is

$$\Psi\left(\boldsymbol{\eta}_\sigma\right) = \begin{pmatrix} \breve{\eta}_{\sigma,1}\breve{\eta}_{\sigma,2}\breve{\eta}_{\sigma,3} \\ \breve{\eta}_{\sigma,1}\breve{\eta}_{\sigma,2}\eta_{\sigma,3} + \breve{\eta}_{\sigma,1}\eta_{\sigma,2}\breve{\eta}_{\sigma,3} + \eta_{\sigma,1}\breve{\eta}_{\sigma,2}\breve{\eta}_{\sigma,3} \\ \breve{\eta}_{\sigma,1}\eta_{\sigma,2}\eta_{\sigma,3} + \eta_{\sigma,1}\breve{\eta}_{\sigma,2}\eta_{\sigma,3} + \eta_{\sigma,1}\eta_{\sigma,2}\breve{\eta}_{\sigma,3} \\ \eta_{\sigma,1}\eta_{\sigma,2}\eta_{\sigma,3} \end{pmatrix},$$

where we set $\boldsymbol{\eta}_{\sigma,i} := (1 - \eta_{\sigma,i})$ and all terms are evaluated at (r,k).

Taking the expected value of eqn. (B.1), i.e., $f^{\mathrm{R}}_{\sigma,i} = E\left(\eta^{\mathrm{R}}_{\sigma,i}(r,k)\right)$, yields

$$\begin{aligned} f^{\mathrm{R}}_{A,i} = f_{A,i} &+ p_c \breve{f}_{A,i}\Big[(f_{A,j}\breve{f}_{A,l} + \breve{f}_{A,j}f_{A,l} + f_{A,j}f_{A,l})\breve{f}_{I,i}\breve{f}_{I,j}\breve{f}_{I,l} \\ &\qquad + f_{A,j}f_{A,l}(f_{I,i}\breve{f}_{I,j}\breve{f}_{I,l} + \breve{f}_{I,i}f_{I,j}\breve{f}_{I,l} + \breve{f}_{I,i}\breve{f}_{I,j}f_{I,l})\Big] \\ &- p_d f_{A,i}\Big[\breve{f}_{A,j}\breve{f}_{A,l}(f_{I,i}f_{I,j}\breve{f}_{I,l} + f_{I,i}\breve{f}_{I,j}f_{I,l} + \breve{f}_{I,i}f_{I,j}f_{I,l}) \\ &\qquad + (\breve{f}_{A,j}\breve{f}_{A,l} + f_{A,j}\breve{f}_{A,l} + \breve{f}_{A,j}f_{A,l})f_{I,i}f_{I,j}f_{I,l}\Big] \end{aligned} \tag{B.2}$$

and

$$
\begin{aligned}
f_{I,i}^{\mathrm{R}} = f_{I,i} + p_c \breve{f}_{I,i} \Big[& \breve{f}_{I,j} \breve{f}_{I,l} (f_{A,i} \breve{f}_{A,j} \breve{f}_{A,l} + \breve{f}_{A,i} f_{A,j} \breve{f}_{A,l} + \breve{f}_{A,i} \breve{f}_{A,j} f_{A,l} \\
& \qquad + f_{A,i} f_{A,j} \breve{f}_{A,l} + f_{A,i} \breve{f}_{A,j} f_{A,l} + \breve{f}_{A,i} f_{A,j} f_{A,l} \\
& \qquad + f_{A,i} f_{A,j} f_{A,l}) \\
& + (f_{I,j} \breve{f}_{I,l} + \breve{f}_{I,j} f_{I,l})(f_{A,i} f_{A,j} \breve{f}_{A,l} + f_{A,i} \breve{f}_{A,j} f_{A,l} \\
& \qquad + \breve{f}_{A,i} f_{A,j} f_{A,l} + f_{A,i} f_{A,j} f_{A,l}) \\
& + f_{I,j} f_{I,l} f_{A,i} f_{A,j} f_{A,l} \Big] \\
- p_d f_{I,i} \Big[& \breve{f}_{I,j} \breve{f}_{I,l} \breve{f}_{A,i} \breve{f}_{A,j} \breve{f}_{A,l} + (f_{I,j} \breve{f}_{I,l} + \breve{f}_{I,j} f_{I,l}) \\
& \cdot (\breve{f}_{A,i} \breve{f}_{A,j} \breve{f}_{A,l} + f_{A,i} \breve{f}_{A,j} \breve{f}_{A,l} + \breve{f}_{A,i} f_{A,j} \breve{f}_{A,l} + \breve{f}_{A,i} \breve{f}_{A,j} f_{A,l}) \\
& + f_{I,j} f_{I,l} (\breve{f}_{A,i} \breve{f}_{A,j} \breve{f}_{A,l} + f_{A,i} \breve{f}_{A,j} \breve{f}_{A,l} + \breve{f}_{A,i} f_{A,j} \breve{f}_{A,l} \\
& \qquad + \breve{f}_{A,i} \breve{f}_{A,j} f_{A,l} + f_{A,i} f_{A,j} \breve{f}_{A,l} + f_{A,i} \breve{f}_{A,j} f_{A,l} \\
& \qquad + \breve{f}_{A,i} f_{A,j} f_{A,l}) \Big],
\end{aligned}
\tag{B.3}
$$

where we set $\breve{f}_{\sigma,i} := (1 - f_{\sigma,i})$; all terms are evaluated at (r, k) and the indices $i, j, l \in \{1, 2, 3\}$ are always chosen to be distinct, i.e., $i \neq j \neq l$.

B.2 Linear Stability Analysis

The matrix elements ω_i, $i = 1, \ldots, 4$, from the matrix $\boldsymbol{I} + \Omega^0$ play an important role in the derivation of the Boltzmann propagator (cp. p. 216). They are explicitly determined by

$$
\begin{aligned}
\omega_1 &= \frac{1}{3}\Big(1 - (\bar{f}_I - 1)^2 \left(2(\bar{f}_I - 1) + 3\bar{f}_A(2 - \bar{f}_A + 4(\bar{f}_A - 1)\bar{f}_I)\right) p_c \\
&\qquad + \bar{f}_I^2 \left(2\bar{f}_I - 3 + 3\bar{f}_A(4 - 3\bar{f}_A + 4(\bar{f}_A - 1)\bar{f}_I)\right) p_d\Big), \\
\omega_2 &= -2(\bar{f}_A - 1)\bar{f}_A \left(1 - \bar{f}_I + \bar{f}_A(2\bar{f}_I - 1)\right) \left((\bar{f}_I - 1)p_c - \bar{f}_I p_d\right), \\
\omega_3 &= -\left((\bar{f}_I - 1)^2 - 2\bar{f}_A(\bar{f}_I - 1)(3\bar{f}_I - 1) + \bar{f}_A^2(1 + 6(\bar{f}_I - 1)\bar{f}_I)\right) \\
&\quad \cdot \left((\bar{f}_I - 1)p_c - \bar{f}_I p_d\right), \\
\omega_4 &= \frac{1}{3}\Big(1 - \bar{f}_A\big(9(\bar{f}_I - 1)^2 - 3\bar{f}_A(\bar{f}_I - 1)(9\bar{f}_I - 5) \\
&\qquad + \bar{f}_A^2(7 + 6\bar{f}_I(3\bar{f}_I - 4))\big)p_c - \big(1 + \bar{f}_A(3 - 3\bar{f}_A \\
&\qquad + \bar{f}_A^2 - 12(\bar{f}_A - 1)^2\bar{f}_I + 9(\bar{f}_A - 1)(2\bar{f}_A - 1)\bar{f}_I^2\big)p_d\Big).
\end{aligned}
\tag{B.4}
$$

For the case $p_c = p_d =: p$, the nontrivial spatially homogeneous steady state is defined by $(\bar{f}_A, \bar{f}_I) = (0.5, 0.5)$, and therefore we obtain

$$\omega_1 = \frac{1}{24}(8+p), \quad \omega_2 = -\frac{1}{4}p, \quad \omega_3 = \frac{3}{8}p \quad \text{and} \quad \omega_4 = \frac{1}{12}(4-7p). \tag{B.5}$$

Using this, the eigenvalues given by (11.21) become

$$\begin{aligned}
&\lambda_{1,2}(q) \\
&= \frac{1}{2}\Bigg(\frac{1}{24}(8+p)u_A(q) - \frac{1}{4}pu_I(q) \\
&\qquad \pm \sqrt{\frac{1}{18}(1-p)(5p-8)u_A(q)u_I(q) + \left(\frac{1}{24}(8+p)u_A(q) - \frac{1}{4}pu_I(q)\right)^2}\Bigg).
\end{aligned} \tag{B.6}$$

C

Excitable Media: Complete Interaction Rule

Recall that the lattice-Boltzmann equation (11.44) for the two-dimensional three-component LGCA model introduced in sec. 11.2 is given by

$$\begin{aligned} f_{\sigma,i}(r + m_\sigma c_i, k+1) - f_{\sigma,i}(r,k) &= \frac{1}{5}\sum_{l=1}^{5} E\left(\eta^{\mathrm{R}}_{\sigma,l}(r,k)\right) - f_{\sigma,i}(r,k) \\ &= \tilde{C}_{\sigma,i}\left(\boldsymbol{f}(r,k)\right). \end{aligned}$$

When particle motion is excluded, i.e., $m_\sigma = 0$, this becomes

$$\varrho_\sigma(r, k+1) = \varrho_\sigma(r,k) + \sum_{i=1}^{5} \tilde{C}_{\sigma,i}\left(\boldsymbol{f}\right),$$

where

$$\begin{aligned} \sum_{i=1}^{5} \tilde{C}_{A,i}\left(\boldsymbol{f}\right) &= \frac{1}{5}\varrho_A\left(\varrho_B(\varrho_C - 5) - 5(\varrho_C - 4)\right) \\ &\quad - \frac{2}{25}\varrho_A^2(\varrho_B - 5)(\varrho_C - 5) - \frac{2}{125}\varrho_A^3(\varrho_B\varrho_C - 25) \\ &\quad + \frac{1}{125}\varrho_A^4\left(-5 + \varrho_B(\varrho_C - 2) - 2\varrho_C\right) \\ &\quad + \frac{1}{15625}\varrho_A^5(\varrho_B(25 - 11\varrho_C) + 25(1 + \varrho_C), \end{aligned} \tag{C.1}$$

$$\begin{aligned} \sum_{i=1}^{5} \tilde{C}_{B,i}\left(\boldsymbol{f}\right) &= \varrho_B(0.1\varrho_C - 1) + \varrho_A\left(0.5 + 0.1(\varrho_B - \varrho_C)\right) \\ &\quad - \frac{2}{125}\varrho_A^3\left(\varrho_B(\varrho_C - 2) - 3\varrho_C\right) \\ &\quad + \frac{1}{625}\varrho_A^4\left(-8\varrho_C + \varrho_B(3\varrho_C - 7)\right) \end{aligned}$$

$$- \frac{3}{15625} \varrho_A^5 \left(-5\varrho_C + \varrho_B(2\varrho_C - 5)\right), \tag{C.2}$$

and

$$\begin{aligned}\sum_{i=1}^{5} \tilde{C}_{C,i}(\boldsymbol{f}) = {} & \varrho_C(0.1\varrho_B - 1) + \varrho_A\left(0.5 + 0.1(\varrho_C - \varrho_B)\right) \\ & - \frac{2}{125}\varrho_A^3\left(\varrho_C(\varrho_B - 2) - 3\varrho_B\right) \\ & + \frac{1}{625}\varrho_A^4\left(-8\varrho_B + \varrho_C(3\varrho_B - 7)\right) \\ & - \frac{3}{15625}\varrho_A^5\left(-5\varrho_B + \varrho_C(2\varrho_B - 5)\right).\end{aligned} \tag{C.3}$$

D

Isotropy, Lattices, and Tensors

To clarify the influence of the underlying lattice for a LGCA model, it is useful to express quantities characterizing the particular dynamics with the help of *tensors* (see sec. D.2), since a tensor representation directly reflects the influence of the lattice. An approporiate lattice should possess sufficient isometries such that the tensors are isotropic. In this section we focus on two-dimensional (square and hexagonal) lattices. The results can be extended to other, particularly higher dimensional, lattices.

D.1 Isotropic Media and Lattices

Consider the problem of particle motion in space. In continuous space, no spatial anisotropy exists, i.e., there are no *a priori* preferred directions of randomly moving particles. The situation is completely different in "discrete media." Cellular automata (CA) defined in this book operate on lattices that can be viewed as discrete media in which particles are forced into directions implied by the underlying lattice, obviously creating a spatial anisotropy. Accordingly, we say a discrete medium (spanned by the lattice together with its dynamics) is *isotropic* if it is invariant by all isometries of the lattice.[98] Having accepted the fact that a lattice is necessarily spatially anisotropic, it is important to characterize the "degree of spatial anisotropy" of a lattice that emerges to different extents depending on the particular dynamics. The basic idea is that we regard a lattice as "sufficiently isotropic" for simulating a particular dynamics if the process is invariant with respect to all isometric transformations of the lattice. We will see that a lattice might offer "sufficient isotropy" for one type of interaction dynamics, but not for another.

Isometries. An *isometry*[99] (orthogonal or symmetry transformation) is a transformation of space that leaves distances invariant. Isometries of d-dimensional space

[98] In this short overview we use the notions of Aris 1989; Ernst 1991; Rothman and Zaleski 1997.

[99] Iso means "same" and metry means "distance" or "measurement," as in "geometry."

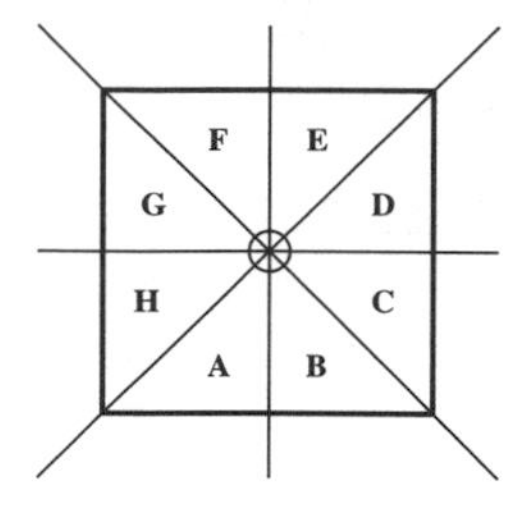

Reflections:

A↦B, H↦C, G↦D, F↦E

A↦C, A↦D, H↦E, G↦F

H↦G, A↦F, B↦E, C↦D

A↦H, B↦G, C↦F, D↦E

Rotation of angle ϕ:

$\phi = \frac{\pi}{2}$: A↦C, B↦D, C↦E, D↦F, E↦G, F↦H, G↦A, H↦B

$\phi = \pi$: A↦E, B↦F, C↦G, D↦H, E↦A, F↦B, G↦C, H↦D

$\phi = \frac{2}{3}\pi$: A↦G, B↦H, C↦A, D↦B, E↦C, F↦D, G↦E, H↦F

Figure D.1. Isometries of a square. Symmetry transformations for equally congruent parts A – H. $A \mapsto B$ means "A is mapped to B."

define the *orthogonal group* $O(d)$ consisting of reflections about a plane and rotations. The rotations form a subgroup $SO(d)$, the *special orthogonal group*. The finite subgroups of $O(d)$ are the symmetry groups of regular polygons, polyhedra, etc. In two dimensions these are the groups that leave polygons invariant. In particular, the symmetry group of the square contains eight isometries: four reflections, three rotations and the identity (cp. fig. D.1). For example, if we assume that the square in fig. D.1 is located at the origin of the Cartesian coordinate system and that the sides of the square have a length of two, then a vector $v := (-1/3, -2/3)$ which is located in region A can be transformed to a vector v' located in region C by a rotation of $\pi/2$, i.e., with

$$R := \begin{pmatrix} \cos(\frac{\pi}{2}) & -\sin(\frac{\pi}{2}) \\ \sin(\frac{\pi}{2}) & \cos(\frac{\pi}{2}) \end{pmatrix} = \begin{pmatrix} 0 & -1 \\ 1 & 0 \end{pmatrix} \qquad \text{we get } Rv^T = \left(\frac{2}{3}, -\frac{1}{3}\right) =: v', \tag{D.1}$$

where the length of the vector is conserved, i.e., $|v| = |v'| = \sqrt{5}/3$. Note that (D.1) can be stated using the *Einstein notation* in which repeated subscripts in a product imply summation over those subscripts, i.e.,

$$v'_\alpha = R_{\alpha\beta} v_\beta = \sum_{\beta=1}^{2} R_{\alpha\beta} v_\beta .$$

The triangle possesses six isometries, and the hexagon twelve. Generally, for a polygon with n edges there exist $2n$ isometries.

D.2 Introduction to Tensors

Typically, dynamical situations in this work can be described in terms of tensors,[100] which are—loosely stated—multiple-indexed objects $T_{\alpha_1,\ldots,\alpha_n}, \alpha_i = 1,\ldots,d$, where n is the *order* of the tensor. A scalar a is a zero-order tensor, a vector v_α is a tensor of order 1, and a second-order tensor $R_{\alpha,\beta}$ can be represented by a matrix.

Medium isotropy requires that the tensorial representation does not depend on the choice of the particular coordinate system. In particular, when space is transformed by an isometry R, scalars are unchanged by this transformation while the coordinates of a vector change as stated in (D.1). More generally, a tensor of order n transforms[101] as

$$T'_{\alpha'\beta'\ldots\gamma'} = R_{\alpha'\alpha}\, R_{\beta'\beta} \ldots R_{\gamma'\gamma}\, T_{\alpha\beta\ldots\gamma}. \tag{D.2}$$

Invariant and Isotropic Tensors. A tensor is *invariant* under a discrete or continuous group of symmetry transformations $R_{\alpha\beta}$ if all its components remain unchanged,

$$T'_{\alpha'\beta'\ldots\gamma'} = T_{\alpha'\beta'\ldots\gamma'},$$

for all transformations in the group. In particular, tensors are said to have *cubic, square, hexagonal, etc. symmetry*, if they are invariant under the symmetry transformations of the corresponding cubic, square, hexagonal, etc. lattices. Obviously, isotropic media are characterized by invariant tensors.

In general, the components of a tensor depend on the orientation of the coordinate system. However, for *isotropic tensors*, the components are always the same; i.e., they are invariant by all isometries in $O(d)$ (rotations and reflections). In particular, scalars are always isotropic, vectors are isotropic if they vanish, isotropic order 2 tensors are *proportional* to the unit matrix, and in order 3 again, only the null tensor is isotropic. That is,

$$T^{iso}_{\alpha} = 0, \quad T^{iso}_{\alpha\beta} = \delta_{\alpha\beta}, \quad T^{iso}_{\alpha\beta\gamma} = 0,$$

where $\delta_{\alpha\beta}$ is the usual Kronecker symbol. In order 4, isotropic tensors are of the form

$$T^{iso}_{\alpha\beta\gamma\delta} = d\left(\delta_{\alpha\beta}\delta_{\gamma\delta} + \delta_{\alpha\gamma}\delta_{\beta\delta} + \delta_{\alpha\delta}\delta_{\beta\gamma}\right)$$

with arbitrary coefficient d (Ernst 1991; Rothman and Zaleski 1997).

Example: As an example for isotropic and anisotropic tensors, consider the set of vectors of the nearest-neighborhood template $\mathcal{N}_b$ (cp. the definition on p. 68), which

[100] Tensors are (multilinear) mappings. Their definition is a natural extension of the vector calculus.

[101] We do not need to distinguish between covariant and contravariant indices because we consider solely orthogonal coordinate systems.

defines a lattice. Tensors up to order 4 generated by these vectors $c_i \in \mathcal{N}_b$ of the two-dimensional lattice are

$$T_\alpha = \sum_{i=1}^{b} c_{i\alpha}, \qquad T_{\alpha\beta} = \sum_{i=1}^{b} c_{i\alpha} c_{i\beta},$$

$$T_{\alpha\beta\gamma} = \sum_{i=1}^{b} c_{i\alpha} c_{i\beta} c_{i\gamma}, \qquad T_{\alpha\beta\gamma\delta} = \sum_{i=1}^{b} c_{i\alpha} c_{i\beta} c_{i\gamma} c_{i\delta}$$

for $\alpha, \beta \in \{1, 2\}$. For the square lattice ($b = 4$),

$$c_i \in \{(1, 0), (0, 1), (-1, 0), (0, -1)\},$$

and hexagonal lattice ($b = 6$),

$$c_i \in \{(1, 0), (0.5, \sqrt{3}/2), (-0.5, \sqrt{3}/2), (-1, 0), (-0.5, -\sqrt{3}/2), (0.5, -\sqrt{3}/2)\},$$

the first- and third-order tensors vanish, i.e., $T_\alpha = T_{\alpha\beta} = 0$, while the second-order tensors are given by

$$T_{\alpha\beta} = \frac{b}{2} \delta_{\alpha\beta} \propto T^{iso}_{\alpha\beta}.$$

Furthermore, it is easy to verify that

$$T_{\alpha\beta\gamma\delta}|_{b=4} = 2\delta_{\alpha\beta\gamma\delta}$$

and

$$T_{\alpha\beta\gamma\delta}|_{b=6} = \frac{3}{4} \left(\delta_{\alpha\beta}\delta_{\gamma\delta} + \delta_{\alpha\gamma}\delta_{\beta\delta} + \delta_{\alpha\delta}\delta_{\beta\gamma} \right) \propto T^{iso}_{\alpha\beta\gamma\delta}.$$

Fourth-order tensors of this type (particularly momentum flux tensors) characterize the microdynamic equation of hydrodynamic lattice gases. Accordingly, here hexagonal lattices are predominantly used since the corresponding fourth-order tensor is isotropic for the hexagonal but not for the square lattice (Doolen et al. 1990; Wolfram 1994).

D.3 LGCA Dynamics and the Influence of the Lattice

In linear approximation and under mean-field assumption the dynamics of a LGCA model can be expressed by the linear lattice-Boltzmann equation (cp. (4.29)), whose temporal evolution is determined by the dominant eigenvalue $\lambda(q)$ of the spectrum of the Boltzmann propagator (cp. (4.42)). In two dimensions the Taylor representation of a given eigenvalue around $q_* = (q_{*1}, q_{*2})$, with $\hat{q}_\alpha := q_\alpha - q_{*\alpha}$, is

$$\begin{aligned}\lambda(q) &= \lambda(q_*) + \hat{q}_1\partial_1\lambda(q_*) + \hat{q}_2\partial_2\lambda(q_*) + \frac{1}{2}\hat{q}_1^2\partial_{11}\lambda(q_*) + \hat{q}_1\hat{q}_2\partial_{12}\lambda(q_*)\\
&+ \frac{1}{2}\hat{q}_2^2\partial_{22}\lambda(q_*) + \frac{1}{6}\hat{q}_1^3\partial_{111}\lambda(q_*) + \frac{1}{2}\hat{q}_1^2\hat{q}_2\partial_{112}\lambda(q_*)\\
&+ \frac{1}{2}\hat{q}_1\hat{q}_2^2\partial_{122}\lambda(q_*) + \frac{1}{6}\hat{q}_2^3\partial_{222}\lambda(q_*) + \frac{1}{24}\hat{q}_1^4\partial_{1111}\lambda(q_*)\\
&+ \frac{1}{6}\hat{q}_1^3\hat{q}_2\partial_{1112}\lambda(q_*) + \frac{1}{4}\hat{q}_1^2\hat{q}_2^2\partial_{1122}\lambda(q_*) + \frac{1}{6}\hat{q}_1\hat{q}_2^3\partial_{2221}\lambda(q_*)\\
&+ \frac{1}{24}\hat{q}_2^4\partial_{2222}\lambda(q_*) + O(|q-q_*|^5).\end{aligned}$$

In Einstein notation this expression reduces to

$$\boxed{\begin{aligned}\lambda(q) = \lambda(q_*) + T_\alpha^A\hat{q}_\alpha + T_{\alpha\beta}^B\hat{q}_\alpha\hat{q}_\beta + T_{\alpha\beta\gamma}^C\hat{q}_\alpha\hat{q}_\beta\hat{q}_\gamma\\ + T_{\alpha\beta\gamma\delta}^D\hat{q}_\alpha\hat{q}_\beta\hat{q}_\gamma\hat{q}_\delta + O(|q-q_*|^5),\end{aligned}} \tag{D.3}$$

where $\alpha, \beta, \gamma, \delta \in \{1, 2\}$, with appropriately defined tensors, for example,

$$T^A = \frac{1}{2}\begin{pmatrix}\partial_{11}\lambda(q_*) & \partial_{12}\lambda(q_*)\\ \partial_{21}\lambda(q_*) & \partial_{22}\lambda(q_*)\end{pmatrix}.$$

The Taylor expansion of the dominant eigenvalue λ may provide an indication of isotropy/anisotropy of the particular LGCA dynamics: if the dominant critical wave number $q_* \approx 0$ (as, e.g., in the diffusive lattice-gas), we expect only weak anisotropy on a square lattice, since in this case the influence of the fourth-order tensor $T_{\alpha\beta\gamma\delta}^D$ is small.[102] But if the LGCA dynamics is characterized by a dominant critical wave number $q_* \neq 0$, low anisotropy on a square lattice can only be expected if $T_{\alpha\beta\gamma\delta}^D \approx 0$; otherwise it is recommended to choose a hexagonal lattice for the LGCA model.

[102] Diffusion processes can be simulated on a square lattice because the tensor in the corresponding diffusive dispersion relation is symmetric with respect to the square lattice (cp. "Further Research Projects," sec. 5.6).

References

Adam, J. A. and N. Bellomo (Eds.) (1996). *A survey of models for tumor immune system dynamics*. Boston: Birkhauser.

Adler, C., D. d'Humiéres, and D. H. Rothman (1994). Surface tension and interface fluctuations in immiscible lattice gases. *J. Phys. I 4*, 29–46.

Agur, Z. (1991). Fixed points of majority rule cellular automata with application to plasticity and precision of the immune system. *Compl. Syst. 2*, 351–357.

Ahmed, E. (1996). Fuzzy cellular automata models in immunobiology. *J. Stat. Phys. 85(1/2)*, 291–294.

Alarcon, T., H. Byrne, and P. K. Maini (2003). A cellular automaton for tumour growth in inhomogeneous environment. *J. Theor. Biol. 225(2)*, 257–274.

Alexander, F. J., I. Edrei, P. L. Garrido, and J. L. Lebowitz (1992). Phase transitions in a probabilistic cellular automaton: growth kinetics and critical properties. *J. Stat. Phys. 68(3/4)*, 497–514.

Alt, W. and G. Hoffmann (Eds.) (1990). *Biological motion*. Berlin, Heidelberg, New York: Springer-Verlag.

Alt, W. and B. Pfistner (1990). A two-dimensional random walk model for swarming behavior. In W. Alt and G. Hoffmann (Eds.), *Biological motion*, pp. 564–565. Berlin: Springer-Verlag.

Ames, W. F. (1977). *Numerical methods for partial differential equations*. New York: Academic Press.

Amoroso, S. and Y. N. Patt (1972). Decision procedures for surjectivity and injectivity of parallel maps for tesselation structures. *J. Comput. Sys. Sci. 6*, 448–464.

Andrecut, M. (1998). A simple three-states cellular automaton for modelling excitable media. *Int. J. Mod. Phys B 12(5)*, 601–607.

Arends, M. J. and A. H. Wyllie (1991). Apoptosis: Mechanisms and roles in pathology. *Int. Rev. Exp. Pathol. 32*, 223–254.

Aris, R. (1989). *Vectors, tensors, and the basic equations of fluid motion*. New York: Dover.

Arlotti, L., A. Deutsch, and M. Lachowicz (2003). *On a discrete Boltzmann-type model of swarming*. Preprint RW03-13(134), Oct. 2003, Inst. of Appl. Math. and Mechanics, Warsaw Uiversity.

Armstrong, P. B. (1985). The control of cell motility during embryogenesis. *Cancer Metast. Rev. 4*, 59–80.

Baer, R. M. and H. M. Martinez (1974). Automata and biology. *Annu. Rev. Biophys. Bio. 3*, 255–291.

Bagnoli, F. (2001). Cellular automata. http://arxiv.org/abs/cond-mat/9810012.

Bagnoli, F. and M. Bezzi (1998). Species formation in simple ecosystems. *Intern. J. Mod. Phys. C 9(4)*, 1–17.

Bagnoli, F. and M. Bezzi (2000). An evolutionary model for simple ecosystems. In D. Stauffer (Ed.), *Annual review of computational physics*. Singapore: World Scientific.

Bagnoli, F., R. Rechtman, and S. Ruffo (1992). Damage spreading and Lyapunov exponents in cellular automata. *Phys. Lett. A 172*, 34–38.

Baldi, P. and S. Brunak (1998). *Bioinformatics. The machine learning approach.* Cambridge, MA: MIT Press.

Balss, H. (1947). *Albertus Magnus, Werk und Ursprung*. Stuttgart: Wiss. Verl. Ges.

Banks, R. B. (1994). *Growth and diffusion phenomena: Mathematical frameworks and applications*. New York: Springer-Verlag.

Bär, M. and L. Brusch (2004). Core breakup of spiral waves caused by radial dynamics: Eckhaus and finite wave number instabilities. *New J. Phys. 6*, 5.

Bär, M. and M. Eiswirth (1993). Turbulence due to spiral breakup in a continuous excitable media. *Phys. Rev. E 48*, 1635.

Barbe, A. M. (1990). A cellular automata ruled by an eccentric conservation law. *Physica D 45*, 49–62.

Bard, J. B. L. (1990). *Morphogenesis: The cellular and molecular processes of developmental anatomy*. Cambridge: Cambridge University Press.

Barkley, D. (1991). A model for fast computer simulation of waves in excitable media. *Physica D 49(61)*, 61–70.

Ben-Avraham, D. and J. Köhler (1992). Mean-field (n,m)-cluster approximation for lattice models. *Phys. Rev. E 45(12)*, 8358–8370.

Ben-Jacob, E., H. Shmueli, O. Shochet, and A. Tenenbaum (1992). Adaptive self-organization during growth of bacterial colonies. *Physica A 187*, 378–424.

Ben-Jacob, E., O. Shochet, A. Tenenbaum, I. Cohen, A. Czirók, and T. Vicsek (1994). Generic modelling of cooperative growth patterns in bacterial colonies. *Nature 368*, 46–49.

Bender, T. (1999). *Entwicklung eines Algorithmus zur konturbasierten Analyse von Videoaufnahmen am Beispiel von Myxobakterien und statistische Auswertung der gewonnenen Daten*. Diploma thesis, University of Bonn.

Berec, L. (2002). Techniques of spatially explicit individual-based models: construction, simulation, and mean-field analysis. *Ecol. Model. 150*, 55–81.

Berec, L., D. S. Boukal, and M. Berec (2001). Linking the Allee effect, sexual reproduction, and temperature-dependent sex determination via spatial dynamics. *Am. Nat. 157(2)*, 217–230.

Bignone, F. A. (1993). Cells-gene interactions simulation on a coupled map lattice. *J. Theor. Biol. 161*, 231–249.

Bleckmann, H. (1993). Role of the lateral line in fish behaviour. In T. J. Pitcher (Ed.), *Behaviour of teleost fishes*, pp. 201–246. New York, London: Chapman & Hall.

Boccara, N. and K. Cheong (1992). Automata network SIR models for the spread of infectious diseases in populations of moving individuals. *J. Phys. A: Math. Gen. 25*, 2447–2461.

Boccara, N. and K. Cheong (1993). Critical behaviour of a probabilistic automata network SIS model for the spread of an infectious disease in a population of moving individuals. *J. Phys. A: Math. Gen. 26*, 3707–3717.

Boccara, N. and H. Fukś (1998). Modeling diffusion of innovations with probabilistic cellular automata. http://xxx.lanl.gov/abs/adap-org/9705004.

Boccara, N., J. Nasser, and M. Roger (1994). Critical behavior of a probabilistic local and nonlocal site-exchange cellular automaton. *Int. J. Mod. Phys. C 5(3)*, 537–545.

Boccara, N., O. Roblin, and M. Roger (1994a). Automata network predator-prey model with pursuit and evasion. *Phys. Rev. E 50(6)*, 4531–4541.

Boccara, N., O. Roblin, and M. Roger (1994b). Route to chaos for a global variable of a two-dimensional "game-of-life type" automata network. *J. Phys. A: Math. Gen. 27*, 8039–8047.

Bodenstein, L. (1986). A dynamic simulation model of tissue growth and cell patterning. *Cell Differ. Dev. 19*, 19–33.

Boerlijst, M. C. (1994). *Selfstructuring: A substrate for evolution*. Ph. D. thesis, University of Utrecht.

Boghosian, B. M., P. V. Coveney, and A. N. Emerton (1996). A lattice-gas model of microemulsions. *Proc. Roy. Soc. Lond. A 452*, 1221–1250.

Bolker, B. and S. W. Pacala (1997). Using moment equations to understand stochastically driven spatial pattern formation in ecological systems. *Theor. Popul. Biol. 52*, 179–197.

Bonner-Fraser, M. and S. E. Fraser (1988). Cell lineage analysis reveals multipotency of some avian neural crest cells. *Nature 335*, 161–164.

Boon, J. P., D. Dab, R. Kapral, and A. T. Lawniczak (1996). Lattice gas automata for reactive systems. *Phys. Rep. 273*, 55–147.

Bornberger, E. (1996). A simple folding model for HP-type lattice proteins. In R. Hofestädt, M. Löffler, T. Lengauer, and D. Schomburg (Eds.), *Computer Science and Biology, Proceedings of the German Conference on Bioinformatics (GCB'96)*. Universität Leipzig. IMISE Report No.1, pp. 72–77.

Börner, U., A. Deutsch, H. Reichenbach, and M. Bär (2002). Rippling patterns in aggregates of myxobacteria arise from cell-cell collisions. *Phys. Rev. Lett. 89*, 078101.

Bossel, H. (1994). *Modeling and simulation*. Wellesley, MA: A K Peters.

Boveri, T. (1910). Die Potenzen der *Ascaris*-Blastomeren bei abgeänderter Furchung: Zugleich ein Beitrag zur Frage qualitativ ungleicher Chromosomenteilung. In *Festschr. R. Hertwig, III*, pp. 133–214. Jena: G. Fischer.

Bowler, P. J. (1975). The changing meaning of "evolution." *J. Hist. Ideas 36*, 95–114.

Brieger, L. and E. Bonomi (1991). A stochastic cellular automaton simulation of the non-linear diffusion equation. *Physica D 47*, 159–168.

Britton, N. F. (1986). *Reaction-diffusion equations and their applications to biology*. London: Academic Press.

Britton, N. F. (2003). *Essential mathematical biology*. New York: Springer-Verlag.

Brown, R. (1828). A brief account of microscopical observations made in the months of June, July and August, 1827, on the particles contained in the pollen of plants; and on the general existence of active molecules in organic and inorganic bodies. *Phil. Mag. (new series) 4*, 161–173.

Burks, A. W. (1970). *Essays on cellular automata*. Urbana IL: University of Illinois Press.

Burks, C. and D. Farmer (1984). Towards modelling DNA sequences as automata. In D. Farmer, T. Toffoli, and S. Wolfram (Eds.), *Cellular automata: Proceedings of an interdisciplinary workshop, New York, 1983*, pp. 157–167. Amsterdam: North-Holland.

Bussemaker, H. (1995). *Pattern formation and correlations in lattice gas automata*. Ph.D. thesis, Instituut voor Theoretische Fysica, Universiteit Utrecht, Holland.

Bussemaker, H. (1996). Analysis of a pattern forming lattice gas automaton: Mean-field theory and beyond. *Phys. Rev. E 53(2)*, 1644–1661.

Bussemaker, H., A. Deutsch, and E. Geigant (1997). Mean-field analysis of a dynamical phase transition in a cellular automaton model for collective motion. *Phys. Rev. Lett. 78*, 5018–5021.

Caginalp, G. and P. C. Fife (1986). Phase-field methods for interfacial boundaries. *Phys. Rev. B 33*, 7792–7794.

Cannas, S. A., S. A. Paez, and D. E. Marco (1999). Modeling plant spread in forest ecology using cellular automata. *Comp. Phys. Communic. 121–122*, 131–135.

Capcarrere, M. S. (2002). *Cellular automata and other cellular systems: design and evolution*. Ph.D. thesis, Swiss Federal Institute of Technology, Lausanne.

Capcarrere, M. S., A. Tettamanzi, and M. Sipper (1998). Statistical study of a class of cellular evolutionary algorithms. *Evol. Comput. 7(3)*, 255–274.

Casciari, J. J., S. V. Sotirchos, and R. M. Sutherland (1992). Variations in tumor cell growth rates and metabolism with oxygen concentration, glucose concentration and extracellular pH. *J. Cell. Physiol. 151*, 386–394.

Casey, A. E. (1934). The experimental alteration of malignancy with a homologous mammalian tumor material I. *Amer. J. Cancer 21*, 760–775.

Castets, V., E. Dulos, J. Boissonade, and P. de Kepper (1990). Experimental evidence of a sustained standing Turing-type nonequilibrium chemical pattern. *Phys. Rev. Lett. 64*, 2953.

Casti, J. L. (1989). *Alternate realities*. New York: Wiley.
Casti, J. L. (2002). Science is a computer program. *Nature 417*, 381–382. Book review.
Chang, Y. Y. (1968). Cyclic 3',5'-adenosine monophosphate phosphodiesterase produced by the slime mold D*ictyostelium discoideum*. *Science 161*, 57–59.
Chaplain, M. A. J. (1996). Avascular growth, angiogenesis and vascular growth in solid tumours: The mathematical modelling of the stages of tumour development. *Math. Comput. Model. 23*, 47–87.
Chaplain, M. A. J., G. D. Singh, and J. C. McLachlan (Eds.) (1999). *On growth and form: Spatio-temporal pattern formation in biology*. Chichester: Wiley.
Chapuis, A. and E. Droz (1958). *Automata: A historical and technological study*. London: B. T. Batsford.
Chaté, H. and P. Manneville (1992). Collective behaviors in spatially extended systems with local interactions and synchronous updating. *Prog. Theor. Phys. 87 (1)*, 1–60.
Chaudhuri, P. P., D. R. Chowdhury, S. Nandi, and S. Chatterjee (1997). *Additive cellular automata—Theory and Applications, Vol. 1*. Los Alamitos, CA: IEEE Computer Society Press.
Chen, K., P. Bak, and C. Tang (1990). A forest-fire model and some thoughts on turbulence. *Phys. Lett. A 147*, 297–300.
Chen, M. Y., R. H. Insall, and P. N. Devreotes (1996). Signaling through chemoattractant receptors in D*ictyostelium*. *Trends Genet. 12*, 52–57.
Chopard, B. and M. Droz (1990). Cellular automata approach to diffusion problems. In P. Manneville, N. Boccara, G. Vichniac, and R. Bidaux (Eds.), *Cellular automata and modeling of complex physical systems, Springer Proceedings in Physics, Vol. 46*. Berlin: Springer-Verlag, p. 130.
Chopard, B. and M. Droz (1998). *Cellular automata modeling of physical systems*. New York: Cambridge University Press.
Civelekoglu, G. and L. Edelstein-Keshet (1994). Modelling the dynamics of F-actin in the cell. *J. Math. Biol. 56(4)*, 587–616.
Codd, E. F. (1968). *Cellular automata*. New York, London: Academic Press.
Comins, H. N., M. P. Hassell, and R. M. May (1992). The spatial dynamics of host-parasitoid systems. *J. Anim. Ecol. 61*, 735–748.
Cook, J. (1995). Waves of alignment in populations of interacting, oriented individuals. *Forma 10*, 171–203.
Coveney, P. V., A. N. Emerton, and B. M. Boghosian (1996). Simulation of self-reproducing micelles using a lattice-gas automaton. *J. Amer. Chem. Soc. 118*, 10719–10724.
Crank, J. (1975). *The mathematics of diffusion*. Oxford: Oxford University Press.
Cross, M. and P. Hohenberg (1993). Pattern formation outside of equilibrium. *Rev. Mod. Phys. 65*, 851.
Crutchfield, J. P. and M. Mitchell (1995). The evolution of emergent computation. *Proc. Nat. Acad. Sci. US 93(23)*, 10742–10746.

Crutchfield, J. P., M. Mitchell, and R. Das (2003). Evolutionary design of collective computation in cellular automata. In J. P. Crutchfield and P. Schuster (Eds.), *Evolutionary dynamics*. New York: Oxford University Press.

Czirók, A., A. Deutsch, and M. Wurzel (2003). Individual-based models of cohort migration in cell cultures. In W. Alt, M. Chaplain, M. Griebel, and J. Lenz (Eds.), *Models of polymer and cell dynamics*. Basel: Birkhäuser, pp. 205–219.

Dallon, J. C., H. G. Othmer, C. v. Oss, A. Panfilov, P. Hogeweg, T. Höfer, and P. K. Maini (1997). Models of *Dictyostelium discoideum* aggregation. In W. Alt, A. Deutsch, and G. Dunn (Eds.), *Dynamics of cell and tissue motion*, pp. 193–202. Basel: Birkhäuser.

Darwin, C. (1859). *On the origin of species by means of natural selection*. London: John Murray.

Das, R., J. P. Crutchfield, M. Mitchell, and J. E. Hanson (1995). Evolving globally synchronized cellular automata. In *Proceedings of the 6th International Conference on Genetic Algorithms (ICGA-95)*, San Fransisco, CA, pp. 336–343. San Mateo, CA: Morgan Kaufmann.

Das, R., M. Mitchell, and J. P. Crutchfield (1994). A genetic algorithm discovers particle based computation in cellular automata. In Y. Davidov, H. P. Schwefel, R. Manner (Eds.), *Parallel problem solving from nature*, pp. 244–353. New York: Springer-Verlag.

Davis, P. J. and R. Hersh (1981). *The mathematical experience*. Boston, Mass.: Birkhäuser. Deutsch: *Erfahrung Mathematik*. Birkhäuser, Basel 1986.

de Boer, R. J., J. van der Laan, and P. Hogeweg (1993). Randomness and pattern scale in the immune network: A cellular automaton approach. In W. D. Stein and F. J. Varela (Eds.), *Thinking about biology*, SFI Studies in the Sciences of Complexity, pp. 231–252. Redwood City, CA: Addison-Wesley.

de la Torre, A. C. and H. O. Mártin (1997). A survey of cellular automata like the "game of life." *Physica A 240*, 560–570.

de Roos, A. M., E. McCauley, and W. G. Wilson (1991). Mobility versus density-limited predator-prey dynamics on different spatial scales. *Proc. Roy. Soc. Lond. B 246*, 117–122.

de Roos, A. M., E. McCauley, and W. G. Wilson (1998). Pattern formation and the spatial scale of interaction between predators and their prey. *Theor. Popul. Biol. 53(2)*, 108–130.

Deutsch, A. (1991). Musterbildung bei dem Schlauchpilz N*eurospora crassa*: Mathematische Modellierung und experimentelle Analyse. Report No. 91/105 of the Sonderforschungsbereich 343, Diskrete Strukturen in der Mathematik, University of Bielefeld.

Deutsch, A. (1993). A novel cellular automaton approach to pattern formation by filamentous fungi. In L. Rensing (Ed.), *Oscillations and morphogenesis*, Chapter 28, pp. 463–480. New York: Marcel Dekker.

Deutsch, A. (Ed.) (1994). *Muster des Lebendigen*. Wiesbaden: Vieweg.

Deutsch, A. (1995). Towards analyzing complex swarming patterns in biological systems with the help of lattice-gas cellular automata. *J. Biol. Syst. 3*, 947–955.

Deutsch, A. (1996). Orientation-induced pattern formation: swarm dynamics in a lattice-gas automaton model. *Int. J. Bifurc. Chaos 6*, 1735–1752.

Deutsch, A. (1999a). *Cellular automata and biological pattern formation*. Habilitationsschrift, University of Bonn.

Deutsch, A. (1999b). Principles of morphogenetic motion: Swarming and aggregation viewed as self-organization phenomena. *J. Biosc. 24(1)*, 115–120.

Deutsch, A. (2000). A new mechanism of aggregation in a lattice-gas cellular automaton model. *Math. Comput. Model. 31*, 35–40.

Deutsch, A. and S. Dormann (2002). Principles and mathematical modeling of biological pattern formation. *Mat. Stos. 3*, 16–38.

Deutsch, A., A. Dress, and L. Rensing (1993). Formation of morphological differentiation patterns in the ascomycete N*eurospora crassa*. *Mech. Develop. 44*, 17–31.

Deutsch, A., M. Falcke, J. Howard, and W. Zimmermann (Eds.) (2004). *Function and regulation of cellular systems: Experiments and models*, Basel. Birkhauser.

Deutsch, A. and A. T. Lawniczak (1999). Probabilistic lattice models of collective motion and aggregation; from individual to collective dynamics. *Mathem. Biosc. 156*, 255–269.

d'Humières, D., Y. H. Qian, and P. Lallemand (1989). Invariants in lattice gas models. In R. Monaco (Ed.), *Proceedings of the Workshop on Discrete Kinetic Theory, Lattice Gas Dynamics and Foundations of Hydrodynamics; Torino, Italy, Sept. 20–24, 1988*. World Scientific, Singapore, pp. 102–113.

Dill, K. A., S. Bromberg, K. Yue, K. M. Fiebig, D. P. Yee, P. D. Thomas, and H. S. Chan (1995). Principles of protein folding—a perspective from simple exact models. *Protein Science 4*, 561–602.

Doolen, G. D., U. Frisch, B. Hasslacher, S. Orszag, and S. Wolfram (Eds.) (1990). *Lattice gas methods for partial differential equations*, Redwood City, CA: Addison-Wesley.

Dorie, M., R. Kallman, and M. Coyne (1986). Effect of cytochalasin B nocodazole and irradiation on migration and internalization of cells and microspheres in tumor cell spheroids. *Exp. Cell Res. 166*, 370–378.

Dorie, M., R. Kallman, D. Rapacchietta, D. van Antwerp, and Y. Huang (1982). Migration and internalization of cells and polystyrene microspheres in tumor cell spheroids. *Exp. Cell Res. 141*, 201–209.

Dormann, S. (2000). *Pattern formation in cellular automaton models—Characterisation, examples and analysis*. Ph.D. thesis, University of Osnabrück, Dept. of Mathematics/Computer Science, Applied Systems Science.

Dormann, S. and A. Deutsch (2002). Modeling of self-organized avascular tumor growth with a hybrid cellular automaton. *In Silico Biol. 2*, 0035. online journal.

Dormann, S., A. Deutsch, and A. T. Lawniczak (2001). Fourier analysis of Turing-like pattern formation in cellular automaton models. *Future Gener. Comp. Sys. 17*, 901–909.

Doucet, P. and P. B. Sloep (1992). *Mathematical modeling in the life sciences.* New York: Ellis Horwood.

Drasdo, D. (1993). *Monte-Carlo-Simulationen in zwei Dimensionen zur Beschreibung von Wachstumskinetik und Strukturbildungsphänomenen in Zellpopulationen.* Ph.D. thesis, University of Göttingen.

Drasdo, D. (2000). A Monte-Carlo approach to growing solid nonvascular tumours. In G. Beysens and G. Forgacs (Eds.), *Networks in biology and medicine*, pp 171–185. Springer, New York.

Drasdo, D. (2003). On selected individual-based approaches to the dynamics in multicellular systems. In W. Alt, M. Chaplain, M. Griebel, and J. Lenz (Eds.), *Models of polymer and cell dynamics.* Basel: Birkhäuser, 169–203.

Drasdo, D. and G. Forgacs (2000). Modeling the interplay of generic and genetic mechanisms in cleavage, blastulation and gastrulation. *Dev. Dynam. 219*, 182–191.

Drasdo, D., S. Höhme, S. Dormann, and A. Deutsch (2004). Cell-based models of avascular tumor growth. In A. Deutsch, M. Falcke, J. Howard, and W. Zimmermann (Eds.), *Function and regulation of cellular systems: Experiments and models.* Basel: Birkhauser, pp. 367–378.

Durrett, R. (1993). Stochastic models of growth and competition. In S. A. Levin, T. M. Powell, and J. H. Steele (Eds.), *Lecture notes in biomathematics 96, Lectures on patch dynamics.* Berlin: Springer-Verlag.

Durrett, R. (1995). Ten lectures on particle systems. In P. Biane and R. Durrett (Eds.), *Lecture notes in mathematics 1608, Lectures on probability theory.* Berlin: Springer-Verlag.

Durrett, R. (1999). Stochastic spatial models. In V. Capasso and O. Dieckmann (Eds.), *Mathematics inspired by biology*, pp. 39–94. Berlin: Springer-Verlag.

Durrett, R. and S. Levin (1994a). The importance of being discrete (and spatial). *Theor. Popul. Biol. 46*, 363–394.

Durrett, R. and S. A. Levin (1994b). Stochastic spatial models: a user's guide to ecological applications. *Phil. Trans. R. Soc. Lond. B 343*, 329–350.

Duryea, M., T. Caraco, G. Gardner, W. Maniatty, and B. K. Szymanski (1999). Population dispersion and equilibrium infection frequency in a spatial epidemic. *Physica D 132*, 511–519.

Dworkin, M. (1993). Cell surfaces and appendages. In M. Dworkin and D. Kaiser (Eds.), *Myxobacteria II*, pp. 63–83. Washington, DC: American Society for Microbiology.

Dworkin, M. (1996). Recent advances in the social and developmental biology of myxobacteria. *Microbiol. Rev. 60*, 70–102.

Dworkin, M. and D. Eide (1983). M*yxococcus xanthus* does not respond chemotactically to moderate concentration gradients. *J. Bacteriol. 154*, 437–442.

Dworkin, M. and D. Kaiser (Eds.) (1993). *Myxobacteria II*, Washington: American Society for Microbiology.

Ebeling, W., A. Engel, and R. Feistel (1990). *Physik der Evolutionsprozesse.* Berlin: Akademie-Verlag.

Edelstein, B. B. (1971). Cell specific diffusion model of morphogenesis. *J. Theor. Biol. 30*, 515–532.

Edelstein-Keshet, L. and B. Ermentrout (1990). Models for contact-mediated pattern formation: Cells that form parallel arrays. *J. Math. Biol. 29*, 33–58.

Eden, M. (1961). A two-dimensional growth process. In J. Neyman (Ed.), *Proceedings of the Fourth Berkeley Symposium on Mathematics Statistics and Probability*, Vol. 4, pp. 223–239. Univ. California Press, Berkeley.

Einstein, A. (1905). Über die von der molekularkinetischen Theorie der Wärme geforderte Bewegung von in ruhenden Flüssigkeiten suspendierten Teilchen. *Ann. Physik 17(4)*, 549–560.

Ellner, S. P. (2001). Pair approximations for lattice models with multiple interaction scales. *J. Theor. Biol. 210*, 435–447.

Eloranta, K. (1997). Critical growth phenomena in cellular automata. *Physica D 103*, 478–484.

Elsdale, T. (1973). The generation and maintenance of parallel arrays in cultures of diploid fibroblasts. In E. Kulonen and J. Pikkarainen (Eds.), *Biology of fibroblasts*. New York: Academic Press.

Emmeche, C. (1994). *Wie die Natur Formen erzeugt*. Hamburg: Rowohlt.

Engelhardt, R. (1994). *Modelling pattern formation in reaction diffusion systems*. Master's thesis, Dept. of Chemistry, University of Copenhagen, Denmark.

Epperlein, H.-H. and J. Löfberg (1990). *The development of the larval pigment patterns in Triturus alpestris and Ambystoma mexicanum*. Advances in Anatomy, Embryology and Cell Biology 118. Berlin, Heidelberg: Springer-Verlag.

Erickson, C. A. (1990). Cell migration in the embryo and adult organism. *Curr. Opin. Cell Biol. 2*, 67–74.

Ermentrout, B., J. Campbell, and G. Oster (1986). A model for shell patterns based on neural activity. *The Veliger 28(4)*, 369–388.

Ermentrout, G. B. and L. Edelstein-Keshet (1993). Cellular automata approaches to biological modeling. *J. Theor. Biol. 160*, 97–133.

Ernst, M. (1991). Statistical mechanics of cellular automata fluids. In J. Hansen, D. Levesque, and J. Zinn-Justin (Eds.), *Liquids, freezing and glass transition*, pp. 43–143. Amsterdam: North Holland.

Evans, K. M. (2001). Larger than life: Digital creatures in a family of two-dimensional cellular automata. In R. Cori, J. Mazoyer, M. Morvan, and R. Mossery (Eds.), *Discrete models: Combinatorics, computation, and geometry, DM-CCG 2001*, Lect. Notes Comput. Sci., pp. 177–192.

Fahse, L., C. Wissel, and V. Grimm (1998). Reconciling classical and individual-based approaches in theoretical population ecology: a protocol for extracting population parameters from individual-based models. *Am. Nat. 152(6)*, 838–852.

Farmer, D., T. Toffoli, and S. Wolfram (Eds.) (1984). *Cellular automata: Proceedings of an interdisciplinary workshop, New York, 1983*, Amsterdam. North-Holland.

Fast, V. G. and I. R. Efimov (1991). Stability of vortex rotation in an excitable cellular medium. *Physica D 49*, 75–81.

Feller, W. (1968). *An introduction to probability theory and its applications*. New York: Wiley.

Feynman, R. P. (1982). Simulating physics with computers. *Int. J. Theor. Phys. 21(6/7)*, 467–488.

Fischer, K. H. and J. A. Hertz (1993). *Spin glasses*. Cambridge: Cambridge University Press.

Fisher, R. A. (1937). The wave of advance of advantageous genes. *Ann. Eugenic. 7*, 355–369.

Folkman, J. and M. Hochberg (1973). Self-regulation of growth in three dimensions. *J. Exp. Med. 138*, 745–753.

Freyer, J. P. (1988). Role of necrosis in regulating the growth saturation of multicellular spheroids. *Cancer Res. 48*, 2432–2439.

Frisch, U., D. d'Humières, B. Hasslacher, P. Lallemand, Y. Pomeau, and J.-P. Rivet (1987). Lattice gas hydrodynamics in two and three dimensions. *Compl. Syst. 1*, 649–707.

Frisch, U., B. Hasslacher, and Y. Pomeau (1986). Lattice-gas automata for the Navier-Stokes equation. *Phys. Rev. Lett. 56*(14), 1505–1508.

Ganguly, N., B. K. Sikdar, A. Deutsch, G. Canright, and P. P. Chaudhuri (2004). *A survey on cellular automata*. submitted.

Gardiner, C. W. (1983). *Handbook of stochastic methods*. Berlin: Springer-Verlag.

Gardner, M. (1983). *Wheels, life and other mathematical amusements*. New York: Freemann.

Gatenby, R. and P. K. Maini (2003). Cancer summed up. *Nature 421*, 321.

Geigant, E. (1999). *Nichtlineare Integro-Differential-Gleichungen zur Modellierung interaktiver Musterbildungsprozesse auf S^1*. Ph. D. thesis, University of Bonn.

Gerhardt, M. and H. Schuster (1989). A cellular automaton describing the formation of spatially ordered structures in chemical systems. *Physica D 36*, 209–221.

Gerhardt, M. and H. Schuster (1995). *Das digitale Universum*. Wiesbaden: Vieweg.

Gerhardt, M., H. Schuster, and J. Tyson (1990a). A cellular automaton model of excitable media. II. curvature, dispersion, rotating waves and meandering waves. *Physica D 46*, 392–415.

Gerhardt, M., H. Schuster, and J. J. Tyson (1990b). A cellular automaton model of excitable media including curvature and dispersion. *Science 247*, 1563–1566.

Gierer, A. and H. Meinhardt (1972). A theory of biological pattern formation. *Kybernetik 12*, 30–39.

Glazier, J. A. and F. Graner (1993). Simulation of the differential adhesion driven rearrangement of biological cells. *Phys. Rev. E 47(3)*, 2128–2154.

Gocho, G., R. Pérez-Pascual, and J. Rius (1987). Discrete systems, cell-cell interactions and color patterns of animals, I and II. *J. Theor. Biol. 125(4)*, 419–435.

Godt, D. and U. Tepass (1998). D*rosophila* oocyte localization is mediated by differential cadherin-based adhesion. *Nature 395*, 387–391.

Goel, N. S. and R. L. Thompson (1988). *Computer simulation of self-organization in biological systems*. Melbourne: Croom.

Gould, S. J. (1976). D'Arcy Thompson and the science of form. *Boston Studies in the Philosophy of Science—Topics in the Philosophy of Biology 27*, 66–97.

Gould, S. J. (1977). *Ontogeny and phylogeny*. Cambridge, MA: The Belknap Press of Harvard University Press.

Gould, S. J. and W. W. Morton (1983). *Hen's teeth and horse's toes. Further reflections in natural history*. New York: Birkhäuser.

Green, P. B. (1996). Transduction to generate plant form and pattern: An essay on cause and effect. *Ann. Bot. 78*, 269–281.

Greenberg, J. M., B. D. Hassard, and S. P. Hastings (1978). Pattern formation and periodic structures in systems modeled by reaction-diffusion equations. *B. Am. Math. Soc. 84*, 1296–1327.

Greenspan, H. P. (1972). Models for the growth of a solid tumor by diffusion. *Stud. Appl. Math. 51*, 317–340.

Grünbaum, D. (1994). Translating stochastic density-dependent individual behavior with sensory constraints to an Eulerian model of animal swarming. *J. Math. Biol. 33*, 139–161.

Grünbaum, D. and A. Okubo (1994). Modeling animal aggregation. In S. A. Levin (Ed.), *Frontiers of theoretical biology*. Lecture Notes in Biomathematics, vol. 100, pp. 296–325.

Grygierzec, W., A. Deutsch, L. Philipsen, M. Friedenberger, and W. Schubert (2004). Modelling tumour cell population dynamics based on molecular adhesion assumptions. *J. Biol. Syst.* In print.

Gueron, S. and S. A. Levin (1993). Self-organization of front patterns in large wildebeest herds. *J. Theor. Biol. 165*, 541–552.

Gueron, S. and S. A. Levin (1995). The dynamics of group formation. *Math. Biosc. 128*, 243–264.

Gueron, S., S. A. Levin, and D. I. Rubenstein (1996). The dynamics of mammalian herds: from individuals to aggregations. *J. Theor. Biol. 182*, 85–98.

Gunji, J. (1990). Pigment colour patterns of molluscs as an autonomous process generated by asynchronous automata. *Biosystems 23*, 317–334.

Günther, H. V. (1994). *Wholeness lost and wholeness regained*. Albany: State University of New York Press.

Gurney, W. S. C. and R. M. Nisbet (1989). *Ecological dynamics*. Oxford: Oxford University Press.

Gutowitz, H. A. (1990). A hierarchical classification of cellular automata. *Physica D 45*, 136–156.

Gutowitz, H. A. and J. D. Victor (1989). Local structure theory: Calculation on hexagonal arrays, and interaction of rule and lattice. *J. Stat. Phys. 54(1/2)*, 495–514.

Gutowitz, H. A., J. D. Victor, and B. W. Knight (1987). Local structure theory for cellular automata. *Physica D 28*, 18–48.

Haberlandt, R., S. Fritzsche, G. Peinel, and K. Heinzinger (1995). *Molekulardynamik—Grundlagen und Anwendungen*. Braunschweig/Wiesbaden: Vieweg.

Haken, H. (1977). *Synergetics. An introduction*. Berlin, Heidelberg, New York: Springer-Verlag.

Haken, H. (1978a). *Cooperative phenomena in multi-component systems*. Stuttgart: Teubner.

Haken, H. (1978b). *Synergetics. An introduction. Nonequilibrium phase transitions and self-organization in physics, chemistry and biology*. Berlin: Springer-Verlag.

Haken, H. and H. Olbrich (1978). Analytical treatment of pattern formation in the Gierer-Meinhardt model of morphogenesis. *J. Math. Biol. 6*, 317–331.

Hall, B. K. and S. Hörstadius (1988). *The neural crest*. New York: Oxford University Press.

Hameroff, S. R., S. A. Smith, and R. C. Watt (1986). Automaton model of dynamic organization in microtubules. *Ann. NY Acad. Sci. 466*, 949–952.

Harrison, R. G. (1907). Observations on the living developing nerve fibre. *Anat. Rec. 1*, 116–118.

Hassell, M. P., H. N. Comins, and R. M. May (1991). Spatial structure and chaos in insect population dynamics. *Nature 353*, 255–258.

Hasslacher, B. (1987). Discrete fluids. Los Alamos Science, pp. 175–217.

Hasslacher, B., R. Kapral, and A. T. Lawniczak (1993). Molecular Turing structures in the biochemistry of the cell. *Chaos 3(1)*, 7–13.

Hastings, A. (1994). Conservation and spatial structure: theoretical approaches. In S. A. Levin (Ed.), *Lecture notes in biomathematics 100, Frontiers in mathematical biology*, pp. 494–503. Berlin: Springer-Verlag.

He, X. and M. Dembo (1997). A dynamical model of cell division. In W. Alt, A. Deutsch, and G. Dunn (Eds.), *Dynamics of cell and tissue motion*, pp. 55–65. Basel: Birkhäuser.

Hegedus, B. (2000). Locomotion and proliferation of glioblastoma cells in vitro: Statistical evaluation of videomicroscopic observations. *J. Neurosurg. 92*, 428–434.

Hegselmann, R. and A. Flache (1998). Understanding complex social dynamics: A plea for cellular automata based modelling. *J. Artif. Soci. and Soc. Simul. (online journal) 1*(3).

Helbing, D. (2001). Traffic and related self-driven many-particle systems. *Rev. Mod. Phys. 73(4)*, 1067–1141.

Hendry, R. J., J. M. McGlade, and J. Weiner (1996). A coupled map lattice model of the growth of plant monocultures. *Ecol. Model. 84(1–3)*, 81–90.

Herz, A. V. M. (1994). Collective phenomena in spatially extended evolutionary games. *J. Theor. Biol. 169*, 65–87.

Hiebeler, D. (1997). Stochastic spatial models: From simulations to mean field and local structure approximations. *J. Theor. Biol. 187*, 307–319.

Hilbert, S. F. (Ed.) (1991). *Developmental biology. A comprehensive synthesis. A conceptual history of modern embryology*, Volume 7. New York: Plenum.

Hildebrandt, S. and A. Tromba (1996). *The parsimonious universe*. New York: Springer.

His, W. (1874). *Unsere Körperform und das physiologische Problem ihrer Entstehung*. Leipzig: Vogel.

Hlatky, L., R. K. Sachs, and E. L. Alpen (1988). Joint oxygen-glucose deprivation as the cause of necrosis on a tumor analog. *J. Cell. Physiol. 134*, 167–178.

Hodgkin, J. and D. Kaiser (1979). Genetics of gliding motility in M*yxococcus xanthus* (Myxobacteriales): Two gene systems control movement. *Mol. Gen. Genet. 172*, 177–191.

Höfer, T., J. A. Sherratt, and P. K. Maini (1995). *Dictyostelium discoideum*: cellular self-organization in an excitable biological medium. *Proc. Roy. Soc. London B 259*, 249–257.

Hogeweg, P. (1989). Local T-T cell and T-B cell interactions: A cellular automaton approach. *Immun. Lett. 22*, 113–122.

Holtfreter, J. (1939). Tissue affinity, a means of embryonic morphogenesis. *Arch. Exptl. Zellforsch. Gewebezücht. 23*, 169–209.

Holtfreter, J. (1943). A study of the mechanics of gastrulation. *J. Exp. Biol. 94*, 261–318.

Holtfreter, J. (1944). A study of the mechanics of gastrulation, Part II. *J. Exp. Zool. 95*, 171–212.

Horder, T. J. (1993). The chicken and the egg. In H. G. Othmer, P. K. Maini, and J. D. Murray (Eds.), *Experimental and theoretical advances in biological pattern formation*, Chapter 14, pp. 121–148. New York: Plenum.

Hurley, M. (1990). Attractors in cellular automata. *Ergod. Th. Dyn. Syst. 10*, 131–140.

Huttenlocher, A., M. Lakonishok, M. Kinder, S. Wu, T. Truong, K. Knudsen, and A. Horwitz (1998). Integrin and cadherin synergy regulates contact inhibition of migration and motile activity. *J. Cell Biol. 141(2),* 515–526.

Igoshin, O., A. Mogilner, R. Welch, D. Kaiser, and G. Oster (2001). Pattern formation and traveling waves in myxobacteria: Theory and modeling. *Proc. Natl. Acad. Sci. US 98*, 14913.

Iwasa, Y. (2000). Lattice models and pair approximation in ecology. In U. Dieckmann, R. Laws, and J. A. J. Metz (Eds.), *The geometry of ecological interactions: simplifying spatial complexity*, pp. 227–251. Cambridge: Cambridge University Press.

Jackson, E. A. (1991). *Perspectives of nonlinear dynamics*, Vol. 1 and 2. Cambridge: Cambridge University Press.

Jantsch, E. (1980). *The self-organizing universe: scientific and human implications of the emerging paradigm of evolution*. Oxford: Pergamon Press. Deutsch: *Die Selbstorganisation des Universums: vom Urknall zum menschlichen Geist,* DTV, München, 1986.

Jen, E. (1990). Aperiodicity in one-dimensional cellular automata. *Physica D 45*, 3–18.

Jepsen, G. L., E. Mayr, and G. G. Simpson (1949). *Genetics, paleontology, and evolution*. Princeton, NJ: Princeton University Press.

Joule, J. (1884). *Matter, living force and heat. The scientific papers of James Prescott Joule, Vol. 1*. London: Taylor and Francis. (quotation, p. 273).

Kadanoff, L. P. (1986). On two levels. *Phys. Today,* Sept., 7–9.

Kaiser, D. (1979). Social gliding is correlated with the presence of pili in *Myxococcus xanthus. Proc. Natl. Acad. Sci. USA 76(11),* 5952–5956.

Kaiser, D. and C. Crosby (1983). Cell movement and its coordination in swarms of *Myxococcus xanthus. Cell Motil. 3*, 227–245.

Kaneko, K. (1986). Attractors, basin structures, and information processing in cellular automata. In: Wolfram, S. (Ed.), *Theory and applications of cellular automata*. World Scientific, Singapore. pp. 367–399.

Kaneko, K. (1993). *Theory and applications of coupled map lattices*. Chichester: Wiley.

Kansal, A. R., S. Torquato, E. A. Chocas, and T. S. Deisboeck (2000). Emergence of a subpopulation in a computational model of tumor growth. *J. Theor. Biol. 207*, 431–441.

Kansal, A. R., S. Torquato, G. R. Harsh, E. A. Chiocca, and T. S. Deisboeck (2000). Simulated brain tumor growth dynamics using a three-dimensional cellular automaton. *J. Theor. Biol. 203*, 367–382.

Kaplan, D. and L. Glass (1995). *Understanding nonlinear dynamics*. New York: Springer-Verlag.

Kapral, R., A. T. Lawniczak, and P. Masiar (1991). Oscillations and waves in a reactive lattice-gas automaton. *Phys. Rev. Lett. 66 (19)*, 2539–2542.

Kari, J. (1990). Reversibility of 2d cellular automata is undecidable. *Physica D 45*, 379–385.

Kari, J. (1994). Reversibility and surjectivity problems of cellular automata. *J. Comput. Syst. Sci. 48(1),* 149–182.

Kearns, D. B. and L. J. Shimkets (1998). Chemotaxis in a gliding bacterium. *Proc. Natl. Acad. Sci. USA 95(20),* 11957–62.

Keller, E. F. and L. A. Segel (1970). Initiation of slime mold aggregation viewed as an instability. *J. Theor. Biol. 26*, 399–415.

Keller, E. F. and L. A. Segel (1971a). A model for chemotaxis. *J. Theor. Biol. 30*, 225–234.

Keller, E. F. and L. A. Segel (1971b). Travelling bands of chemotactic bacteria: A theoretical analysis. *J. Theor. Biol. 30*, 235–248.

Kelley, W. G. and A. C. Peterson (1991). *Difference equations*. Boston: Academic Press.

Kikuchi, R. (1966). The path probability method. *Prog. Theor. Phys. Suppl. 35*, 1–64.

Killich, T., P. J. Plath, E. C. Haß, W. Xiang, H. Bultmann, L. Rensing, and M. G. Vicker (1994). Cell movement and shape are non-random and determined by intracellular, oscillatory, rotating waves in *Dictyostelium* amoebae. *Biosystems 33*, 75–87.

Kirschner, M. and T. Mitchison (1986). Beyond self-assembly: From microtubules to morphogenesis. *Cell 45*, 329–342.

Klevecz, R. R. (1998). Phenotypic heterogeneity and genotypic instability in coupled cellular arrays. *Physica D 124*, 1–10.

Kruse, K. (2002). A dynamic model for determining the middle of *Escherichia coli*. *Biophys. J. 82*, 618.

Kubica, K., A. Fogt, and J. Kuczera (1994). Influence of amphiphilic anionic and cationic mixture on calcium ion desorption from lecithin liposomes. *Polish J. Environm. Stud. 3*(4), 37–41.

Kumar, K. (1984). The physics of swarms and some basic questions of kinetic theory. *Phys. Rev. 112*, 319–375.

Kumar, K., H. R. Skullerud, and R. E. Robson (1980). Kinetic theory of charged particle swarms in neutral gases. *Aust. J. Phys. 33*, 343–448.

Kuner, J. M. and D. Kaiser (1982). Fruiting body morphogenesis in submerged cultures of M*yxococcus xanthus*. *J. Bacteriol. 151*, 458–461.

Küng, H. (1996). *Große christliche Denker*. München: Piper.

Kurka, P. (1997). Languages, equicontinuity and attractors in cellular automata. *Ergod. Theor. Dyn. Syst. 17*, 229–254.

Kusch, I. and M. Markus (1996). Cellular automaton simulations and evidence for undecidability. *J. Theor. Biol. 178*, 333–340.

Landau, L. D. and E. M. Lifshitz (1979). *Fluid mechanics*. Oxford: Pergamon Press. (Section 132.)

Langer, J. S. (1980). Instabilities and pattern formation in crystal growth. *Nature 343*, 523.

Langton, C. G. (1984). Self-reproduction in cellular automata. *Physica D 10,* 135–144.

Langton, C. G. (1989). *Artificial life*. Redwood, CA: Addison-Wesley.

Langton, C. G. (1990). Computation at the edge of chaos. *Physica D 42,* 12–37.

Lawniczak, A. T. (1997). Lattice gas automata for diffusive-convective transport dynamics. *Center for Nonlinear Studies, Newsletter No. 136*, LALP–97–010.

Lawniczak, A. T., D. Dab, R. Kapral, and J. Boon (1991). Reactive lattice-gas automata. *Physica D 47*, 132–158.

LeDouarin, N. M. (1982). *The neural crest*. Cambridge: Cambridge University Press.

Lendowski, V. (1997). *Modellierung und Simulation der Bewegung eines Körpers in reaktiven Zweiphasenflüssigkeiten.* Doctoral thesis, University of Bonn.

Levin, S. A. (1974). Dispersion and population interactions. *Am. Nat. 108*, 207–228.

Levin, S. A. (1992). The problem of pattern and scale. *Ecology 73(6),* 1943–1967.

Li, W., N. H. Packard, and C. G. Langton (1990). Transition phenomena in cellular automata rule space. *Physica D 45*, 77–94.

Liggett, T. M. (1985). *Interacting particle systems*. New York: Springer-Verlag.

Liggett, T. M. (1999). *Stochastic interacting systems: Contact, voter and exclusion processes*. New York: Springer-Verlag.

Lindenmayer, A. (1982). Developmental algorithms: Lineage versus interactive control mechanisms. In S. Subtelny and P. Green (Eds.), *Developmental order:*

Its origin and regulation, Volume 40, New York, pp. 219–245. 40th Symp. Soc. Dev. Biol.: A. R. Liss, New York.

Lindenmayer, A. and G. Rozenberg (Eds.) (1976). *Automata, languages, development*, Amsterdam, New York. North-Holland.

Liu, F. and N. Goldenfeld (1991). Deterministic lattice model for diffusion-controlled crystal growth. *Physica D 47*, 124–131.

Löfberg, J. and K. Ahlfors (1980). Neural crest cell migration in relation to extracellular matrix organization in the embryonic axolotl trunk. *Dev. Biol. 75*, 148–167.

Löfberg, J., H. H. Epperlein, R. Perris, and M. Stigson (1989). Neural crest cell migration. In J. B. Armstrong and G. M. Malacinski (Eds.), *Developmental biology of the axolotl*. New York, Oxford: Oxford University Press, pp. 83–101.

Lutscher, F. (2002). Modeling alignment and movement of animals and cells. *J. Math. Biol. 45*, 234–260.

Lutscher, F. and A. Stevens (2002). Emerging patterns in a hyperbolic model for locally interacting cell systems. *J. Nonlin. Sci. 12(6)*, 619–640.

Magnus, A. (1867). *De vegetabilibus libri VII. Editionem criticam ab Ernesto Meyero coeptam absolvit Carolus Jessen*. Berlin: G. Reimer. cited after H. Balss: Albertus Magnus, Werk und Ursprung, Wiss. Verl. Ges. 1947, Stuttgart.

Maini, P. K. (1999). Mathematical models in morphogenesis. In V. Capasso and O. Dieckmann (Eds.), *Mathematics inspired by biology*, pp. 151–189. Berlin: Springer-Verlag.

Maini, P. K., K. J. Painter, and H. N. P. Chau (1997). Spatial pattern formation in chemical and biological systems. *J. Chem. Soc., Faraday Trans. 93(20)*, 3601–3610.

Manak, J. R. and M. P. Scott (1994). A class act: Conservation of homeodomain protein functions. *Development Suppl.*, 61–77.

Marée, A. F. M. and P. Hogeweg (2001). How amoeboids self-organize into a fruiting body: Multicellular coordination in *Dictyostelium discoideum*. *Proc. Natl. Acad. Sci. USA 98*, 3879–3883.

Markus, M. and B. Hess (1990). Isotropic cellular automaton for modeling excitable media. *Nature 347(6288)*, 56–58.

Markus, M. and I. Kusch (1995). Cellular automata for modelling shell pigmentation of molluscs. *J. Biol. Syst. 4*, 999–1011.

Markus, M. and H. Schepers (1993). Turing structures in a semi-random cellular automaton. In J. Demongeot and V. Capasso (Eds.), *Mathematics applied to biology and medicine*, pp. 473–481. Winnipeg: Wuerz Publishing. (Proc. of the 1st European Conference on Mathematics Applied to Biology and Medicine.)

Mayr, E. (1982). *The growth of biological thought: diversity, evolution, and inheritance*. Cambridge, MA: Belknap Press.

McCauley, E., W. G. Wilson, and A. M. de Roos (1996). Dynamics of age-structured predator-prey populations in space: Asymmetrical effects of mobility in juvenile and adult predators. *OIKOS 76*, 485–497.

McIntosh, H. V. (1990). Wolfram's class IV automata and a good life. *Physica D 45*, 105–121.

McNamara, G. and G. Zanetti (1988). Use of the Boltzmann equation to simulate lattice-gas automata. *Phys. Rev. Lett. 61*, 2332–2335.

Meakin, P. (1986). A new model of biological pattern formation. *J. Theor. Biol. 118*, 101–113.

Medvinskii, A. B., D. A. Tikhonov, J. Enderlein, and H. Malchow (2000). Fish and plankton interplay determines both plankton spatio-temporal pattern formation and fish school walks. A theoretical study. *Nonlinear Dynamics, Psychology and Life Sciences 4(2)*, 135–152.

Mehr, R. and Z. Agur (1992). Bone marrow regeneration under cytotoxic drug regimens. *Biosystems 26*, 231–237.

Meinhardt, H. (1982). *Models of biological pattern formation*. London: Academic Press.

Meinhardt, H. (1992). Pattern formation in biology: a comparison of models and experiments. *Rep. Prog. Phys. 55*, 797–849.

Meinhardt, H. and P. A. J. de Boer (2001). Pattern formation in E*scherichia coli*: A model for the pole-to-pole oscillations of Min proteins and the localization of the division site. *Proc. Nat. Acad. Sci. 98(25),* 14202–14207.

Metropolis, N., M. Rosenbluth, A. Teller, and E. Teller (1953). Equation of state calculations by fast computing machines. *J. Chem. Phys. 21(6),* 1087–1092.

Meyer-Hermann, M. (2002). A mathematical model for the germinal center morphology and affinity maturation. *J. Theor. Biol. 216*, 273–300.

Meyer-Hermann, M., A. Deutsch, and M. Or-Guil (2001). Recycling probability and dynamical properties of germinal center reactions. *J. Theor. Biol. 210*, 265–285.

Mikhailov, A. S. (1994). *Foundations of synergetics I*. Berlin: Springer-Verlag.

Mimura, M. (1981). Stationary pattern of some density-dependent diffusion system with competitive dynamics. *Hiroshima Math. J. 11*, 621–635.

Mitchell, M. (2002). Is the universe a universal computer? *Science 298*, 65–68. Book review.

Mitchell, M., P. T. Hraber, and J. P. Crutchfield (1993). Revisiting the edge of chaos: Evolving cellular automata to perform computations. *Compl. Syst. 7*, 89–130.

Mochizuki, A., Y. Iwasa, and Y. Takeda (1996). A stochastic model for cell sorting and measuring cell-cell adhesion. *J. Theor. Biol. 179*, 129–146.

Mochizuki, A., N. Wada, H. Ide, and Y. Iwasa (1998). Cell-cell adhesion in limb-formation, estimated from photographs of cell sorting experiments based on a spatial stochastic model. *Dev. Dynam. 211*, 204–214.

Mogilner, A., A. Deutsch, and J. Cook (1997). Models for spatio-angular self-organization in cell biology. In W. Alt, A. Deutsch, and G. Dunn (Eds.), *Dynamics of cell and tissue motion*, Chapter III.1, pp. 173–182. Basel: Birkhäuser.

Mogilner, A. and L. Edelstein-Keshet (1995). Selecting a common direction. I. how orientational order can arise from simple contact responses between interacting cells. *J. Math. Biol. 33*(6), 619–660.

Mogilner, A. and L. Edelstein-Keshet (1996a). Selecting a common direction. II. peak-like solutions representing total alignment of cell clusters. *J. Math. Biol. 34*, 811–842.

Mogilner, A. and L. Edelstein-Keshet (1996b). Spatio-angular order in populations of self-aligning objects: Formation of oriented patches. *Physica D 89*, 346–367.

Moore, E. F. (1962). Machine models of self-reproduction. In *Mathematical problems in the biological sciences (Proceedings of Symposia in Applied Mathematics 14)*, pp. 17–33. Providence, RI: American Mathematical Society.

Moore, J. H. and L. W. Hahn (2000). A cellular automata-based pattern recognition approach for identifying gene-gene and gene-environment interactions. *Am. J. Hum. Genet. 67*, 52.

Moore, J. H. and L. W. Hahn (2001). Multilocus pattern recognition using cellular automata and parallel genetic algorithms. In *Proc. of the Genetic and Evolutionary Computation Conference (GECCO-2001)*, p. 1452.

Morale, D. (2000). Cellular automata and many-particle systems modeling aggregation behaviour among populations. *Int. J. Appl. Math. Comp. Sci. 10*, 157–173.

Moreira, J. and A. Deutsch (2002). Cellular automaton models of tumour development—A critical review. *Adv. Compl. Syst. (ACS) 5(2),* 1–21.

Moreira, J. and A. Deutsch (2004). Zebrafish pigment pattern formation driven by local cellular interactions. *Dev. Dynam.* In print.

Moscona, A. (1961). Rotation-mediated histogenetic aggregation of dissociated cells. *Exp. Cell Res. 22*, 455–475.

Moscona, A. A. (1962). Analysis of cell recombinations in experimental synthesis of tissues in vitro. *J. Cell Comp. Physiol. 60* (Suppl. 1), 65–80.

Mostow, G. D. (Ed.) (1975). *Mathematical models for cell rearrangement*, New Haven, CT. Yale University Press.

Murray, J. D. (1981). A pre-pattern formation mechanism for animal coat markings. *J. Theor. Biol. 88*, 161–199.

Murray, J. D. (1988). How the leopard gets its spots. *Sci. Am. 258*, 62–69.

Murray, J. D. (1989). *Mathematical biology*. New York: Springer-Verlag.

Murray, J. D. (2002). *Mathematical biology* (3rd ed.). New York: Springer-Verlag.

Murray, J. D. and G. F. Oster (1984). Generation of biological pattern and form. *IMA J. Math. Appl. Med. Biol. 1*, 51–75.

Myhill, J. (1963). The converse of Moore's Garden of Eden theorem. *Proc. Amer. Math. Soc. 14*, 685–686.

Newell, P. C. (1983). Attraction and adhesion in the slime mold *Dictyostelium*. In J. E. Smith (Ed.), *Fungal differentiation: A contemporary synthesis. Mycology series 43*, pp. 43–71. New York: Marcel Decker.

Nicola, E., M. Or-Guil, W. Wolf, and M. Bär (2002). Drifting pattern domains in reaction-diffusion systems with nonlocal coupling. *Phys. Rev. E 65*, 055101.

Nicolis, G. and I. Prigogine (1977). *Self-organization in nonequilibrium systems.* New York: Wiley.

Nijhout, H. F., G. A. Wray, C. Krema, and C. Teragawa (1986). Ontogeny, phylogeny and evolution of form: An algorithmic approach. *Syst. Zool. 35*, 445–457.

Nüsslein-Volhard, C. (1991). Determination of the embryonic axes of *Drosophila. Development 112 Suppl.*, 1–10.

Oakley, R. A. and K. W. Tosney (1993). Contact-mediated mechanisms of motor axon segmentation. *J. Neurosci. 13(9),* 3773–3792.

O'Connor, K. and D. R. Zusman (1989). Patterns of cellular interactions during fruiting-body formation in M*yxococcus xanthus. J. Bacteriol. 171(11),* 6013–6024.

Oelschläger, K. (1989). *Many-particle systems and the continuum description of their dynamics.* Habilitationsschrift, Universität Heidelberg.

Okubo, A. (1980). *Diffusion and ecological problems: Mathematical models.* Berlin: Springer-Verlag.

Okubo, A. (1986). Dynamical aspects of animal grouping: Swarms, schools, flocks, and herds. *Adv. Biophys. 22*, 1–94.

Okubo, A. and S. Levin (2002). *Diffusion and ecological problems: Mathematical models* (2nd ed.). New York: Springer-Verlag.

Olsson, L. and J. Löfberg (1993). Pigment cell migration and pattern formation in salamander larvae. In L. Rensing (Ed.), *Oscillations and morphogenesis*, Chapter 27, pp. 453–462. New York: Marcel Dekker.

Omohundro, S. (1984). Modelling cellular automata with partial differential equations. *Physica D 10*, 128–134. (Proceedings of an interdisciplinary workshop, "Cellular Automata" D. Farmer, T. Toffoli and S. Wolfram (eds.), Los Alamos.)

Onsager, L. (1931). Reciprocal relations in irreversible processes. *Phys. Rev. 37*, 405.

Oster, G. F., J. D. Murray, and A. K. Harris (1983). Mechanical aspects of mesenchymal morphogenesis. *J. Embryol. Exp. Morphol. 78*, 83–125.

Oster, G. F., J. D. Murray, and P. K. Maini (1985). A model for chondrogenic condensations in the developing limb: The role of extracellular matrix and cell tractions. *J. Embryol. Exp. Morphol. 89*, 93–112.

Othmer, H. G., P. K. Maini, and J. D. Murray (Eds.) (1993). *Experimental and theoretical advances in biological pattern formation*, New York. Plenum Press.

Othmer, H. G. and A. Stevens (1997). Aggregation, blowup and collapse: the ABCs of taxis in reinforced random walks. *SIAM J. Appl. Math. 57*, 1044–1082.

Ouyang, Q. and H. L. Swinney (1991). Transition from a uniform state to hexagonal and striped Turing patterns. *Nature 352*, 610.

Packard, N. H. (1988). Adaptation towards the edge of chaos. In J. A. S. Kelso, A. J. Mandell, and M. F. Shlesinger (Eds.), *Dynamic patterns in complex systems*, pp. 293–301. Singapore: World Scientific.

Packard, N. H. and S. Wolfram (1985). Two-dimensional cellular automata. *J. Stat. Phys. 38*, 901–946.

Pearson, H. (2001). Two become one. *Nature 413*, 244–246.

Pelce, P. and J. Sun (1993). Geometrical models for the growth of unicellular algae. *J. Theor. Biol. 160*, 375–386.

Pennisi, E. and W. Roush (1997). Developing a new view of evolution. *Science 277*, 34–37.

Pfeifer, M. (1998). Developmental biology: Birds of a feather flock together. *Nature 395*, 324–325.

Pfistner, B. (1990). A one-dimensional model for the swarming behaviour of myxobacteria. In W. Alt and G. Hoffmann (Eds.), *Biological motion*, pp. 556–563 Heidelberg, Berlin, New York: Springer-Verlag.

Pfistner, B. (1995). Simulation of the dynamics of myxobacteria swarms based on a one-dimensional interaction model. *J. Biol. Syst. 3(2),* 579–588.

Phipps, M. (1992). From local to global: the lesson of cellular automata. In D. L. DeAngelis and L. J. Gross (Eds.), *Individual-based models and applications in ecology*, pp. 165–187. New York, London: Chapman & Hall.

Plath, P. and J. Schwietering (1992). Improbable events in deterministically growing patterns. In J. Encarnacao, H.-O. Peitgen, A. Sakas, and B. Englert (Eds.), *Fractal dynamics and computer graphics*, Chapter 28, pp. 162–172. Heidelberg: Springer-Verlag.

Potten, C. S. and M. Löffler (1987). A comprehensive model of the crypts of the small intestine of the mouse provides insight into the mechanisms of cell migration and the proliferation hierarchy. *J. Theor. Biol. 127*, 381–391.

Press, W. H., B. P. Flannery, S. A. Teukolsky, and W. T. Vetterling (1988). *Numerical recipes in C*. Cambridge: Cambridge University Press.

Preziosi, L. (Ed.) (2003). *Cancer modelling and simulation*. Boca Raton, FL, USA: Chapman and Hall/CRC Press.

Prigogine, I. and I. Stengers (1984). *Order out of chaos: Man's new dialogue with nature*. New York: Bantam Books.

Pulvirenti, M. and N. Bellomo (Eds.) (2000). *Modelling in applied sciences: A kinetic theory approach*, Boston: Birkhäuser.

Qi, A.-S., X. Zheng, C.-Y. Du, and B.-S. An (1993). A cellular automaton model of cancerous growth. *J. Theor. Biol. 161*, 1–12.

Qian, Y., D. d'Humières, and P. Lallemand (1992). Diffusion simulation with a deterministic one-dimensional lattice-gas model. *J. Stat. Phys. 68 (3/4)*, 563–573.

Rand, D. A. (1999). Correlation equations and pair approximations for spatial ecologies. In J. McGlade (Ed.), *Advanced ecological theory: Principles and applications*, pp. 100–142. Oxford: Blackwell Science.

Rand, D. A. and H. B. Wilson (1995). Using spatio-temporal chaos and intermediate-scale determinism to quantify spatially extended ecosystems. *Proc. R. Soc. Lond. B 259*, 111–117.

Rasmussen, S., H. Karampurwala, R. Vaidyanath, K. S. Jensen, and S. Hameroff (1990). Computational connectionism within neurons: A model of cytoskeletal automata subserving neural networks. *Physica D 42*, 428–449.

Rensing, L. (Ed.) (1993). *Oscillations and morphogenesis*, New York: Marcel Dekker.

Résibois, P. and M. de Leener (1977). *Classical kinetic theory of fluids*. New York: Wiley.

Richards, F. C., P. M. Thomas, and N. H. Packard (1990). Extracting CA rules directly from experimental data. *Physica D 45*, 189–202.

Richardson, D. (1973). Random growth in a tesselation. *Proc. Camb. Phil. Soc. 74*, 515–528.

Richter, P. and H. Dullin (1994). Harmonie der Proportionen: Der Goldene Schnitt in der Blattstellung höherer Pflanzen. In A. Deutsch (Ed.), *Muster des Lebendigen*, pp. 55–70. Wiesbaden: Vieweg.

Rieger, H., A. Schadschneider, and M. Schreckenberg (1994). Reentrant behavior in the Domany-Kinzel cellular automaton. *J. Phys. A 27*, L423–L430.

Ross, J., S. C. Müller, and C. Vidal (1988). Chemical waves. *Science 240*, 460.

Rothman, D. H. (1989). Negative-viscosity lattice gases. *J. Stat. Phys. 56(3/4)*, 517–524.

Rothman, D. H. and S. Zaleski (1997). *Lattice-gas cellular automata: Simple models of complex hydrodynamics*. Cambridge: Cambridge University Press.

Rozenfeld, A. F. and E. V. Albano (1999). Study of a lattice-gas model for a prey-predator system. *Physica A 266*, 322–329.

Rueff, J. (1554). *De conceptu et generatione hominis, et iis quae circa hec potissimum consyderantur, libri sex*. Zurich: Christopher Froschauer.

Russell, B. (1993). *Wisdom of the west: A historical survey of western philosophy in its social and political setting*, Volume 9, impr., reprint. London: Routledge.

Sager, B. and D. Kaiser (1993). Spatial restriction of cellular differentiation. *Gene. Dev. 7*, 1645–1653.

Savill, N. J. and P. Hogeweg (1997). Modeling morphogenesis: From single cells to crawling slugs. *J. Theor. Biol. 184*, 229–235.

Schaller, G. and M. Meyer-Hermann (2004). Kinetic and dynamic Delaunay tetrahedralizations in three dimensions. *Comput. Phys. Commun.* In print.

Schatz, M. F., S. J. VanHook, J. B. Swift, W. D. McCormick, and H. L. Swinney (1995). Onset of surface-tension-driven Bénard convection. *Phys. Rev. Lett. 75*, 1938.

Schenk, C. P., M. Or-Guil, M. Bode, and H. G. Purwins (1997). Interacting pulses in three-component reaction-diffusion-systems on two-dimensional domains. *Phys. Rev. Lett. 78*, 3781–3783.

Schimansky-Geier, L., F. Schweitzer, and M. Mieth (1997). Interactive structure formation with Brownian particles. In F. Schweitzer (Ed.), *Self-organization of complex structures—from individual to collective dynamics*, pp. 101–118. Amsterdam: Gordon and Breach.

Schönfisch, B. (1993). *Zelluläre Automaten und Modelle für Epidemien*. Ph.D. thesis, Fakultät für Biologie, Universität Tübingen.

Schönfisch, B. (1995). Propagation of fronts in cellular automata. *Physica D 80*, 433–450.

Schönfisch, B. (1996). Cellular automata and differential equations: an example. In M. Martelli (Ed.), *Proceedings of the International Conference on Differential Equations and Applications to Biology and to Industry*, pp. 431–438. Singapore: World Scientific.

Schönfisch, B. (1997). Anisotropy in cellular automata. *Biosystems 41*, 29–41.

Schönfisch, B. and A. de Roos (1999). Synchronous and asynchronous updating in cellular automata. *Biosystems 51*, 123–143.

Schönfisch, B. and K. P. Hadeler (1996). Dimer automata and cellular automata. *Physica D 94*, 188–204.

Schreckenberg, M., A. Schadschneider, K. Nagel, and N. Ito (1995). Discrete stochastic models for traffic flow. *Phys. Rev. E 51(4)*, 2939–2949.

Schubert, W. (1998). Molecular semiotic structures in the cellular immune system: Key to dynamics and spatial patterning. In J. Parisi, S. C. Müller, and W. Zimmermann (Eds.), *A perspective look at nonlinear physics; from physics to biology and social sciences*, pp. 197–206. Heidelberg: Springer-Verlag.

Schubert, W., C. L. Masters, and K. Beyreuther (1993). APP^+ T-lymphocytes selectively sorted to endomysial tubes in polymyositis displace NCAM-expressing muscle fibers. *Eur. J. Cell Biol. 62*, 333–342.

Schulman, L. S. and P. E. Seiden (1978). Statistical mechanics of a dynamical system based on Conway's game of life. *J. Stat. Phys. 19(3)*, 293–314.

Scott, M. P., J. W. Tamkun, and G. Hartzell (1989). The structure and function of the homeodomain. *Biochim. Biophys. Acta 989*, 25–48.

Segel, L. A. (1984). *Modeling dynamic phenomena in molecular and cellular biology*. Cambridge: Cambridge University Press.

Sepulchre, J. A. and A. Babloyantz (1995). Spiral and target waves in finite and discontinuous media. In R. Kapral and K. Showalter (Eds.), *Chemical waves and patterns*. Cambridge: Kluwer Academic Publishers.

Sigmund, K. (1993). *Games of life—explorations in ecology, evolution, and behaviour*. Oxford: Oxford University Press.

Skellam, J. G. (1951). Random dispersal in theoretical populations. *Biometrika 38*, 196–218.

Skellam, J. G. (1973). The formulation and interpretation of mathematical models of diffusionary processes in population biology. In M. S. Bartlett and R. W. Hiorns (Eds.), *The mathematical theory of the dynamics of biological populations*. New York: Academic Press.

Smith, A. R. (1971). Simple computation-universal cellular spaces. *J. Assoc. Comput. Mach. 18*, 339–353.

Smith, S. A., R. C. Watt, and R. Hameroff (1984). Cellular automata in cytoskeletal lattices. *Physica D 10*, 168–174.

Smoluchowski, M. (1916). Drei Vorträge über Diffusion, Brownsche Bewegung und Koagulation von Kolloidteilchen. *Physik. Z. 17*, 557–585.

Solé, R. V., O. Miramontes, and B. C. Goodwin (1993). Oscillations and chaos in ant societies. *J. Theor. Biol. 161*, 343–357.

Spemann, H. (1938). *Embryonic development and induction.* New Haven, CT: Yale University Press.

Starruss, J. and A. Deutsch (2004). *An extended Potts model formulation for the description and analysis of cell-shape-based pattern formation of myxobacteria*, in preparation.

Stein, S., R. Fritsch, L. Lemaire, and M. Kessel (1996). Checklist: Vertebrate homeobox genes. *Mech. Develop. 55*, 91–108.

Steinberg, M. S. (1958). On the chemical bonds between animal cells. A mechanism for type-specific association. *Am. Nat. 92(863),* 65–81.

Steinberg, M. S. (1963). Reconstruction of tissues by dissociated cells. *Science 141(3579),* 401–408.

Steinberg, M. S. (1964). The problem of adhesive selectivity in cellular interactions. In: Locke, M. (Ed.), *Cellular Membranes in Development.* Academic Press, New York, pp. 321-366.

Steinberg, M. S. (1970). Does differential adhesiveness govern self-assembly processes in histogenesis? Equilibrium configurations and the emergence of a hierarchy among populations of embryonic cells. *J. Exp. Zool. 173*, 395.

Stevens, A. (1992). *Mathematical modeling and simulations of the aggregation of myxobacteria. Chemotaxis equations as limit dynamics of moderately interacting stochastic processes.* Ph.D. thesis, University of Heidelberg, Heidelberg.

Stevens, A. and F. Schweitzer (1997). Aggregation induced by diffusing and nondiffusing media. In W. Alt, A. Deutsch, and G. Dunn (Eds.), *Dynamics of cell and tissue motion*, Chapter III.2, pp. 183–192. Basel: Birkhäuser.

Stowe, K. (1984). *Introduction to statistical mechanics and thermodynamics.* New York: Wiley.

Strassburger, E. (1978). *Lehrbuch der Botanik.* Stuttgart, New York: Gustav Fischer.

Swindale, N. V. (1980). A model for the formation of ocular dominance stripes. *Proc. Roy. Soc. Lond. B Bio. 208*, 243–264.

Takeichi, M. (1991). Cadherin cell adhesion receptors as a morphogenetic regulator. *Science 251*, 1451–1455.

Tanaka, E. M. (2003). Regeneration: If they can do it, why can't we? *Cell 113*, 559–562.

Technau, U. and T. W. Holstein (1992). Cell sorting during the regeneration of *Hydra* from reaggregated cells. *Dev. Biol. 151*, 117–127.

Theraulaz, G. and E. Bonabeau (1995a). Coordination in distributed building. *Science 269*, 686–688.

Theraulaz, G. and E. Bonabeau (1995b). Modelling the collective building of complex architectures in social insects with lattice swarms. *J. Theor. Biol. 177*, 381–400.

Thom, R. (1972). *Stabilité structurelle et morphogénèse.* New York: Benjamin.

Thompson, D. W. (1917). *On growth and form.* Cambridge: Cambridge University Press.

Thurman, R. A. F. (1996). Tibet—sein Buddhismus und seine Kunst. In M. M. Rhie and R. A. F. Thurman (Eds.), *Weisheit und Liebe—1000 Jahre Kunst des tibetischen Buddhismus*, pp. 20–38. Bonn: Kunst- und Ausstellungshalle der Bundesrepublik Deutschland.

Toffoli, T. (1984). Cellular automata as an alternative to (rather than an approximation of) differential equations in modeling physics. *Physica D 10*, 117–127. (Proceedings of an interdisciplinary workshop "Cellular Automata" D. Farmer, T. Toffoli and S. Wolfram (Eds.), Los Alamos.)

Toffoli, T. and N. Margolus (1987). *Cellular automata machines: A new environment for modeling*. Cambridge, MA: MIT Press.

Townes, P. L. and J. Holtfreter (1955). Directed movements and selective adhesion of embryonic amphibian cells. *J. Exp. Zool. 128*, 53–120.

Tucker, G. C., J. L. Duband, S. Dufour, and J. P. Thiery (1988). Cell-adhesion and substrate-adhesion molecules: Their instructive roles in neural crest cell migration. *Development 103*, 82–94.

Turing, A. M. (1952). The chemical basis of morphogenesis. *Phil. Trans. R. Soc. Lond. B 237*, 37–72.

Tyson, J. J. and J. P. Keener (1988). Singular perturbation theory of travelling waves in excitable media (a review). *Physica D 32*, 327–361.

Umeo, H., M. Maeda, and N. Fujiwara (2003). An efficient mapping scheme for embedding any one-dimensional firing squad synchronization algorithm onto two-dimensional arrays. In *Lecture notes in computer science: Proceedings of the 5th International Conference on Cellular Automata for Research and Industry, ACRI 2002, Geneva, Switzerland, October 9–11, 2002*, Volume 2493/2002, Heidelberg, pp. 69–81. Springer-Verlag.

van Laarhoven, P. and E. Aarts (1987). *Simulated annealing: Theory and applications*. Dordrecht: Reidel.

Varela, F., H. Maturana, and R. Uribe (1974). Autopoesis: The organization of living systems, its characterization and a model. *Biosystems 5*, 187–196.

Vicsek, T., A. Czirók, E. Ben-Jacob, I. Cohen, and O. Shochet (1995). Novel type of phase transition in a system of self-driven particles. *Phys. Rev. Lett. 75*, 1226–1229.

Vincent, J. V. V. (1986). Cellular automata: A model for the formation of colour patterns in molluscs. *J. Mollus. Stud. 52*, 97–105.

Vitanyi, P. M. B. (1973). Sexually reproducing cellular automata. *Math. Biosc. 18*, 23–54.

Vollmar, R. (1979). *Algorithmen in Zellularautomaten*. Stuttgart: Teubner.

von Neumann, J. (1966). *The theory of self-reproducing automata*. Urbana, IL: University of Illinois Press. Edited by A. W. Burks.

Voorhees, B. (1990). Nearest neighbour cellular automata over Z_2 with periodic boundary conditions. *Physica D 45*, 26–35.

Voss-Böhme, A. and A. Deutsch (2004). *An interacting particle approach to models of differential adhesion*, in preparation.

Waddington, C. H. (1940). *Organisers and genes*. New York: Cambridge University Press.

Waddington, C. H. (1962). *New patterns in genetics and development.* New York: Columbia University Press.

Walker, C. C. (1990). Attractor dominance patterns in sparsely connected Boolean nets. *Physica D 45*, 441–451.

Walther, T., A. Grosse, K. Ostermann, A. Deutsch, and T. Bley (2004). Mathematical modeling of regulatory mechanisms in yeast colony development. *J. Theor. Biol. 229(3)*, 327–338.

Ward, J. P. and J. R. King (1997). Mathematical modelling of avascular tumor growth. *IMA J. Math. Appl. Medic. Biol. 14*, 39–69.

Watson, J. D. and F. H. C. Crick (1953a). Genetical implications of the structure of deoxyribonucleic acid. *Nature 171*, 964–967.

Watson, J. D. and F. H. C. Crick (1953b). Molecular structure of nucleic acids: A structure for desoxyribose nucleic acid. *Nature 171*, 737–738.

Watzl, M. and A. Münster (1995). Turing-like spatial patterns in a polyacrylamide-methylene blue sulfide-oxygen system. *Chem. Phys. Lett. 242*, 273–278.

Weimar, J. R. (1995). *Cellular automata for reactive systems.* Ph.D. thesis, Université Libre de Bruxelles, Faculté des Sciences Service de Chimie Physique.

Weimar, J. R. (2001). Coupling microscopic and macroscopic cellular automata. *Parall. Comput. 27*, 601–611.

Weimar, J. R., D. Dab, J.-P. Boon, and S. Succi (1992). Fluctuation correlations in reaction-diffusion systems: Reactive lattice gas automata approach. *Europhys. Lett. 20(7)*, 627–632.

Weimar, J. R., J. J. Tyson, and L. T. Watson (1992). Third generation cellular automaton for modeling excitable media. *Physica D 55*, 328–339.

Weisheipl, J. A. (Ed.) (1980). Albertus Magnus and the sciences: Commemorative essays. Institute of Mediaeval Studies, Toronto (incl. bibliography).

Weliky, M., S. Minsuk, R. Keller, and G. Oster (1991). Notochord morphogenesis in X*enopus laevis*: Simulation of cell behavior underlying tissue convergence and extension. *Development 113*, 1231–1244.

Weliky, M. and G. Oster (1990). The mechanical basis of cell rearrangement. I. Epithelial morphogenesis during F*undulus* epiboly. *Development 109*, 373–386.

Whitehead, A. N. (1969). *Process and reality. An essay in cosmology.* New York: The Freedom Press.

Wiener, N. and A. Rosenbluth (1946). The mathematical formulation of the problem of conduction of impulses in a network of connected excitable elements, specifically in cardiac muscle. *Arch. Inst. Cardiol. Mexico 16*, 205–265.

Williams, H. T., S. G. Desjardins, and F. T. Billings (1998). Two-dimensional growth models. *Phys. Lett. A 250*, 105–110.

Wilson, H. V. (1907). On some phenomena of coalescence and regeneration in sponges. *J. Exp. Zool. 5*, 245–258.

Winfree, A. T. (1972). Spiral waves of chemical activity. *Science 175*, 634–636.

Winfree, A. T. (1987). *When time breaks down.* Princeton, NJ: Princeton University Press.

Witten, T. A. and L. M. Sander (1981). Diffusion-limited aggregation, a kinetic critical phenomenon. *Phys. Rev. Lett. 47(19),* 1400–1403.

Witten, T. A. and L. M. Sander (1983). Diffusion-limited aggregation. *Phys. Rev. B 27(9)*, 5686–5697.

Wolff, C. F. (1966). *Theorie von der Generation in zwei Abhandlungen erklärt und bewiesen (Berlin 1764)/Theoria Generationis (Halle 1759)*. Hildesheim: Olms. with an introduction by R. Herrlinger (in German).

Wolfram, S. (1983). Statistical mechanics of cellular automata. *Rev. Mod. Phys. 55*, 601–644.

Wolfram, S. (1984). Universality and complexity. *Physica D 10*, 1–35.

Wolfram, S. (1985). Twenty problems in the theory of cellular automata. *Physica Scripta T9*, 170–183. Proceedings of the Fifty-Ninth Nobel Symposium.

Wolfram, S. (1986a). Cellular automaton fluids 1: Basic theory. *J. Stat. Phys. 45(3/4)*, 471–526.

Wolfram, S. (Ed.) (1986b). *Theory and applications of cellular automata*, Singapore: World Publishing.

Wolfram, S. (1994). *Cellular automata and complexity—collected papers*. Reading: Addison-Wesley.

Wolfram, S. (2002). *A new kind of science*. Champaign, IL: Wolfram Media, Inc.

Wolpert, L. (1981). Positional information and pattern formation. *Phil. Trans. Roy. Soc. Lond. B 295*, 441–450.

Wootters, W. W. and C. G. Langton (1990). Is there a sharp phase transition for deterministic CA? *Physica D 45*, 95–104.

Wuensche, A. (1996). The emergence of memory; categorization far from equilibrum. In: S. R. Hammeroff, A. W. Kaszniak, A. C. Scott (Eds.), *Towards a Science of Consciousness: The First Tuscon Discussions and Debates*. Cambridge, MA: MIT Press.

Wuensche, A. and M. Lesser (1992). *The global dynamics of cellular automata*, Volume 1. Reading, MA: Addison-Wesley.

Wyatt, T. (1973). The biology of *Oikopleura dioica* and *Fritillaria borealis* in the Southern Bight. *Mar. Biol. 22*, 137–158.

Yotsumoto, A. (1993). A diffusion model for phyllotaxis. *J. Theor. Biol. 162*, 131–151.

Young, D. A. (1984). A local activator-inhibitor model of vertebrate skin patterns. *Math. Biosci. 72*, 51–58.

Zhabotinskii, A. M. and A. N. Zaikin (1970). Concentration wave propagation in a two-dimensional liquid-phase self-oscillating system. *Nature 225*, 535–537.

Zipori, D. (1992). The renewal and differentiation of hemopoietic stem cells. *FASEB J. 6*, 2691–2697.

Zwick, M. and H. Shu (1995). Set theoretic reconstructibility of elementary cellular automata. *Adv. Syst. Sci. Applic., Spec. Issue 1*, 31–36.

Index